Springer Series in Computational Mechanics

Edited by S. N. Atluri

S. N. Atluri, A. K. Amos

T.M

Large Space Structures: Dynamics and Control

With 166 Figures

Springer-Verlag Berlin Heidelberg NewYork
London Paris Tokyo

Editor of the series:

Prof. S. N. Atluri
Georgia Institute of Technology
Center for the Advancement of Computational Mechanics
School of Civil Engineering
Atlanta, GA 30332
USA

Editors of this volume:

Prof. S. N. Atluri
Georgia Institute of Technology
Center for the Advancement of Computational Mechanics
School of Civil Engineering
Atlanta, GA 30332
USA

Dr. A. K. Amos
Air Force Office of Scientific Research
Bolling Air Force Base
Washington, D.C. 20332
USA

ISBN 3-540-18900-9 Springer-Verlag Berlin Heidelberg New York
ISBN 0-387-18900-9 Springer-Verlag New York Heidelberg Berlin

Offsetprinting: Color-Druck, G. Baucke, Berlin
Bookbinding: Lüderitz & Bauer, Berlin

2161/3020 543210

Preface

This monograph is intended to provide a snapshot of the status and opportunities for advancement in the technologies of dynamics and control of large flexible spacecraft structures. It is a reflection of the serious dialog and assessments going on all over the world, across a wide variety of scientific and technical disciplines, as we contemplate the next major milestone in mankind's romance with space: the transition from exploration and experimentation to commercial and defense exploitation.

This exploitation is already in full swing in the space communications area. Both military and civilian objectives are being pursued with increasingly more sophisticated systems such as large antenna reflectors with active shape control. Both the NATO and Warsaw pact alliances are pursuing permanent space stations in orbit: large structural systems whose development calls for in-situ fabrication and/or assembly and whose operation will demand innovations in controls technology.

The last ten years have witnessed a fairly brisk research activity in the dynamics and control of large space structures in order to establish a technology base for the development of advanced spacecraft systems envisioned for the future. They have spanned a wide spectrum of activity from fundamental methods development to systems concept studies and laboratory experimentation and demonstrations. Some flight experiments have also been conducted for various purposes such as the characterization of the space enviroment, durability of materials and devices in that environment, assembly and repair operations, and the dynamic behavior of flexible structures. It is this last area that has prompted this monogram. The emphasis is clearly on the basic analytical and experimental methods development aspects of the technology.

The principal aims of this monopgrah are to bring together the view points of the structural dynamicists and the control theoreticians and, through this interdisciplinary dialogue, to facilitate further coordinated efforts in resolving outstanding technical problems in the nonlinear dynamics and control of highly flexible space structures.

In Chapter 1, Noor and Mikulas deal with several issues pertinent to the development of equivalent-continuum models for beam-like and plate-like lattices. These issues include: (i) a definition of the equivalent continuum, (ii) characterization of the continuum model and (ii) different approaches for the generation of stiffness, inertia, and thermal properties of the equivalent continuum. In Chapter 2, Atluri and Iura discuss computational methods to treat nonlinerarities of the structural, the inertial, and of the damping (due to flexible hysteritic joints) type that arise in the context of dynamics and control of LSS. Both semi-discrete type and space-time type methods to analyse the transient nonlinear response are discussed. Algorithms for implementing control on nonlinear semi-discrete type coupled systems of ordinary differential equations, are discussed. Reduced-order structural modeling techniques for both the equivalent-continuum models of LSS, as well as for truss and frame type lattice structures are discussed. Simple finite element methods for beam type LSS undergoing large rotational deformations, and field-boundary-element methods for shell-type LSS undergoing large deformations are discussed. Exact and explicit expressions for tangent stiffness operators of truss and frame type LSS are given. Finally, Atluri and Iura discuss the mechanical coupling between structural members and piezo-electric-film type control actuators, and give expressions for the actuator forces as functions of the excitation voltages and the magnitudes of generalized internal forces in the member.

In Chapter 3, Hu, Skelton, and Yang discuss the question of the contribution of a specific mode of linear vibration to the norm of the nodal response vector of a structure, in a situation wherein the promary consideration is the accuracy of dynamic response of a structure at specific locations. This question is of importance in deciding which modes are to be retained in creating a reduced-order structural model for the purpose of designing the controllers. In Chapter 4, Modi and Ibrahim discuss a general formulation, applicable to a class of space platforms with flexible extendable members, for analyzing transient dynamic situations involving complex interactions between deployment, attitude dynamics, and flexural rigidity. In Chapter 5, Park surveys partitioned solution procedures for large-scale simulation of dynamics and control of space structures which involve structural elements capable of large overall as well as flexible motions.

In Chapter 6, Zak discusses the case of pulse excitations of structures. In such cases, if a finite dimensional discrete model of the structure is employed, a loss of contribution of the high-frequency modes to the dynamic response may result. Zak shows that the thus unmodelled part of the response can be represented by a system of thin pulses and discusses the associated fundamental dynamical properties of the system. In Chapter 7, Bainum reviews the topics of mathematical models for orbital dynamics of large flexible structures, numerical techniques for synthesizing shape and attitude control laws, and the modelling of environmental disturbance torques due to the interaction of solar pressure on vibrating and thermally deformated structures.

In Chapter 8, Srinivasan discusses the phenomenon of friction between contacting surfaces, and addresses related issues such as quantifying the nature and magnitude of friction forces, quantifying the nature and magnitude of vibratory motion at contacting interface, and predicting the extent of damping that may be present. These topics are of interest in the passive control of space structures. In Chapter 9, Meirovitch discusses the active control concept of the independent-modal-space-method, and the concept of direct feddback control. He also discusses the related issues of deciding suitable control gains, the presence of damping, etc. In Chapter 10, von Flotow discusses the limiting case, labelled the acoustic limit, wherein the control bandwidth includes a very large number of natural modes. He argues that, in this limit, the modal analysis approach to control design is of limited value, and discusses alternate approaches involving wave-propagation formalisms applicable to flexible lattice-type structures.

In Chapter 11, Lynch and Banda present the control design techniques of Linear Quadratic Gaussian with Loop Transfer Recovery, for large space structures, wherein the high-frequency modeling uncertainties necessitate a robust control design. In Chapter 12, Bernstein and Hyland review the machinery of Optimal Projection for Uncertain Systems, for active control of flexible structures, and demonstrate its practical value. In Chapter 13, Kosut presents an approach to the problem of designing a robust control using on-line measurements, emlpoying the methods of parameter identification to obtain a nominal estimate of the plant-transfer function. Non-parametric spectral methods are then used to obtain a freqeuency domain expression for modal uncertainty. If the modal uncertainty exceeds a specified frequency bound, data filters used in the system-identification are modified, and the procedure is repeated. He also presents an analysis which establishes conditions under which the procedure converges to a satisfacotry robust design.

In Chapter 14, Junkins and Rew address the question whether a feed-back control law, designed based on a linear finite dimensional discrete mathematical model of a flexible structure, will stabilize and near-optimally control the real system. They emphasize the development of robust eigenstructure assignment methods and summarize optimization methods in which both the controller and selected strucural parameters are redesigned to improve the robustness. In the final Chapter 15, Khot presents two approaches for the optimum design of a strucutre and its control system with the objective of modifying the structural stiffness in order to achieve both a minimum weight structure, and a desired spectrum of closed-loop eigenvalues and structural frequencies.

The editors believe that these fifteen chapters collectively form a sound foundation for the subject of dynamics and control of flexible structures, wherein rapid scientific advances are to be expected in the next decade or so. It is towards this objective that the editors hope that this monograph would serve as a catalyst.

It is a great pleasure to thank all the authors for their kind cooperation through a timely preparation of their manuscripts. The editors also thank the Harris Corporation, Government Aerospace Systems Division, of Melbourne, Florida for their kind permission to use the illustration that appears on the cover of this manograph. A note of thanks to Ms. Deanna Winkler is also recorded here, for her assistance in the various editoral tasks.

Atlanta and Washington, D. C., July 1987 Satya N. Atluri, Anthony K. Amos

Contents

Continuum Modeling of Large Lattice Structures: Status and Projections

Ahmed K. Noor and Martin M. Mikulas
NASA Langley Research Center
Hampton, Virginia 23665

SUMMARY

The status and some recent developments of continuum modeling for large repetitive lattice structures are summarized. Discussion focuses on a number of aspects including definition of an effective substitute continuum; characterization of the continuum model; and the different approaches for the generation of the properties of the continuum, namely, the constitutive matrix, the matrix of mass densities, and the matrix of thermal coefficients. Also, a simple approach is presented which can be used to generate analytic expressions and/or numerical values of the continuum properties.

Application of the proposed approach to some beamlike and double-layered platelike lattices, currently considered as candidates of large space structures, is described. Future directions of research on continuum modeling are identified. These include needed extensions and applications of continuum modeling as well as computational strategies and modeling techniques.

SYMBOLS

$A^{(k)}$	cross-sectional area of member k of the repeating cell
$[C]$	matrix of stiffness coefficients of the simplified continuum
$C_{11}, C_{12}, C_{13}, \dots C_{88}$	stiffness coefficients of the simplified continuum (see Tables 1 and 2 and Figs. 4 and 5)
$\{C_T\}$	thermal load vector of the simplified continuum
$d_1, d_2, \dots d_6$	generalized displacements (see Figs. 4 and 5)
$E^{(k)}$	elastic modulus of the material of member k of the repeating cell
$F_1, F_2, \dots F_6$	generalized internal forces in the continuum beam model (see Fig. 4)
G_{11}, G_{12}, G_{13}	partitions of the matrix $[G]_c$, see Eqs. 37
$[G]_c$	geometric stiffness matrix of the continuum
$[g]$	geometric stiffness matrix of the simplified continuum
$[g^{(k)}]$	geometric stiffness matrix of member k of the repeating cell
$[K]$	stiffness matrix of the repeating cell
$[K]_c$	stiffness matrix of continuum
$K_{11}, K_{12}, K_{13}, K_{22}, K_{23}, K_{33}$	partitions of the matrix $[K]_c$, see Eqs. 14

Springer Series in Computational Mechanics
S. N. Atluri, A. K. Amos (Eds.)
Large Space Structures: Dynamics and Control

K	kinetic energy density of the continuum
$L^{(k)}$	length of member k of the repeating cell
M_2, M_3, M_t	bending and twisting moments in continuum beam model (see Fig. 4)
M_{11}, M_{22}, M_{12}	bending and twisting stress resultants in continuum plate model (see Fig. 5)
$\mathcal{M}_{11}$, $\mathcal{M}_{12}$, $\mathcal{M}_{22}$	partitions of the matrix $[\mathcal{M}]_c$, see Eqs. 15
$[\mathcal{M}]_c$	matrix of density parameters of the continuum
$[m]$	matrix of density parameters of the simplified continuum model
m_{11}, m_{12}, m_{13}, ... m_{66}	density parameters of the simplified continuum (see Tables 1 and 2 and Figs. 4 and 5)
$[\mathfrak{m}^{(k)}]$	consistent mass matrix of member k of the repeating cell
$\bar{N}$	axial force in beamlike lattices (see Fig. 4)
N_{11}, N_{22}, N_{12}	extensional stress resultants in continuum plate model (see Fig. 5)
$\{P_T\}$	thermal load vector of the repeating cell
$\{P_T\}_c$	thermal load vector of the continuum
$\{P_{T1}\}$, $\{P_{T2}\}$, $\{P_{T3}\}$	partitions of the vector $\{P_T\}_c$, see Eqs. 16
Q_{12}, Q_{13}	transverse shearing forces in the y and z directions in continuum beam model
Q_1, Q_2	transverse shear stress resultants in continuum plate model
$[R^{(k)}]$	transformation matrix whose entries are products of direction cosines of member k
S_1, S_2, ... S_8	stress resultants in the continuum plate model (see Fig. 5)
T^o	temperature parameter (see Eqs. 26 and 32)
$T^{(k)}$	temperature of member k of the repeating cell
$\{T\}_c$	vector of temperature parameters used in describing the continuum
U	thermoelastic strain energy density of the continuum
U_1, U_2	contributions to the strain energy of the linear and quadratic terms in the temperature parameters
U_o	isothermal strain energy
u, v, w	displacement components in the coordinate directions
$\bar{u}$, $\bar{v}$, $\bar{w}$	displacement parameters characterizing warping and cross-sectional distortions
u^o, v^o, w^o	displacement parameters in the coordinate directions
$\{u\}^{(k)}$	vector of nodal displacements of member k of the repeating cell
$\{u\}_c$	vector of displacement parameters used in describing the continuum
$\{u_{c1}\}$, $\{u_{c2}\}$	partitions of the vector $\{u_c\}$, see Eqs. 13
x, y, z	Cartesian coordinates
$\alpha^{(k)}$	coefficient of thermal expansion of member k of the repeating cell
$[\Gamma_u^{(k)}]$, $[\Gamma_\varepsilon^{(k)}]$, $[\Gamma_T^{(k)}]$ $[\bar{\Gamma}_\Delta^{(k)}]$	transformation matrices (see Eqs. 6, 7, 8 and 34)

$\gamma^o_{xy}, \gamma^o_{xz}, \gamma^o_{yz}$ — shearing strain parameters in the coordinate planes

$\gamma_{12}, \gamma_{13}, \gamma_{23}$ — shearing strains in the coordinate planes

$\{\Delta\}_c$ — vector of displacement parameters and their spatial derivatives for the continuum

$\varepsilon^{(k)}$ — axial strain of member k of the repeating cell

$\varepsilon_{11}, \varepsilon_{22}, \varepsilon_{33}$ — axial strains in the coordinate directions

$\varepsilon^o_x, \varepsilon^o_y, \varepsilon^o_z$ — extensional strain parameters in the coordinate directions

$\{\varepsilon\}_c$ — vector of strain parameters used in describing the continuum

$\{\varepsilon\}^{(k)}$ — vector of strain components in the coordinate directions used in the expansion of $\varepsilon^{(k)}$

$\{\varepsilon_{c1}\}, \{\varepsilon_{c2}\}, \{\varepsilon_{c3}\}$ — partitions of the vector $\{\varepsilon\}_c$, see Eqs. 12

$\bar{\kappa}$ — strain parameter (see Eqs. 25)

$\kappa^o_x, \kappa^o_y, \kappa^o_z, \kappa^o_t, 2\kappa^o_{xy}$ — curvature changes and twist parameters

ρ — mass density of the material (see Fig. 6)

ϕ_x, ϕ_y, ϕ_z — rotation components

ψ^o — strain parameter (see Eqs. 25)

Ω — characteristic geometric property of the repeating cell of the lattice (length of repeating cell for beamlike lattices and planform area of repeating cell for platelike lattices)

ω — frequency of vibration (see Figs. 8-11)

$(\ell, m, n)^{(k)}$ — direction cosines of member k

$\partial_x \equiv \partial/\partial x$, $\partial_y \equiv \partial/\partial y$, $\partial_z \equiv \partial/\partial z$

Superscript t denotes transposition.

1. INTRODUCTION

Lattice structures have been used for many years in spanning large areas with few intermediate supports. These structures can combine low cost with light weight and an esthetically pleasing appearance. Also, due to their ease of packaging, transporting, and assembling in space, lattice structures have attracted considerable attention for use in large-area space structures such as the space station, large space mirrors, antennas, multipurpose platforms, and power systems for supporting space operations. A main feature of the large-area lattice structures considered for space applications is that the basic pattern or configuration is repeated many times.

A review of the state-of-the-art in the analysis, design and construction of lattice structures until 1976 is given in [14 and 31]. The currently-used approaches for analyzing large repetitive lattices can be grouped into four classes; namely:

1) direct method
2) discrete field methods
3) periodic structure approaches
4) substitute continuum approaches.

In the first approach (direct method) the structure is analyzed as a system of discrete finite elements, and the methods of solving structural framework problems are applied. It has the obvious drawback of being computationally expensive for large lattices. This is particularly true when a buckling, vibration, or a nonlinear analysis is required.

The second approach (discrete field methods) takes advantage of the regularity of the structure and involves writing the equilibrium and compatibility equations at a typical joint of the lattice and either solving the resulting difference equations directly, or using truncated Taylor series expansions to replace the difference equations by differential equations (see, for example, [15, 16, 50, 51 and 62]). This approach works well for simple lattice configurations, but becomes quite involved for lattices with complex geometry.

The third group of methods are referred to as periodic structure approach, and are based on either: a) the combined use of finite elements and transfer matrix methods, which is efficient only for rotationally periodic (i.e., cyclically symmetric) structures or lattices with simple geometries [33, 64 and 65], or b) the exact representation of the stiffness of an individual member from which the analysis of beamlike lattices with simply supported edges can be performed [4, 5 and 6].

The fourth approach is based on replacing the actual lattice structure by a substitute continuum model which is equivalent to the original structure in some sense, such as the constitutive relations, strain energy and/or kinetic energy (see, for example, [3, 17, 18, 19, 21, 24, 29, 34, 35, 37, 42, 43, 54, 55, 56, 59, 63 and 66]). The use of continuum models to simulate the behavior of planar lattice beams dates back to the previous century [61, p. 483]. It has gained popularity only in recent years and has been applied to a variety of other discrete systems and phenomena including solid and liquid crystals, dislocations and defects, composite materials and biological systems.

The number of publications on continuum modeling of repetitive lattice structures has been steadily increasing. Therefore, there is a need to broaden awareness among practicing engineers and research workers about the recent developments in various aspects of continuum modeling for large lattice structures. The present paper is a modest attempt to fill this void. Specifically, the objectives of this paper are:

1) to assess the effectiveness of the currently used approaches for continuum modeling;

2) to present a simple and rational approach for development of continuum models for large repetitive lattice structures; and

3) to identify the future directions of research which have high potential for realizing the advantages of continuum modeling.

The scope of the present study includes thermoelastic stress analysis, buckling, free vibration, and geometrically nonlinear problems of large lattice structures. Beamlike and platelike repetitive lattices with pin and rigid joints are considered. Continuum modeling of lattices with flexible joints will also be discussed.

2. ADVANTAGES OF CONTINUUM MODELING

Before an assessment is made of the different approaches for developing continuum models, the following three advantages of using the continuum modeling approach for analyzing repetitive lattice structures are identified. First, it offers a practical and efficient approach for analyzing large lattice structures. This is particularly true for beamlike and platelike lattices, wherein a dimensionality reduction can result in a substantial reduction in the number of degrees of freedom. Second, it provides a simple means of comparing structural, thermal, and dynamic characteristics of lattices with different configurations and assessing the sensitivity of their responses to variations in material and geometric properties; and third, it provides an effective tool for parameter/system identification and feedback control system design of lattice structures.

3. DEFINITION AND KEY ELEMENTS OF A SUBSTITUTE CONTINUUM MODEL

A number of definitions have been given for the substitute continuum model. Herein an *effective continuum model* is defined to be a continuum which has the following characteristics:

1) the same amount of thermoelastic strain and kinetic energies are stored in it as those of the original lattice structure when both are deformed identically;

2) the temperature distribution, loading and boundary conditions of the continuum simulate those of the original lattice structure being modeled;

3) for beamlike and platelike lattices the continuum models are one-dimensional beams and two-dimensional plates, respectively (see Fig. 1);

4) local deformations are accounted for; and

5) lattices with pin joints are modeled as classical continua, and lattices with rigid (and/or flexible) joints are generally modeled as micropolar continua.

The last two characteristics are perhaps the most important in terms of recent developments and are discussed subsequently.

3.1 Local Deformations

The local deformations of two axially loaded planar trusses are shown in Fig. 2. The first truss has double lacing and a single-bay repeating cell. The second truss has single lacing and a double bay repeating cell. The cord members of the first truss remain straight as shown on the top sketch. On the other hand, the actual deformation of the single-laced truss has the zig-zag pattern shown on the top right sketch. On the average, however, the cord members remain straight. Early continuum models averaged these deformations, thereby substantially overestimating the axial stiffness. Recent continuum models, for lattices with more than one bay in their repeating cells, do account for the local deformations [37, 38 and 42].

3.2 Ordinary Versus Micropolar Continua

A contrast between the ordinary and micropolar continua is made in Fig. 3. For an

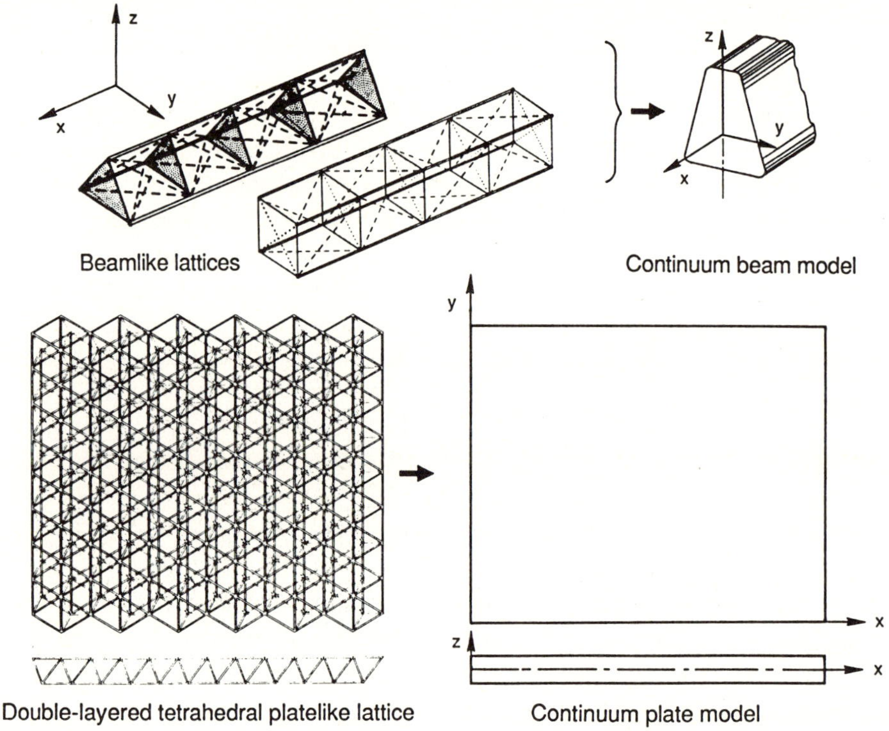

Figure 1 - Continuum models for beamlike and platelike lattice structures.

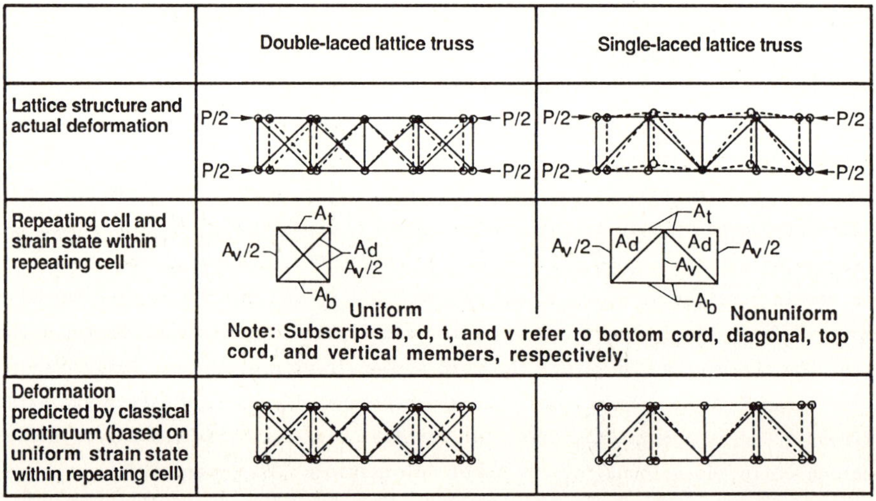

Figure 2 - Local deformations in planar lattice trusses subjected to axial loading.

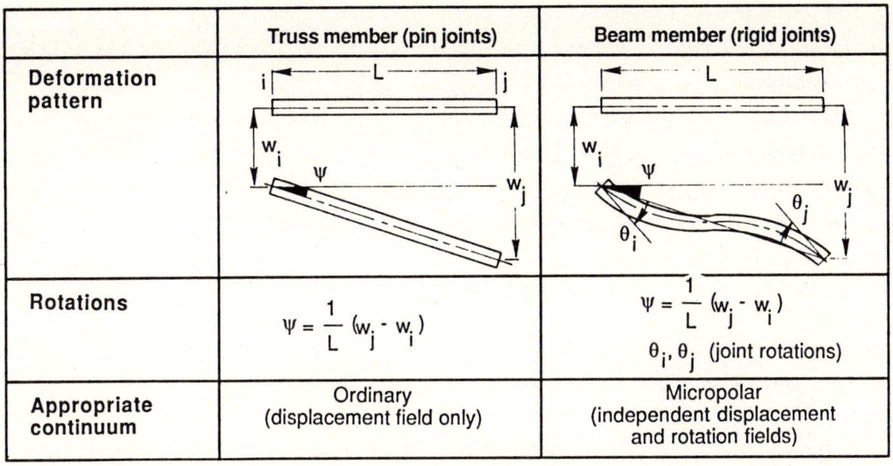

Figure 3 - Deformation patterns for pin-jointed and rigid-jointed one-dimensional members

axially loaded pin-jointed truss member the transverse motion is completely characterized by the joint displacements. The member rotation ψ is related to the joint displacements w_i and w_j. Therefore, the appropriate continuum to use in modeling pin-jointed trusses is the ordinary continuum for which the displacement field completely characterizes the motion of the structure.

On the other hand, for a rigid-jointed member, the transverse motion is characterized by both the joint displacements w_i, w_j as well as the joint rotations θ_i, θ_j which are independent degrees of freedom. Therefore, the appropriate continuum to use in modeling rigid-jointed flexural members is one whose motion is characterized by both a displacement field and an independent rotation field (referred to as microrotation field). The micropolar continuum is such a continuum.

3.3 Characterization of the Substitute Continuum Model

The substitute continuum model is characterized by the thermoelastic constitutive relations and density parameters which are determined in terms of the geometric and material properties of the original lattice structure. The thermoelastic constitutive relations and density parameters of the continuum can then be used to determine: a) the thermoelastic strain and kinetic energies; b) the governing differential equations; and whenever appropriate c) equivalent discrete finite element models.

3.4 Comments on Continuum Models

The following three comments regarding continuum models seem to be in order:

1. For some lattices the substitute continuum models may not have much resemblance to the continuum theories commonly used in engineering practice. Also, for complicated lattices the continuum models may be fairly complicated, and therefore, not useful for practical applications.

2. The accuracy of the predictions of the continuum approximation increases with the increase in the number of repeating cells (or modules) constituting the original lattice structure.

3. The response of the substitute continuum model (which simulates that of the original lattice structure) can be generated through: a) exact (or analytic) solution of the governing differential equations, or b) application of a discretization technique such as Rayleigh-Ritz technique or the finite element method.

4. DEVELOPMENT OF SUBSTITUTE CONTINUUM MODELS FOR STRESS ANALYSIS AND FREE VIBRATION PROBLEMS

A number of approaches have been proposed for developing continuum models, and for determining the appropriate constitutive relations and density parameters. These approaches include:

a) relating the force or deformation characteristics (or both) of a small segment of the lattice to those of a small segment of the continuum [20, 21, 22, 25, 26, 27, 28, 29 and 53];

b) using the discrete field method to obtain the governing difference equations of the lattice and either solving them directly or converting them to approximate differential equations [16, 50 and 51];

c) applying homogenization techniques based on using multiple-scale asymptotic expansions (see [7, 8, 11 and 32]); and

d) using energy equivalence concepts. The potential and kinetic energies of a typical (repeating) cell of the lattice are equated to those of the continuum, after expanding the nodal displacements of the lattice in a Taylor series.

The latter approach has been applied to a number of beamlike and platelike lattices. Computerized symbolic manipulation was used to generate analytic expressions for the stiffness and density parameters of the continuum (see [37, 38, 39, 41 and 42]). More recently, an equivalent approach was proposed for generating the properties of simplified one- and two-dimensional continuum models of beamlike and platelike lattice trusses with pin joints, which does not require the use of computerized symbolic manipulation (see [46]). Rather, numerical values of the stiffness and mass coefficients can be obtained by using a small Fortran program on an IBM PC (see [49]). A modified version of this approach is described subsequently.

The three key elements of the foregoing approach are:

1) introduction of kinematic and temperature assumptions to reduce the dimensionality of the continuum;

2) expansion of each of the nodal displacements, strain components, and temperature in a Taylor series; and

3) generation of four transformation matrices which relate nodal displacements, axial strains and temperatures of individual members of the repeating cell to the displacements, strain and temperature parameters of the continuum.

The procedure consists of the three major phases which are discussed subsequently for the case of lattices with pin joints.

Phase 1 - Generation of the Thermoelastic Stiffnesses of a Repeating Cell

1) A repeating cell (or module) is isolated from the lattice grid. The axial strain, temperature, and consistent mass matrix of a typical member, k, of the repeating cell are given by $\varepsilon^{(k)}$, $T^{(k)}$ and $[m^{(k)}]$, respectively.

2) The axial strain $\varepsilon^{(k)}$ of member k is expressed in terms of the vector of strain components in the coordinate directions through the following matrix equation:

$$\varepsilon^{(k)} = [R^{(k)}]\{\varepsilon\}^{(k)} \tag{1}$$

where

$$\{\varepsilon\}^{(k)} = \begin{Bmatrix} \varepsilon_{11} \\ \varepsilon_{22} \\ \varepsilon_{33} \\ \gamma_{12} \\ \gamma_{13} \\ \gamma_{23} \end{Bmatrix} \tag{2}$$

$$[R^{(k)}] = [\ell^2 \quad m^2 \quad n^2 \quad \ell m \quad \ell n \quad mn]^{(k)} \tag{3}$$

ε_{11}, ε_{22}, ε_{33} are the axial strains in the coordinate directions; γ_{12}, γ_{13}, γ_{23} are the shearing strains; and (ℓ, m, n) are the direction cosines of the member.

For simplicity, in the present study the strain state is assumed to be uniform within each repeating cell. Variation of the strain state within the repeating cell can be accounted for by expanding $\{\varepsilon\}^{(k)}$ in a Taylor series about the center of the repeating cell. The number of terms in the Taylor series expansion is equal to the number of independent deformation modes of the repeating cell.

3) The stiffness matrix and the thermal load vector of the repeating cell are generated using the following equations:

$$[K] = \sum_{\text{members}} (EAL)^{(k)} [R^{(k)}]^t [R^{(k)}] \tag{4}$$

$$\{P_T\} = \sum_{\text{members}} (\alpha EAL)^{(k)} [R^{(k)}]^t T^{(k)} \tag{5}$$

where E, A, L, α are the elastic modulus, cross-sectional area, length and coefficient of thermal expansion of member k; and superscript t denotes transposition.

The thermoelastic stiffnesses of the equivalent three-dimensional classical continuum are obtained by dividing the right-hand sides of Eqs. 4 and 5 by the volume of the repeating cell. Note that for members shared by n repeating cells, their cross sectional areas in Eqs. 4 and 5 are divided by n.

Phase 2 - Generation of Thermoelastic Stiffnesses and Density Parameters of the Beam and Plate Continua

4) Reduction in dimensionality is achieved by introducing kinematic and temperature assumptions regarding the variation of the displacement components, strain components and temperature in the plane of the cross-section (for beamlike lattices), or in the thickness direction (for platelike lattices).

The vectors of nodal displacements and strain components, and the temperature of member k are expressed in terms of the corresponding continuum parameters by means of the following three transformation matrices:

$$\{u\}^{(k)} = [\Gamma_u^{(k)}]\{u\}_c \tag{6}$$

$$\{\varepsilon\}^{(k)} = [\Gamma_\varepsilon^{(k)}]\{\varepsilon\}_c \tag{7}$$

$$T^{(k)} = [\Gamma_T^{(k)}]\{T\}_c \tag{8}$$

where $\{u\}_c$, $\{\varepsilon\}_c$ and $\{T\}_c$ are the vectors of displacement parameters, strain parameters, and temperature parameters used in describing the beam (or plate) continuum; $[\Gamma_u^{(k)}]$, $[\Gamma_\varepsilon^{(k)}]$ and $[\Gamma_T^{(k)}]$ are transformation matrices.

5) The stiffness matrix, matrix of density parameters and thermal load vector of the continuum are given by:

$$[K]_c = \frac{1}{\Omega} \sum_{\text{members}} (EAL)^{(k)}[\Gamma_\varepsilon^{(k)}]^t[R^{(k)}]^t[R^{(k)}][\Gamma_\varepsilon^{(k)}] \tag{9}$$

$$[M]_c = \frac{1}{\Omega} \sum_{\text{members}} [\Gamma_u^{(k)}][m^{(k)}][\Gamma_u^{(k)}] \tag{10}$$

$$\{P_T\}_c = \frac{1}{\Omega} \sum_{\text{members}} (\alpha EAL)^{(k)}[R^{(k)}]^t[\Gamma_u^{(k)}]\{T\}_c \tag{11}$$

where Ω is a characteristic geometric property of the repeating cell (length of the repeating cell for beamlike lattices, and planform area of the repeating cell for platelike lattices).

Phase 3 - Generation of Properties of Simplified Beam and Plate Continua

6) Simplified continuum models are obtained by partitioning the displacement and strain parameters of the continuum as follows:

$$\{\varepsilon\}_c = \begin{Bmatrix} \varepsilon_{c1} \\ \varepsilon_{c2} \\ \varepsilon_{c3} \end{Bmatrix} \tag{12}$$

$$\{u\}_c = \begin{Bmatrix} u_{c1} \\ u_{c2} \end{Bmatrix} \tag{13}$$

where $\{\varepsilon_{c1}\}$ are the strain parameters to be retained in the simplified model; $\{\varepsilon_{c2}\}$ are the strain parameters associated with the neglected stress resultants (or internal forces); $\{\varepsilon_{c3}\}$ are the strain parameters to be omitted; $\{u_{c1}\}$, $\{u_{c2}\}$ are the displacement parameters associated with the retained and neglected mass coefficients, respectively. The stiffness matrix, matrix of density parameters, and thermal load vector of the continuum are partitioned accordingly, i.e.,

$$[K]_c = \begin{bmatrix} K_{11} & K_{12} & K_{13} \\ & K_{22} & K_{23} \\ \text{Symm.} & & K_{33} \end{bmatrix} \tag{14}$$

$$[M]_c = \begin{bmatrix} M_{11} & M_{12} \\ \text{Symm.} & M_{22} \end{bmatrix} \tag{15}$$

$$\{P_T\}_c = \begin{Bmatrix} P_{T1} \\ P_{T2} \\ P_{T3} \end{Bmatrix} \tag{16}$$

The effective thermoelastic coefficients of the simplified continuum models are obtained by deleting the rows and columns associated with $\{\varepsilon_{c3}\}$, and by expressing $\{\varepsilon_{c2}\}$ in terms of $\{\varepsilon_{c1}\}$ (using static condensation). The resulting constitutive matrix and thermal load vector have the following forms:

$$[C] = [K_{11}] - [K_{12}][K_{22}]^{-1}[K_{21}] \tag{17}$$

$$\{C_T\} = \{P_{T1}\} - [K_{12}][K_{22}]^{-1}\{P_{T2}\} \tag{18}$$

The density parameters of the simplified continuum are obtained by neglecting the terms associated with $\{u_{c2}\}$, i.e.

$$[m] = [M_{11}] \tag{19}$$

7) The thermoelastic strain energy density of the continuum can be written in the following form:

$$U = U_o - U_1 - U_2 \tag{20}$$

where U_o is the isothermal strain energy density, and U_1 and U_2 are the contributions to the strain energy of the linear and quadratic terms in the temperature parameters. The expressions for U_o and U_1 are:

$$U_o = \frac{1}{2}\{\varepsilon_{c1}\}^t[C]\{\varepsilon_{c1}\} \quad , \quad U_1 = \{\varepsilon_{c1}\}^t\{C_T\} \tag{21}$$

The expression for U_2 is not presented herein since U_2 is inconsequential in the development of the properties of the continuum models.

8) The kinetic energy density of the continuum is given by:

$$K = \frac{1}{2} \{\dot{u}_{cl}\}^t [m] \{\dot{u}_{cl}\} \tag{22}$$

where a dot ($\dot{}$) refers to derivative with respect to time.

The sign convention for the internal forces and generalized displacements, along with the associated direct stiffnesses and mass coefficients, are given in Figs. 4 and 5 for the simplified beam and plate continua.

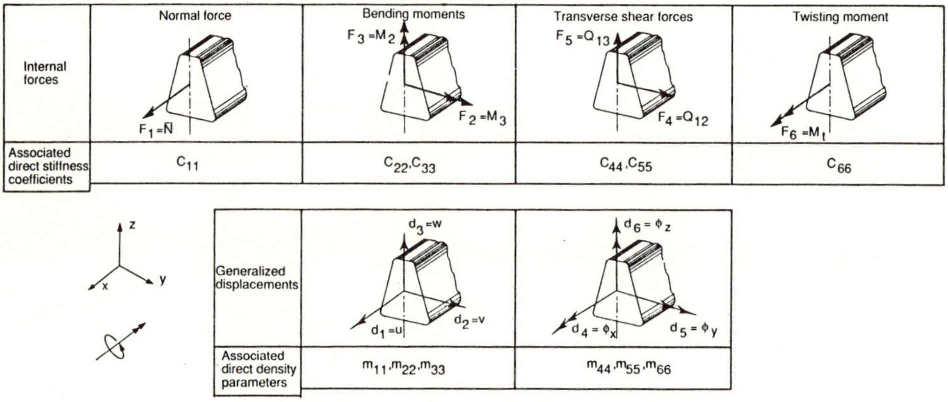

Figure 4 - Simplified continuum beam element, sign convention for internal forces and generalized displacements, and associated stiffness and density parameters.

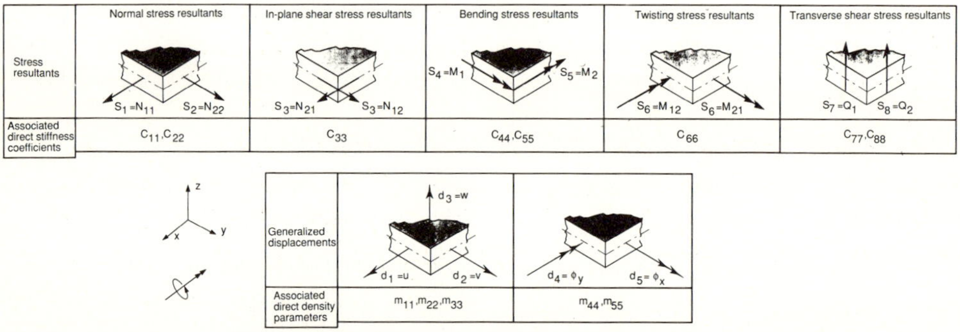

Figure 5 - Simplified continuum plate element, sign convention for stress resultants and generalized displacements, and associated stiffness and density parameters.

The foregoing approach is applied in the succeeding sections to the development of continuum models for beamlike and platelike lattice trusses with pin joints.

5. APPLICATION TO BEAMLIKE LATTICES

The foregoing approach is applied to the beamlike lattices having four longerons and orthogonal tetrahedral (unsymmetric) configuration (see Fig. 6). These trusses

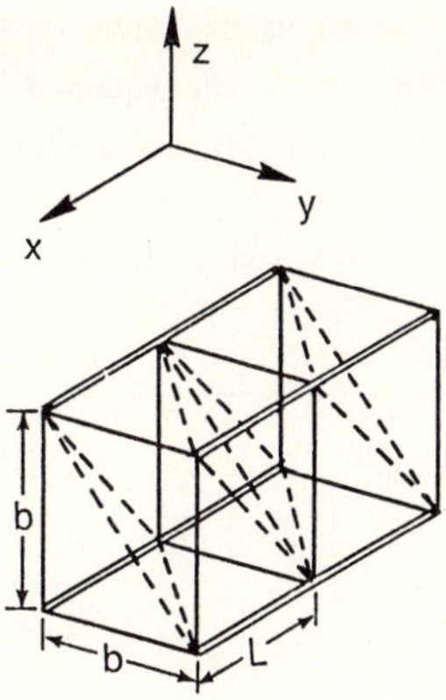

	C. Sec Area	Length	Designation
Longitudinal bars	A_1	L	═══
Diagonal bars	A_d	d	- - - -
Battens	A_b	b	———

b=L=5m.

$A=2.359 \times 10^{-4} m^2$.

ρ =1743.8 kg/m^3

$E=2.482 \times 10^{11}$ Pa (Longerons and battens)

$=2.557 \times 10^{11}$ Pa (diagonals)

Figure 6 - Orthogonal tetrahedral beamlike lattice.

are primary candidates for use in the keel beam design of the space station. For the lattice shown in Fig. 6, the smallest possible repeating cell, which can be isolated from the grid extends over one bay of the original structure.

5.1 Kinematic and Temperature Assumptions

The deformed position of any cross-section of the lattice is assumed to be specified by twelve displacement parameters. A Timoshenko-type beam theory is obtained by eliminating six of these parameters. The following expressions are used for the displacement field in the plane of the cross-section (plane y-z):

$$\begin{Bmatrix} u(x,y,z) \\ v(x,y,z) \\ w(x,y,z) \end{Bmatrix} = \begin{Bmatrix} u^o \\ v^o \\ w^o \end{Bmatrix} + \begin{bmatrix} \cdot & z & -y \\ -z & \cdot & \cdot \\ y & \cdot & \cdot \end{bmatrix} \begin{Bmatrix} \phi_x \\ \phi_y \\ \phi_z \end{Bmatrix} + \begin{bmatrix} \cdot & \cdot & \cdot \\ y & \cdot & \frac{1}{2}z \\ \cdot & z & \frac{1}{2}y \end{bmatrix} \begin{Bmatrix} \varepsilon_y^o \\ \varepsilon_z^o \\ \gamma_{yz}^o \end{Bmatrix} + yz \begin{Bmatrix} \bar{u} \\ \bar{v} \\ \bar{w} \end{Bmatrix} \tag{23}$$

where u^o, v^o and w^o are the displacement components at y=z=0 (chosen to be at the center of the repeating cell); ε_y^o and ε_z^o are the extensional strains in the y and z directions; γ_{yz}^o is the shearing strain in the plane of the cross section (plane yz); $\bar{u}$, $\bar{v}$ and $\bar{w}$ represent warping and distortion of the cross section; and the dots refer to zero terms. The sign convention for the displacement and rotation components is shown in Fig. 4. The twelve parameters u^o, v^o, w^o, ϕ_x, ϕ_y, ϕ_z, ε_y^o, ε_z^o, γ_{yz}^o, $\bar{u}$, $\bar{v}$

and $\bar{w}$ are functions of x only. Note that Eqs. 23 provide an exact representation of the displacement field in the plane of the cross section. As a consequence of the kinematic assumptions, Eqs. 23, the strain components have a bilinear variation in the plane of the cross section, y-z plane, as follows:

$$
\begin{aligned}
\varepsilon_{11} &= \varepsilon_x^o - y\,\kappa_y^o + z\,\kappa_z^o + yz\,\psi^o \\
\varepsilon_{22} &= \varepsilon_y^o + z\,\bar{v} \\
\varepsilon_{33} &= \varepsilon_z^o + z\,\bar{w} \\
\gamma_{12} &= \gamma_{xy}^o + y\,\partial\varepsilon_y^o + z\,(-\kappa_t^o + \bar{\kappa}) + yz\,\partial\bar{v} \\
\gamma_{13} &= \gamma_{xz}^o + y\,(\kappa_t^o + \bar{\kappa}) + z\,\partial\varepsilon_z^o + yz\,\partial\bar{w} \\
\gamma_{23} &= \gamma_{yz}^o + y\,\bar{v} + z\,\bar{w}
\end{aligned}
\tag{24}
$$

where ε_x^o is the extensional strain of the centerline; κ_y^o, κ_z^o are the curvature changes in the y and z directions; κ_t^o is the twist; γ_{xy}^o and γ_{xz}^o are the transverse shear strains; and $\partial \equiv d/dx$. The strain parameters ε_x^o, κ_y^o, κ_z^o, γ_{xz}^o, γ_{xy}^o, κ_t^o, ψ^o and $\bar{\kappa}$ are functions of x only, and can be expressed in terms of the displacement and rotation components as follows:

$$
\begin{aligned}
\varepsilon_x^o &= \partial u^o\,, & \gamma_{xz}^o &= \partial w^o + \phi_y \\
\kappa_y^o &= \partial\phi_z\,, & \kappa_t^o &= \partial\phi_x \\
\kappa_z^o &= \partial\phi_y\,, & \psi^o &= \partial\bar{u} \\
\gamma_{xy}^o &= \partial v^o - \phi_z\,, & \bar{\kappa} &= \frac{1}{2}\,\partial\gamma_{yz}^o + \bar{u}
\end{aligned}
\tag{25}
$$

The temperature distribution is also assumed to be bilinear in the plane of the cross section, i.e.,

$$
T(x,y,z) = T^o + y\,\partial_y T^o + z\,\partial_z T^o + yz\,\partial_y\partial_z T^o \tag{26}
$$

where T^o is the temperature at y=z=0; and

$\partial_y T^o \equiv \dfrac{\partial T^o}{\partial y}$, $\partial_z T^o \equiv \dfrac{\partial T^o}{\partial z}$, and $\partial_y\partial_z T^o = \dfrac{\partial^2 T^o}{\partial y\,\partial z}$ are the temperature gradients.

5.2 Transformation Matrices and Simplified Beam Models

The kinematic assumptions, Eqs. 23, are used to generate the transformation matrix $[\Gamma_u^{(k)}]$ of the kth member. Equations 24 and 26 are used to generate the transformation matrices $[\Gamma_\varepsilon^{(k)}]$ and $[\Gamma_T^{(k)}]$, with y and z in these equations denoting the coordinates of the center of the kth member.

A simplified (Timoshenko-type) continuum beam model is obtained by retaining six displacement parameters in Eqs. 23 and six strain parameters in Eqs. 24. This is accomplished by defining the partitions of $\{\varepsilon\}_c$ and $\{u\}_c$ as follows:

$$
\begin{aligned}
\{\varepsilon_{c1}\}^t &= [\varepsilon_x^o, \kappa_y^o, \kappa_z^o, \gamma_{xy}^o, \gamma_{xz}^o, \kappa_t^o] \\
\{\varepsilon_{c2}\}^t &= [\bar{\kappa}, \gamma_{yz}^o, \psi^o, \varepsilon_y^o, \varepsilon_z^o, \bar{v}, \bar{w}] \\
\{\varepsilon_{c3}\}^t &= [\partial\varepsilon_y^o, \partial\varepsilon_z^o, \partial\gamma_{yz}^o, \partial\bar{v}, \partial\bar{w}] \\
\{u_{c1}\}^t &= [u^o, v^o, w^o, \phi_x, \phi_y, \phi_z] \\
\{u_{c2}\}^t &= [\bar{u}, \gamma_{yz}^o, \varepsilon_y^o, \varepsilon_z^o, \bar{v}, \bar{w}]
\end{aligned}
\tag{27}
$$

The constitutive matrix and the matrix of material densities are obtained by following the procedure described in the preceding section. The numerical values for the coefficients of these matrices are given in Table 1 for the lattice structure shown in Fig. 6. Stiffness coefficients and density parameters for the simplified and higher-order continuum models are presented in [46] for orthogonal tetrahedral lattices with more than four longerons.

Table 1 - Numerical Values of the Stiffness Coefficients and Density Parameters for the Continuum Model of the Beamlike Lattice Structure Shown in Fig. 6

Coefficient	Value
C_{11}	2.967×10^8 N
$C_{14}=C_{15}=C_{44}=C_{55}$	3.126×10^7 N
$C_{22}=C_{33}$	1.561×10^9 N.m^2
C_{23}	9.770×10^7 N.m^2
$C_{26}=C_{36} = -1/2\ C_{66}$	-1.954×10^8 N.m^2
$m_{11}=m_{22}=m_{33}$	6.200 kg/m
$m_{44}=2m_{55}=2m_{66}$	56.096 kg.m
m_{56}	1.212 kg.m

Note: C_{11} is the extensional stiffness; C_{22}, C_{33} are the bending stiffnesses; C_{44}, C_{55} are the transverse shear stiffnesses; C_{66} is the torsional stiffness; and C_{14}, C_{15}, C_{23}, C_{26}, C_{36} are coupling coefficients.

6. APPLICATION TO PLATELIKE LATTICES

In this section applications of the foregoing approach to the tetrahedral and hexahedral grids shown in Figs. 2 and 7, respectively, are outlined. The tetrahedral grid is chosen because it has many attractive features for application to space structures. The hexahedral grid is chosen because it contains more redundancies and has vertical members that present difficulties in some continuum modeling approaches. In each case, a typical repeating element is isolated from the grid. The continuum model in this case is a two-dimensional plate continuum.

6.1 Kinematic and Temperature Assumptions

The three displacement components u, v and w, are assumed to have a linear variation in the thickness coordinate z, i.e.,

$$\begin{Bmatrix} u(x,y,z) \\ v(x,y,z) \\ w(x,y,z) \end{Bmatrix} = \begin{Bmatrix} u^o \\ v^o \\ w^o \end{Bmatrix} + z \begin{Bmatrix} \phi_x \\ \phi_y \\ \varepsilon_z^o \end{Bmatrix} \tag{28}$$

where u^o, v^o and w^o are the displacement components at z=0 (chosen to be at the center of the repeating cell); ϕ_x, ϕ_y are the rotation components; and ε_z^o is the transverse normal strain in the z direction. The six parameters u^o, v^o, w^o, ϕ_x, ϕ_y and ε_z^o are functions of x and y only. The sign convention for the displacement and rotation components is shown in Fig. 5. Equations 28 represent the exact displacement variation in the thickness direction provided no internal nodes are present.

As a consequence of the displacement assumptions, Eqs. 28, the strain components have a linear variation across the thickness of the plate as follows:

$$\begin{Bmatrix} \varepsilon_{11} \\ \varepsilon_{22} \\ \varepsilon_{33} \\ \gamma_{12} \\ \gamma_{13} \\ \gamma_{23} \end{Bmatrix} = \begin{Bmatrix} \varepsilon_x^o \\ \varepsilon_y^o \\ \varepsilon_z^o \\ \gamma_{xy}^o \\ \gamma_{xz}^o \\ \gamma_{yz}^o \end{Bmatrix} + z \begin{Bmatrix} \kappa_x^o \\ \kappa_y^o \\ \cdot \\ 2\kappa_{xy}^o \\ \partial_x \varepsilon_z^o \\ \partial_y \varepsilon_z^o \end{Bmatrix} . \tag{29}$$

where ε_x^o, ε_y^o, γ_{xy}^o are the extensional and shearing strains of the middle plane; ε_z^o is the transverse normal strain; κ_x^o, κ_y^o, $2\kappa_{xy}^o$ are the curvature changes and twist; γ_{xz}^o, γ_{yz}^o are the transverse shear strains; and a dot refers to zero. The strain measures ε_x^o, ε_y^o, γ_{xy}^o, κ_x^o, κ_y^o, $2\kappa_{xy}^o$, γ_{xz}^o and γ_{yz}^o are independent of z; and can be expressed in terms of the displacement and rotation components of the middle plane as follows:

$$\varepsilon_x^o = \partial_x u^o \quad , \qquad \kappa_y^o = \partial_y \phi_y$$

$$\varepsilon_y^o = \partial_y v^o \quad , \qquad 2\kappa_{xy}^o = \partial_x \phi_y + \partial_y \phi_x$$

$$\gamma^o_{xy} = \partial_x v^o + \partial_y u^o \quad , \qquad \gamma^o_{xz} = \partial_x w^o + \phi_x \tag{30}$$

$$\kappa^o_x = \partial_x \phi_x \quad , \qquad \gamma^o_{yz} = \partial_y w^o + \phi_y$$

For the platelike lattices considered in the present study, in order to account for the local transverse deformation of the repeating cell, it is necessary to augment the transverse shear strain expressions by their derivatives with respect to x and y as follows:

$$\gamma_{13} = \gamma^o_{xz} + z\, \partial_x \varepsilon^o_z + x \frac{\partial \gamma^o_{xz}}{\partial x} + y \frac{\partial \gamma^o_{xz}}{\partial y}$$

$$\gamma_{23} = \gamma^o_{yz} + z\, \partial_y \varepsilon^o_z + x \frac{\partial \gamma^o_{yz}}{\partial x} + y \frac{\partial \gamma^o_{yz}}{\partial y} \tag{31}$$

The temperature distribution is also assumed to be linear in the z direction, i.e.,

$$T(x,y,z) = T^o + z\, \partial_z T^o \tag{32}$$

where T^o is the temperature at z=0, and $\partial_z T^o = \partial T^o/\partial z$ is the temperature gradient in the z direction. Both T^o and $\partial_z T^o$ are functions of x and y only.

6.2 Transformation Matrices and Simplified Plate Models

Equations 28, 29, 31 and 32 are used to generate the three transformation matrices $[\Gamma^{(k)}_u]$, $[\Gamma^{(k)}_\varepsilon]$ and $[\Gamma^{(k)}_T]$ for the kth member. A simplified shear-deformation type plate theory is obtained by defining the partitions of $\{\varepsilon_c\}$ and $\{u_c\}$ as follows:

$$\{\varepsilon_{c1}\}^t = [\varepsilon^o_x, \varepsilon^o_y, \gamma^o_{xy}, \kappa^o_x, \kappa^o_y, 2\kappa^o_{xy}, \gamma^o_{xz}, \gamma^o_{yz}]$$

$$\{\varepsilon_{c2}\}^t = [\frac{\partial \gamma^o_{xz}}{\partial x}, \frac{\partial \gamma^o_{xz}}{\partial y}, \frac{\partial \gamma^o_{yz}}{\partial x}, \frac{\partial \gamma^o_{yz}}{\partial y}, \varepsilon^o_z]$$

$$\{\varepsilon_{c3}\}^t = [\partial_x \varepsilon^o_z , \partial_y \varepsilon^o_z] \tag{33}$$

$$\{u_{c1}\}^t = [u^o, v^o, w^o, \phi_x, \phi_y]$$

$$\{u_{c2}\}^t = [\varepsilon^o_z]$$

The constitutive matrix and the matrix of material densities are obtained by following the procedure outlined in the previous section. The numerical values for the coefficients of these matrices are given in Table 2 for the double-layered hexahedral lattice configuration shown in Fig. 7. Analytical expressions for the stiffness coefficients and density parameters of simplified plate models are given in [37] for the double-layered tetrahedral and hexahedral grids.

7. COMMENTS ON THE FOREGOING APPROACH FOR DEVELOPING CONTINUUM MODELS

The following comments regarding the procedure for developing continuum models

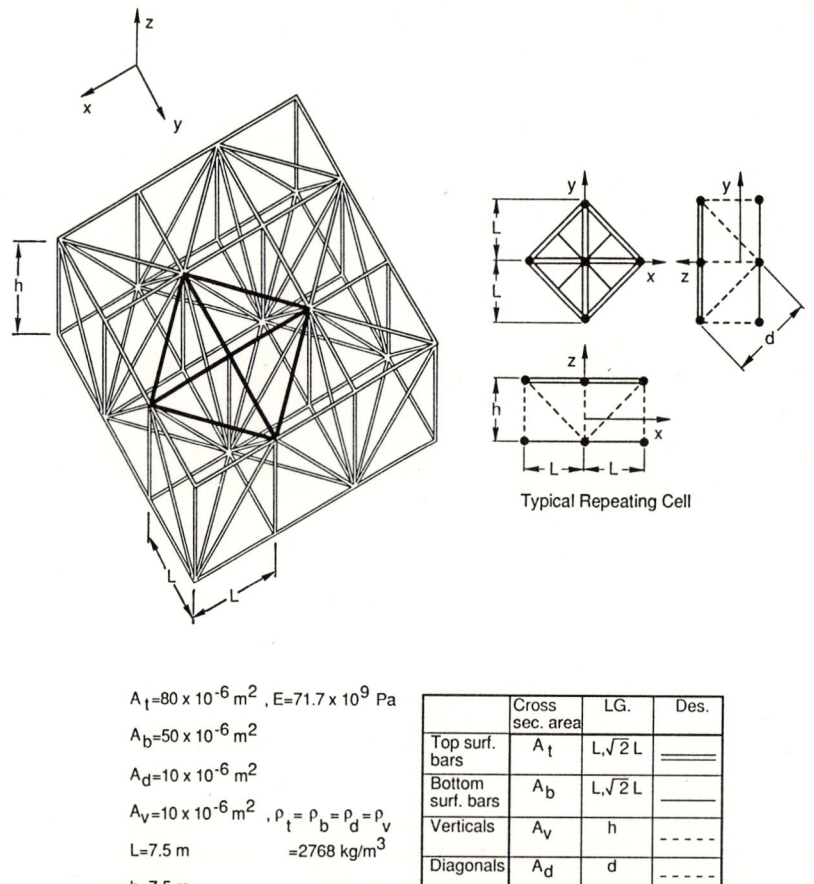

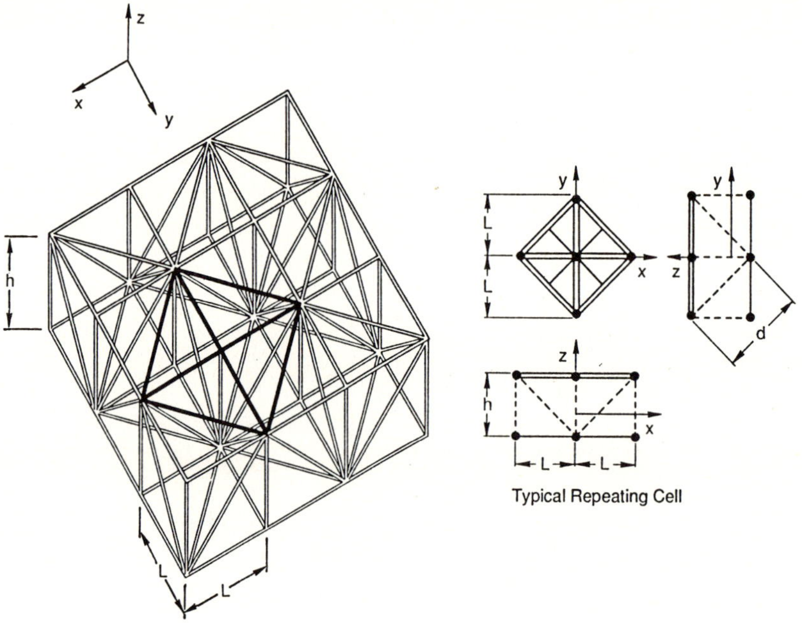

$A_t = 80 \times 10^{-6}\ m^2$, $E = 71.7 \times 10^9$ Pa

$A_b = 50 \times 10^{-6}\ m^2$

$A_d = 10 \times 10^{-6}\ m^2$

$A_v = 10 \times 10^{-6}\ m^2$, $\rho_t = \rho_b = \rho_d = \rho_v$

L=7.5 m =2768 kg/m³

h=7.5 m

	Cross sec. area	LG.	Des.
Top surf. bars	A_t	$L, \sqrt{2} L$	═══
Bottom surf. bars	A_b	$L, \sqrt{2} L$	———
Verticals	A_v	h	- - - - -
Diagonals	A_d	d	- - - - -

Figure 7 - Double-layered hexahedral platelike lattice truss.

seem to be in order:

1. The selection of the strain parameters $\{\varepsilon_{c1}\}$, $\{\varepsilon_{c2}\}$ and $\{\varepsilon_{c3}\}$ can be based on a sensitivity analysis of the response with respect to each of these parameters.

2. For simple lattice configurations, the constitutive matrix [C], the matrix of density parameters [m], and the vector of thermal loads $\{C_T\}$ can be obtained in symbolic form using a computerized symbolic manipulation language such as MACSYMA (see [39]). However, for more complicated configurations, it is more convenient to generate the matrices [C], [m], and $\{C_T\}$ numerically using the Fortran program given in [49].

3. Higher-order continuum models can be obtained by using more strain and displacement parameters than $\{\varepsilon_{c1}\}$ and $\{u_{c1}\}$. The additional parameters may be needed for the accurate prediction of the detailed displacement and stress distribution in the lattice (for example, localized displacements and stresses near a loaded edge).

4. A similar procedure for generating the characteristics of the continuum model was presented in [34 and 35]. In the cited references the lattice structure is considered as the sum of several arrays of parallel elements and thermoelastic stiffness coefficients of the continuum model are obtained by applying the tensor transformation relations to the unidirectional stiffness and thermal coefficients of each of the individual members constituting the repeating cell. The three-dimensional continuum coefficients generated are equal to the stiffness matrix and thermal load vector, [K], $\{P_T\}$ of Eqs. 4 and 5, each divided by the volume of the repeating cell.

Table 2 - Numerical Values of the Stiffness Coefficients and Density Parameters for the Continuum Model of the Platelike Lattice Shown in Fig. 7

Coefficient	Value
$C_{11}=C_{22}$	1.682×10^6 N/m
$C_{12}=C_{33}$	0.4394×10^6 N/m
$C_{14}=C_{25}$	1.456×10^6 N
$C_{15}=C_{24}=C_{36}$	0.3802×10^6 N
$C_{44}=C_{55}$	2.366×10^7 N
$C_{45}=C_{66}$	0.6179×10^7 N.m
$C_{77}=C_{88}$	3.380×10^4 N/m
$m_{11}=m_{22}=m_{33}$	0.1779 kg/m^2
$m_{14}=m_{25}$	0.1418 kg/m
$m_{44}=m_{55}$	2.370 kg

Note: C_{11}, C_{22} are the extensional stiffnesses in the x and y directions, C_{33} is the in-plane shear stiffness, C_{44}, C_{55} are the bending stiffnesses in the x and y directions, C_{66} is the torsional stiffness, C_{77}, C_{88} are the transverse shear stiffnesses; and C_{14}, C_{15}, C_{24}, C_{25}, C_{36} are bending-extensional coupling stiffnesses.

8. NUMERICAL STUDIES

To test and evaluate the accuracy of the predictions of the continuum models developed, a number of thermoelastic stress analysis, free vibration, and bifurcation buckling problems were solved using these models. Comparisons were made with exact solutions based on direct solution of the actual lattice structure. Two problem sets are presented herein: 1) stress and free vibration analysis of anisotropic beamlike lattice; and 2) free vibration analysis of platelike lattice. These problems are discussed subsequently.

8.1 Stress and Free Vibration Analysis of Anisotropic Beamlike Lattices

The first problem set considered is that of the orthogonal tetrahedral truss configuration shown in Fig. 6 which is a candidate for the primary truss support structure for the space station. Because of the unsymmetry of the basic configuration, the continuum model is anisotropic in the sense that both the extensional and shear effects, as well as the bending and twisting effects, are coupled. The elastic and dynamic characteristics of the continuum model are listed in Table 1. For stress analysis problems, the structure was subjected to longitudinal, transverse loadings and twisting moments at its free end. Typical results are shown in Figs. 8 and 9 and in Tables 3 and 4.

Table 3 - Comparison of Maximum Displacements Obtained by the Continuum Beam Models with Exact Solutions for the Cantilevered Lattice Structure Shown in Fig. 6. Number of Repeating Cells = 10.

10^7 × Normalized Displacements	Continuum Model		Exact
	Coupling Terms Neglected	Coupling Terms Included	
$u^o/\bar{N}$ (m/N)	1.685	2.135	2.135
w^o/Q_{13} (m/N)	282.8	302.8	296.9
v^o/Q_{12} (m/N)	282.8	302.8	296.9
w^o/M_t (N^{-1})	0	4.270	4.276
w^o/M_2 (N^{-1})	8.005	8.540	8.442
v^o/M_3 (N^{-1})	8.005	8.540	8.442

Note: $\bar{N}$ is the axial force; Q_{13}, Q_{12} are the transverse shearing forces; M_t is the twisting moment; and M_2, M_3 are the bending moments (see Fig. 4).

Table 3 gives the maximum displacements u^o, v^o and w^o at the free end obtained by the beam models, along with the exact solutions obtained by the direct analysis of the actual structure for lattices with ten bays. For the sake of comparison, the predictions of the foregoing continuum model are given along with those of the continuum model in which all the coupling stiffness and density parameters are neglected. Table 4 and Figure 8 give an indication of the accuracy of the lowest ten frequencies obtained by the two continuum models. Then vibration mode shapes associated with the lowest ten

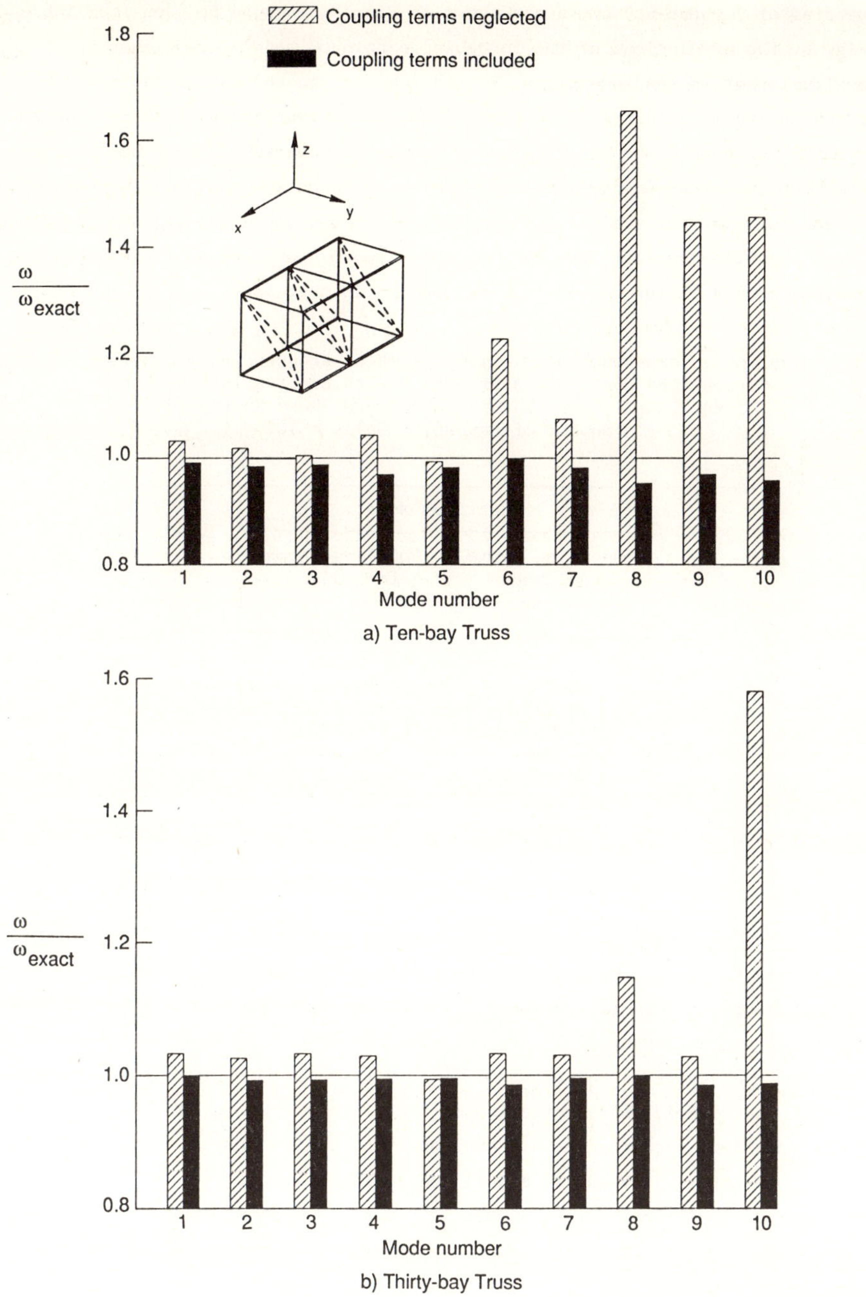

Figure 8 - Accuracy of minimum frequencies obtained by continuum beam models for the cantilevered beamlike lattice structure shown in Fig. 6.

frequencies of the ten-bay truss are shown in Fig. 9. As can be seen from Table 4 and Fig. 8, the predictions of the foregoing continuum models are highly accurate. The maximum error in the lowest ten frequencies for the ten bay truss is 4.73%. The error reduces to less than 1.5%, when the number of bays increases to 30. On the other hand, neglecting the coupling terms results in overestimating the stiffness of the structure, thereby increasing the vibration frequencies and reducing the maximum displacements. This effect is very pronounced for the vibration frequencies associated with the higher modes. For ten-bay truss, when the coupling terms are neglected, the maximum errors in the lowest ten frequencies is over 65%.

Table 4 - Comparison of Minimum Vibration Frequencies (in Hertz) Obtained by Continuum Beam Models with Exact Solutions for the Cantilevered Lattice Structure Shown in Fig. 6. Number of Repeating Cells = 10.

Mode	Continuum Model		Exact
	Coupling Terms Neglected	Coupling Terms Included	
1	3.385	3.252	3.280
2	3.385	3.274	3.325
3	13.197	12.977	13.136
4	17.059	15.853	16.353
5	17.059	16.877	17.172
6	38.794	31.608	31.641
7	38.794	35.468	36.116
8	62.185	35.817	37.596
9	62.185	41.704	42.996
10	86.209	56.732	59.198

8.2 Free Vibration Analysis of Platelike Lattice Grid

The second problem considered is that of the free vibration analysis of a cantilevered hexahedral double-layered grid. The characteristics of the grid are shown in Fig. 7. The boundary nodes at x=0 are completely restrained. The continuum model is taken to be a square, shear-flexible plate with one edge completely restrained. The elastic and dynamic characteristics of the plate are listed in Table 2. The continuum solutions presented herein are converged finite element solutions. Typical results are shown in Figs. 10 and 11 and in Table 5.

An indication of the accuracy of the lowest ten vibration frequencies is given in Table 5 and Fig. 10. The standard of comparison is taken to be the vibration frequencies obtained by the direct finite element solutions of the actual grid. Figure 11 shows the vibration mode shapes associated with the minimum ten frequencies of the 8L × 8L grid. As can be seen from Table 5, for an 8L × 8L grid, the maximum error in the lowest ten vibration frequencies is 1.1%. The error reduces to less than 0.55% when a 16L × 16L grid is used.

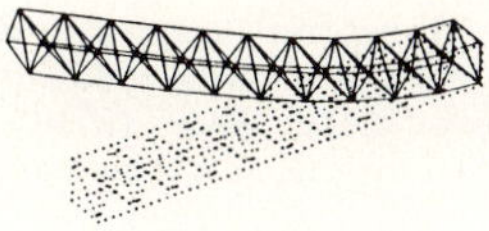

ω_1=3.252 (3.280)

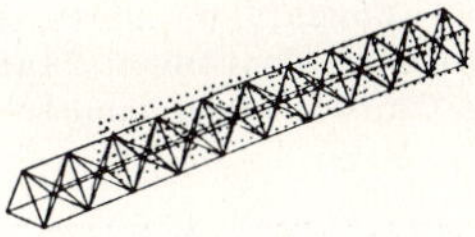

ω_2=3.274 (3.325)

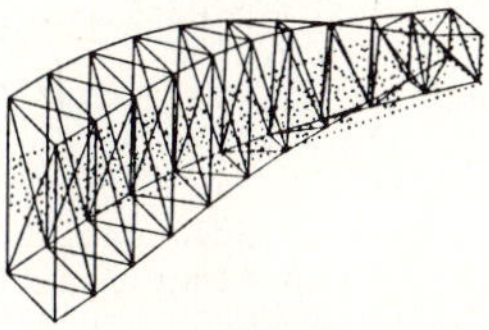

ω_3=12.98 (13.14)

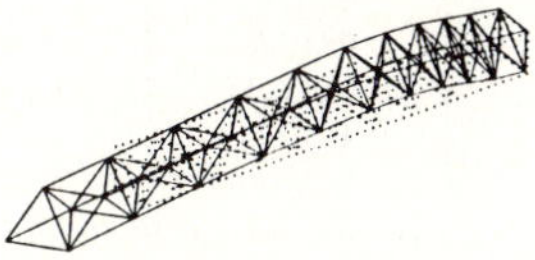

ω_4=15.85 (16.35)

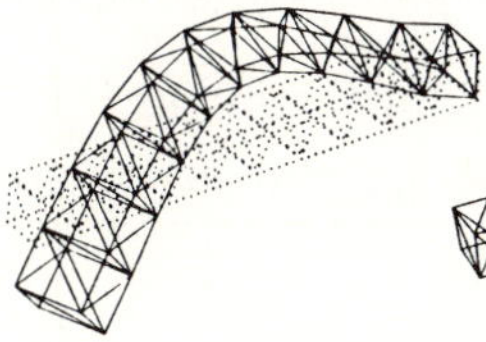

ω_5=16.88 (17.17)

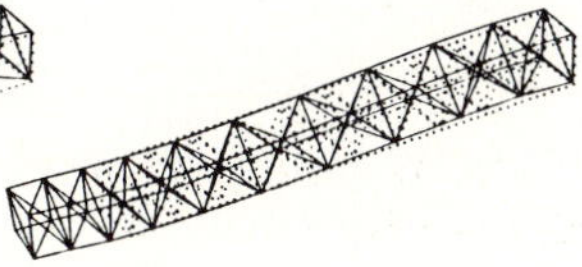

ω_6=31.61 (31.64)

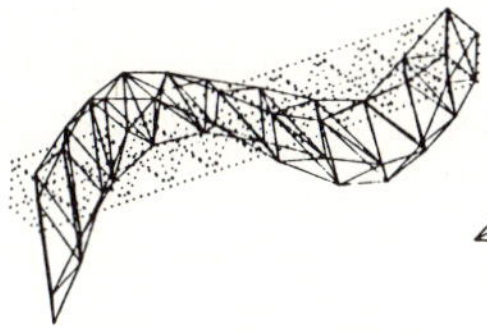

ω_7=35.47 (36.12)

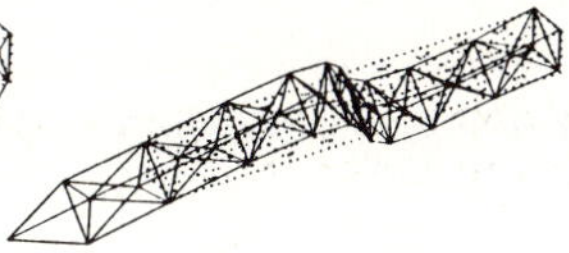

ω_8=35.82 (37.60)

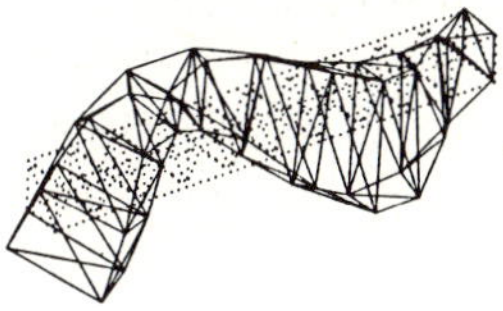

ω_9=41.70 (43.00)

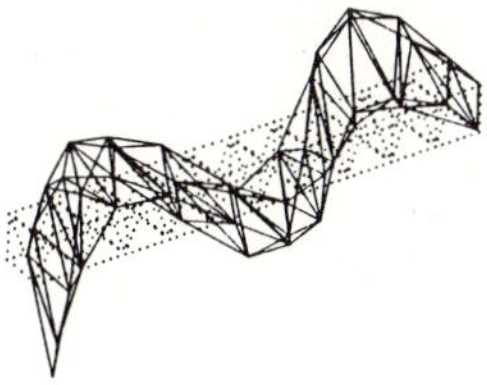

ω_{10}=56.73 (59.20)

Note: Numbers between parentheses refer to exact frequencies (in Hertz) ; other numbers are continuum beam frequencies.

Figure 9 - Vibration mode shapes for the cantilevered lattice truss shown in Fig. 6; number of bays = 10.

Table 5 - Comparison of Minimum Vibration Frequencies (in Hertz) Obtained by Continuum Plate Models with Exact Solutions for the Cantilevered Double-Layered Grid Shown in Fig. 7.

Mode	8L × 8L Grid		16L × 16L Grid	
	Continuum	Exact	Continuum	Exact
1	1.253	1.248	0.3887	0.3878
2	2.064	2.087	0.7152	0.7178
3	4.260	4.249	1.703	1.700
4	4.768	4.792	2.182	2.189
5	5.527	5.573	2.346	2.345
6	6.303	6.298	2.387	2.391
7	8.273	8.347	3.436	3.455
8	8.377	8.396	3.670	3.664
9	9.090	9.144	3.979	3.984
10	9.856	9.809	4.360	4.354

9. APPLICATION OF CONTINUUM MODELS TO STABILITY PROBLEMS

An important criterion in the design of lightweight lattice trusses is the onset of elastic instability. For finely divided lattice trusses the failure mode is usually that of general (system) instability rather than local snap-through of a single joint or member buckling. The foregoing continuum models can be extended to the general (global) stability analysis of lattice structures. This is accomplished by equating the potential energy due to initial stresses of the original lattice to that of the continuum, from which expressions for the geometric stiffness coefficients of the continuum are obtained in terms of the geometric and elastic properties of the original lattice structure. The procedure is outlined subsequently.

1. A repeating cell is isolated from the lattice grid. The geometric stiffness matrix of a typical member, k, of the repeating cell referred to the coordinates of the repeating element is given by $[g^{(k)}]$.

2. The nodal displacements of member k are expressed in terms of the continuum displacement parameters and their derivatives with respect to the spatial coordinates by means of the following transformation:

$$\{u\}^{(k)} = [\bar{\Gamma}_{\Delta}^{(k)}]\{\Delta\}_c \tag{34}$$

where $\{\Delta\}_c$ is the vector of displacement parameters and its spatial derivatives; and $[\bar{\Gamma}_{\Delta}^{(k)}]$ is a transformation matrix. Equations 34 are obtained by expanding the nodal displacements of member k in a Taylor series about the center of the repeating cell.

3. The geometric stiffness matrix of the continuum is given by:

$$[G]_c = \frac{1}{\Omega} \sum_{\text{members}} [\bar{\Gamma}_{\Delta}^{(k)}]^t [g^{(k)}][\bar{\Gamma}_{\Delta}^{(k)}] \tag{35}$$

4. Simplified continuum models are obtained by partitioning the vector $\{\Delta\}_c$ and

the matrix $[G]_c$ as follows:

$$\{\Delta\}_c = \begin{Bmatrix} \Delta_{c1} \\ \Delta_{c2} \end{Bmatrix} \tag{36}$$

$$[G]_c = \begin{bmatrix} G_{11} & G_{12} \\ G_{21} & G_{22} \end{bmatrix} \tag{37}$$

where $\{\Delta_{c1}\}$ and $\{\Delta_{c2}\}$ are the displacement parameters associated with the retained and neglected geometric stiffness coefficients, respectively. The geometric stiffness coefficients of the simplified continuum are given by:

$$[g] = [G_{11}] \tag{38}$$

Note that the foregoing continuum approach for stability analysis assumes that the individual members of the lattice remain straight and stable during buckling. Local member instability and snap-through buckling of a single joint are not predicted by the present theory.

10. OTHER REPORTED APPLICATIONS OF CONTINUUM MODELING

Other reported applications of continuum modeling which have not been discussed in the preceding sections are described subsequently.

10.1 Geometrically Nonlinear Problems of Beamlike Lattices with Pin Joints

In [1 and 13] an incremental stiffness matrix of the equivalent continuum is used to account for the large-displacement nonlinear effects in beamlike lattices. Applications of this approach are made to static, large deflection analysis, large-amplitude free vibration analysis, as well as buckling and postbuckling analysis.

10.2 Beamlike Lattices with Material Damping

Planar beamlike lattices with members having viscous damping are modeled as continuum Timoshenko beams in [2 and 60]. The global damping characteristics are determined in terms of the damping coefficients and dimensions of the truss members.

10.3 Stress, Free Vibrations and Buckling Problems of Beamlike Lattices with Rigid and Flexible Joints

Micropolar continuum models are developed in [10, 40 and 41] for the static, free vibrations and buckling problems of planar and spatial beamlike lattices with rigid joints. The study made in [41] shows that ordinary shear-flexible continuum beam models are found to be adequate for predicting the global response characteristics of lattices which do not need to have rigid joints for their kinematic stability. An exception to this is the case when diagonal members, which have very small cross-sectional area, are used. The incorporation of joint flexibility in the continuum model is discussed in [52].

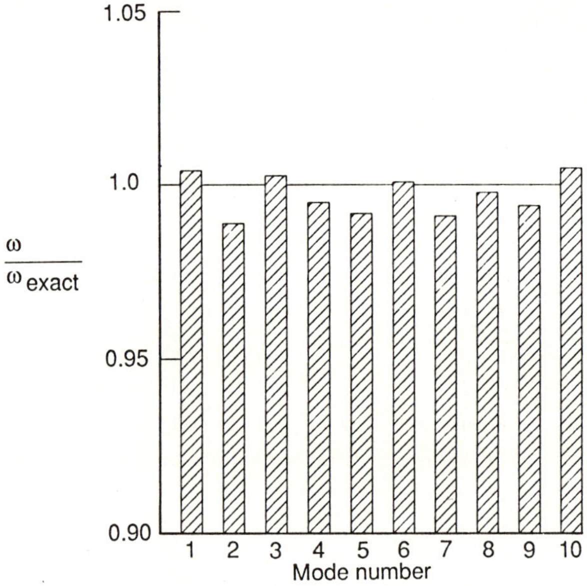

a) 8L x 8L grid

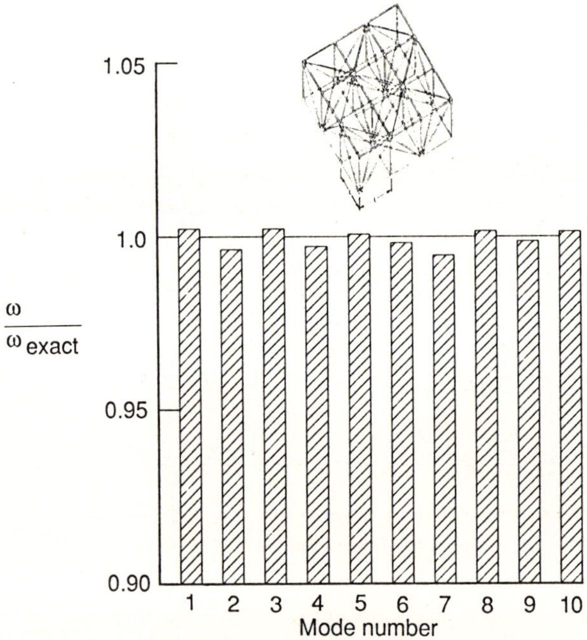

b) 16L x 16L grid

Figure 10 - Accuracy of minimum frequencies obtained by continuum plate models for the cantilevered platelike lattice truss shown in Fig. 7.

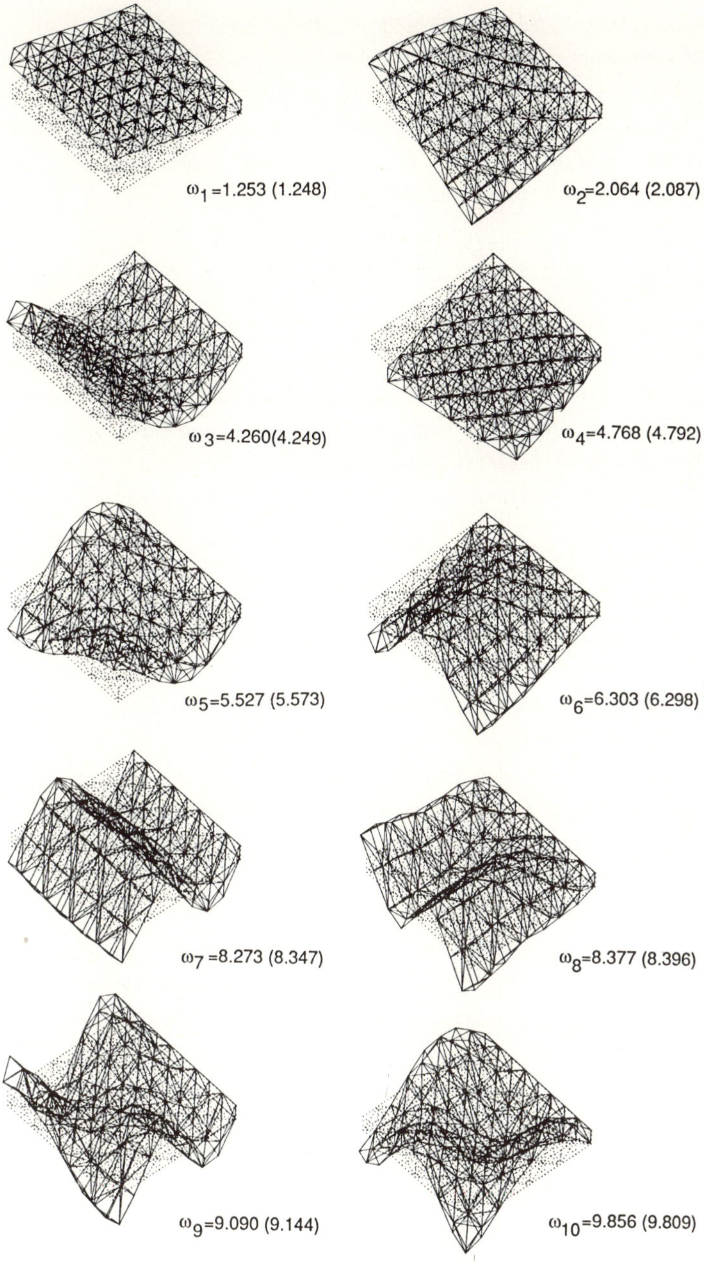

Note: Numbers between parentheses refer to exact frequencies (in Hertz) ; other numbers are continuum plate frequencies.

Figure 11 - Vibration mode shapes for the cantilevered hexahedral grid shown in Fig. 7.

10.4 Stress and Free Vibration Problems of Single-Layer Grids

Ordinary continuum plate models are developed in [21 and 36] for predicting the response of single-layered grids with rigid joints. The transverse motion of the grids are predicted accurately by these models. However, the accurate prediction of the in-plane motion of single-layer grids with rigid joints requires the use of micropolar continuum (see [10, 57 and 58]).

10.5 Use of Continuum Models for System Identification and Control Law Design of Lattice Structures

Because of the significant reduction in the number of material and structural parameters obtained by replacing the actual lattice structure by a continuum model, a number of recent studies have employed continuum modeling in the control design and/or system identification analysis. In the cited references use was made of either the governing partial differential equations of the continuum, or the systems of ordinary differential equations resulting from discretizations of the continuum model (see [9, 12, 23 and 30]).

11. FUTURE DIRECTIONS OF RESEARCH ON CONTINUUM MODELING

Continuum modeling is likely to play a significant role in the analysis, design and control of large space structures, and in order for this to happen a number of extensions and computational strategies need to be developed.

11.1 Needed Extensions and Future Applications

The needed extensions include incorporating effects of joint flexibility and damping into the continuum model.

The new applications include:

- o random periodic structures (i.e., structures with random variations in material properties, geometric characteristics, or periodicity)
- o nonlinear structural dynamics and wave propagation problems
- o Multidisciplinary optimization problems (e.g., simultaneous optimization of structures and control systems).

11.2 Computational Strategies and Modeling Techniques

As far as computational strategies and modeling techniques are concerned, the following two areas appear to have high potential for research:

a) Hybrid Modeling Approaches

These include combination of:

i) continuum/repetitive structure approach for handling lattice structures with arbitrary boundary conditions;

ii) continuum/discrete structure approach for specialized problems such as stress concentration.

b) Application of Operator Splitting/Reduction Technique for Analyzing Complicated Continuum Models

These are very effective techniques for generating the response of a complex structure (e.g., anisotropic continuum model) using large perturbations from the response of a simpler structure (e.g., corresponding orthotropic model). Application of these techniques to eigenvalue and nonlinear problems are presented in [48 and 49].

Reduction techniques have been shown to be related to the preconditioned conjugate gradient iterative method. The preconditioning matrix is taken to be the matrix of the simpler structure; and the preconditioned residuals provide sensitivity information of the response with respect to the complicating factors (e.g., anisotropy).

12. CONCLUDING REMARKS

The status and some recent developments in continuum modeling of large repetitive lattice structures are summarized. Discussion focuses on a number of aspects including definition of an effective substitute continuum, characterization of the continuum model, and the different approaches for generating the constitutive matrix and the matrix of mass densities of the continuum. Also, a simple approach is presented for generating the continuum properties. Also, a simple approach is presented for generating the continuum properties. The approach can be used in conjunction with computerized symbolic manipulation to generate analytic expressions for the continuum properties, or in a small Fortran program to determine the numerical values of these properties. Applications of the proposed approach to the generation of continuum properties for beamlike and platelike lattices is described.

The status of continuum modeling for repetitive lattice structures can be summarized in the following:

Pin-jointed beamlike and platelike lattices. Effective and verified ordinary continuum models exist for linear thermoelastic stress and free vibration problems. Bifurcation buckling loads associated with general instability (global buckling modes) can also be predicted accurately by continuum models. Applications of continuum modeling to nonlinear static and dynamic problems is limited.

Simplified continuum models (associated with known engineering theories such as Timoshenko-type beam theory and Reissner-Mindlin type plate theory) are adequate for many practical problems. Continuum models associated with higher-order theories may be required for predicting localized effects.

Lattices with rigid joints. Ordinary continuum models are available for the accurate prediction of the global response of lattices with rigid joints, provided the response is not dominated by local member deformation (e.g., transverse motion of single-layered grids). Micropolar continuum models exist for predicting the in-plane motions of lattices with rigid joints.

Needed extensions of continuum modeling include incorporation of joint flexibility and damping. Also, application of continuum modeling to multidisciplinary optimization problems needs more development. These extensions are recommended as future directions for research. New computational and modeling strategies that have high potential for use in conjunction with continuum modeling include hybrid continuum/repetitive structure; hybrid continuum/discrete structures approaches; and the application of operator splitting/reduction technique for analyzing complicated continuum models.

ACKNOWLEDGEMENT

The work of the first author is supported by a NASA Grant No. NAG1-740.

REFERENCES

1. Abrate, S. and Sun, C. T., "Dynamic Analysis of Geometrically Nonlinear Truss Structures," Computers and Structures, Vol. 17, No. 4, 1983, pp. 491-497.

2. Abrate, S. and Sun, C. T., "Continuum Modeling of Damping in Large Space Structures," Second International Conference on Recent Advances in Structural Dynamics, Southampton, England, 1984.

3. Abrate, S., "Continuum Modeling of Latticed Structures," The Shock and Vibration Digest, Vol. 17, No. 1, 1985, pp. 15-21.

4. Anderson, M. S., "Buckling of Periodic Lattice Structures," AIAA Journal, Vol. 19, No. 6, June 1981, pp. 782-788.

5. Anderson, M. S., "Vibration of Prestressed Periodic Lattice Structures," AIAA Journal, Vol. 20, April 1982, pp. 551-556.

6. Anderson, M. S. and Williams, F. W., "Natural Vibration and Buckling of General Periodic Lattice Structures," AIAA Journal, Vol. 24, No. 1, January 1986, pp. 163-169.

7. Artola, M. and Duvaut, G., "Homogenesation d'une plaque renforcee," C.R. Acad. Sc., Paris, Serie A 284, March 1977, pp. 707-710.

8. Babuska, I., "Homogenization and Its Application. Mathematical and Computational Problems," in Numerical Solution of Partial Differential Equations - III, SYNSPADE 1975, ed. by B. Hubbard, Academic Press, New York, 1976, pp. 89-116.

9. Banks, H. T. and Crowley, J. M., "Parameter Identification in Continuum Models," NASA CR-172132, May 1983.

10. Bazant, Z. P. and Christensen, M., "Analogy Between Micropolar Continuum and Grid Frameworks Under Initial Stress," International Journal of Solids and Structures, Vol. 8, 1972, pp. 327-346.

11. Bensoussan, A., Lions, J. L. and Papanicolau, G., Asymptotic Analysis for Periodic Structures, North Holland, Amsterdam, 1978.

12. Berry, D. T., Yang, T. Y., and Skelton, R. E., "Dynamics and Control of Lattice Beams Using Simplified Finite Element Models," Journal of Guidance, Vol. 8, No. 5, Sept.-Oct. 1985, pp. 612-619.

13. Berry, D. T. and Yang, T. Y., "Simplified Lattice Beam Elements for Geometrically Nonlinear Static, Dynamic and Postbuckling Analysis," AIAA Journal, Vol. 24, Aug. 1986, pp. 1346-1347.

14. Bibliography on Latticed Structures, Journal of the Structural Division, ASCE, Vol. 98, No. ST7, July 1972, pp. 1545-1566.

15. Dean, D. L. and Avent, R. R., "State-of-the-Art of Discrete Field Analysis of Space Structures," Proceedings of the Second International Conference on Space Structures, ed. by W. J. Supple, University of Surrey, Guildford, England, Sept. 1975, pp. 7-16.

16. Dean, D. L., "Discrete Field Analysis of Structural Systems," Course No. 203, International Center for Mechanical Sciences, Udine, Italy, Pergamon Press, 1976.

17. Dow, J. O., Su, Z. W., Feng, C. C. and Bodley, C. S., "Equivalent Continuum Representation of Structures Composed of Repeated Elements," AIAA Journal, Vol. 23, No. 10, Oct. 1985, pp. 1564-1569.

18. Dow, J. O. and Huyer, S. A., "An Equivalent Continuum Analysis Procedure for Space Station Lattice Structures," Proceedings of the AIAA/ASME/ASCE/AHS 28th Structures, Structural Dynamics and Materials Conference, April 6-8, 1987, Monterey, CA, Part 1, pp. 110-122.

19. Flower, W. R. and Schmidt, L. C., "Analysis of Space Truss as Equivalent Plate," ASCE Journal of the Structural Division, Vol. 97, No. ST12, Dec. 1971, pp. 2777-2789.

20. Heki, K. and Fujitani, Y., "The Stress Analysis of Grids Under the Action of Bending and Shear," in Space Structures, ed. by R. M. Davies, Blackwell Scientific Publications, Oxford and Edinburgh, 1967.

21. Heki, K., "On the Effective Rigidities of Lattice Plates," in Recent Researches of Structural Mechanics, ed. by H. Tanaka and S. Kawamata, Uno Shoten, Tokyo, 1968.

22. Heki, K., "The Effect of Shear Deformation on Double Layer Lattice Plates and Shells," Second International Conference on Space Structures, Guildford, England, Sept. 1975, pp. 189-198.

23. Juang, J. N. and Sun, C. T., "System Identification of Large Flexible Structures by Using Simple Continuum Models," Journal of Astronautical Sciences, Vol. 31, 1983, pp. 77-98.

24. Kleiber, M., "The Approximate Methods in the Theory of Elastic Lattice-Type Shells," Archives of Mechanics, Vol. 25, No. 2, 1973, pp. 195-211.

25. Kollar, L. and Hegedus, I., "Solution of Double-Layer Space Trusses of General Triangular Grid by the Equivalent Continuum Method," Acta Techn. Acad. Sci. Hung., Vol. 74, 1972, pp. 363-381.

26. Kollar, L., "Continuum Method of Analysis for Double-Layer Space Trusses with Upper and Lower Cord Planes of Different Rigidities," Acta Techn. Acad. Sci. Hung., Vol. 76, No. 1-2, 1974, pp. 53-63.

27. Kollar, L., "Analysis of Double-Layer Space Trusses with Diagonally Square Mesh by the Continuum Method," Acta Techn. Acad. Sci. Hung., Vol. 76, No. 3-4, 1974, pp. 273-292.

28. Kollar, L., "Continuum Method of Analysis for Double Layer Space Trusses of 'Hexagonal Over Triangular' Mesh," Acta Techn. Acad. Sci. Hung., Vol. 86, 1978, pp. 55-77.

29. Kollar, L. and Hegedus, I., "Analysis and Design of Space Frames by the Continuum Method," Elsevier, Amsterdam, 1985.

30. Lamberson, S. E. and Yang, T. Y., "Integrated Design of Space Structures Using Lattice Plate Finite Elements," Journal of Guidance, Control and Dynamics, Vol. 9, No. 4, 1986, pp. 478-484.

31. Latticed Structures: State-of-the-Art Report, Journal of the Structural Division, ASCE, Vol. 102, No. ST11, Nov. 1976, pp. 2197-2230.

32. Lutoborski, A., "Homogenization of Linear Elastic Shells," Journal of Elasticity, Vol. 15, 1985, pp. 69-87.

33. McDaniel, T. J. and Chang, K. J., "Dynamics of Rotationally Periodic Large Space Structures," Journal of Sound and Vibration, Vol. 68, No. 3, 1980, pp. 351-368.

34. Nayfeh, A. H. and Hefzy, M. S., "Continuum Modeling of Three-Dimensional Truss-like Space Structures," AIAA Journal, Vol. 16, No. 8, Aug. 1978, pp. 779-787.

35. Nayfeh, A. H. and Hefzy, M. S., "Continuum Modeling of the Mechanical and Thermal Behavior of Discrete Large Structures," AIAA Journal, Vol. 19, No. 6, June 1981, pp. 766-773.

36. Nemeth, M. P., "Continuum Models for Repetitive Lattice Structures with Rigid Joints," M.S. Thesis, George Washington University, Sept. 1979.

37. Noor, A. K., Anderson, M. S. and Greene, W. H., "Continuum Models for Beam- and Platelike Lattice Structures," AIAA Journal, Vol. 16, No. 12, Dec. 1978, pp. 1219-1228.

38. Noor, A. K. and Andersen, C. M., "Analysis of Beamlike Lattice Trusses," Computer Methods in Applied Mechanics and Engineering, Vol. 20, 1979, pp. 53-70.

39. Noor, A. K. and Andersen, C. M., "Computerized Symbolic Manipulation in Structural Mechanics - Progress and Potential," Computers and Structures, Vol. 10, No. 1/2, 1979, pp. 95-118.

40. Noor, A. K. and Nemeth, M. P., "Micropolar Beam Models for Lattice Grids with Rigid Joints," Computer Methods in Applied Mechanics and Engineering, Vol. 21, 1980, pp. 249-263.

41. Noor, A. K. and Nemeth, M. P., "Analysis of Spatial Beamlike Lattices with Rigid Joints," Computer Methods in Applied Mechanics and Engineering, Vol. 24, 1980, pp. 35-59.

42. Noor, A. K. and Weisstein, L. S., "Stability of Beamlike Lattice Trusses," Computer Methods in Applied Mechanics and Engineering, Vol. 25, 1981, pp. 179-193.

43. Noor, A. K., "Assessment of Current State-of-the-Art in Modeling Techniques and Analysis Methods for Large Space Structures," in Modeling, Analysis and Optimization Issues for Large Space Structures, NASA CP-2258, 1982, pp. 5-31.

44. Noor, A. K., "On Making Large Nonlinear Problems Small," Computer Methods in Applied Mechanics and Engineering, Vol. 34, 1982, pp. 955-985.

45. Noor, A. K. and Peters, J. M., "Recent Advances in Reduction Methods for Instability Analysis of Structures," Computers and Structures, Vol. 16, No. 1-4, Jan. 1982, pp. 67-80.

46. Noor, A. K. and Russell, W. C., "Anisotropic Continuum Models for Beamlike Lattice Trusses," Computer Methods in Applied Mechanics and Engineering, Vol. 57, 1986, pp. 257-277.

47. Noor, A. K. and Peters, J. M., "Nonlinear Analysis of Anisotropic Panels," AIAA Journal, Vol. 24, No. 9, Sept. 1986, pp. 1545-1553.

48. Noor, A. K. and Whitworth, S. L., "Model-Size Reduction for the Buckling and Vibration Analyses of Anisotropic Panels," Journal of the Engineering Mechanics Division, ASCE, Vol. 113, No. 2, Feb. 1987, pp. 170-185.

49. Noor, A. K. and Mikulas, M. M., "Continuum Modeling of Large Lattice Structures - Status and Projections," NASA TP (to appear).

50. Renton, J. D., "The Related Behavior of Plane Grids, Space Grids and Plates," Space Structures, ed. by R. M. Davies, Blackwell Scientific Publications, Oxford and Edinburgh, 1967, pp. 19-32.

51. Renton, J. D., "The Beamlike Behavior of Space Trusses," AIAA Journal, Vol. 22, No. 2, Feb. 1984, pp. 273-280.

52. Russell, W. C., "Continuum Modeling Theories for Repetitive Lattice Space Structures," M.S. Thesis, George Washington University, Oct. 1985.

53. Saka, T. and Heki, K., "On the Effective Rigidities of Barlike Plane Trusses," in Memoirs of the Faculty of Engineering, Osaka City University, Osaka, Japan, Vol. 22, Dec. 1981, pp. 167-173.

54. Soare, M., "Application of the Equivalent Continuum Method to the Analysis of Double-Layer Parallel Square Mesh Grids," Buletinul Stiintific al Institutului de Constructii Bucuresti, Vol. 14, 1971, pp. 251-269.

55. Soare, M., "Statica si Dinamica Retelelor Spatiale Planar Patrate Simple Prin Aplicarea Metodei Mediului Continuu Echivalent," St. Cerc. Mec. Apl., Tom. 31, NR.3, 1972, pp. 673-702.

56. Sun, C. T. and Yang, T. Y., "A Continuum Approach Towards Dynamics of Gridworks," Journal of Applied Mechanics, Vol. 40, No. 1, 1973, pp. 186-192.

57. Sun, C. T. and Yang, T. Y., "A Couple-Stress Theory for Gridwork-Reinforced Media," Journal of Elasticity, Vol. 5, No. 1, March 1975, pp. 45-58.

58. Sun, C. T., Lo, H., Cheng, N. C. and Bogdanoff, J. L., "A Simple Continuum Model for Dynamic Analysis of Complex Plane Frame Structures," Report on NSF Grant GI-41897, June 3, 1976.

59. Sun, C. T., Kim, B. J. and Bogdanoff, J. A., "On the Derivation of Equivalent Simple Models for Beam- and Platelike Structures in Dynamic Analysis," 22nd AIAA/ASME/ASCE/AHS Structures, Structural Dynamics and Materials Conference, Atlanta, GA, 1981, pp. 523-532.

60. Sun, C. T. and Juang, J. N., "Modeling Global Structural Damping in Trusses Using Simple Continuum Models," 24th AIAA/ASME/ASCE/AHS Structures, Structural Dynamics and Materials Conference, Lake Tahoe, NV, Part 1, May 1983, pp. 722-729.

61. Timoshenko, S. P. and Gere, J. M., Theory of Elastic Stability, McGraw-Hill, New York, 1961.

62. Wah, T. and Calcote, L. R., Structural Analysis by Finite Difference Calculus, Von Nostrand Reinhold Co., New York, 1970.

63. Weisstein, L. S., "Introduction and Survey on Continuum Models for Repetitive Lattice Structures," in JPL, Proceedings of the Workshop on Application of Distributed System Theory to the Control of Large Space Structures, July 1983, pp. 63-70.

64. Williams, F. W., "An Algorithm for Exact Eigenvalue Calculations for Rotationally Periodic Structures," International Journal for Numerical Methods in Engineering, Vol. 23, 1986, pp. 609-622.

65. Williams, F. W., "Exact Eigenvalue Calculations for Structures with Rotationally Periodic Substructures," International Journal for Numerical Methods in Engineering, Vol. 23, 1986, pp. 695-706.

66. Wright, D. T., "A Continuum Analysis for Double-Layer Space Frame Shells," IABSE Publ., Zurich, Vol. 26, 1966.

Nonlinearities in the Dynamics and Control of Space Structures: Some Issues for Computational Mechanics

S. N. Atluri and M. Iura
Center for the Advancement of Computational Mechanics
Mail Code 0356
Georgia Institute of Technology, Atlanta, Georgia 30332

Introduction

This article deals with nonlinearities that arise in the study of dynamics and control of highly flexible large-space-structures. Broadly speaking, these nonlinearities have various origins: (i) geometrical: due to large deformations and large rotations of these structures and their members; (ii) inertia: depending on the coordinate systems used in characterizing the overall dynamic motion as well as elastic deformations; (iii) damping: due to nonlinear hysterisis in flexible joints; viscoelastic coatings etc. , and (iv) material: due to the nonlinear behavior of the structural material. The geometrical and material nonlinearities affect the "tangent stiffness operator" of the structure; the inertia nonlinearities affect the "tangent inertia operator".

To study the nonlinear transient dynamic response and control of flexible space-structures, one may think of: (i) semi-discrete approximation methods, and (ii) space-time methods. In the former class of methods, an appropriate spatial discretization is employed through weak-formulations (finite-element and field/boundary element) in space, and thus a set of coupled nonlinear ordinary differential equations (O. D. E.) is derived. These O. D. E.'s are solved often through temporal integration techniques of the finite difference-type. The semi-discrete methods are not ideally suited for travelling-wave type propagating disturbances. The second category of methods, viz. ,the space-time methods, wherein weak formulations in both space and time are employed, are somewhat better suited for wave-propagation type problems. In this article, attention is primarily focused on semi-discrete methods, while some results recently obtained on space-time methods are deferred to a later publication.

Depending on the scale of the response that is required to be studied, a large-space-structure may either be modeled as an equivalent continuum, or as a lattice structure with the details of each member being accounted for. The spatial discretization in either case is required to be of the least-order possible so that the control algorithms may be meaningfully implemented. The reduced-order-modelling of the "tangent stiffness" operator of either a continuum model, or a lattice-model of a space-structure is treated in some detail in this chapter, for structures undergoing large dynamic deformations.

The control of dynamic motion of space-structures is currently envisaged to be through either active processes, passive processes, or some combinations there of. One of the concepts of active control that is considered in details here, and by other authors, is the use of piezo-ceramic actuators that are bonded to the truss and frame members of the space-structure in various locations. The controlling shear stress transmitted by the actuator to the truss by frame member depends on the axial force, transverse shear forces, and bending and twisting moments, in the member itself, as well as the excitation voltage applied to the piezo-actuator. This problem of

Springer Series in Computational Mechanics
S. N. Atluri, A. K. Amos (Eds.)
Large Space Structures: Dynamics and Control

mechanical coupling between the structural member, and actuators, is discussed in some detail in this work.

The problem of control of nonlinear dynamic motion is addressed in this article. The problem is posed in the form of determining the feed-back gain matrix and the attendant control force vector, such that the response as predicted by a semi-discrete system of coupled nonlinear ordinary differential equations, subject to a set of arbitrary initial conditions, is damped out in a pre-set time.

In the first part of the article, continuum models of space-structures are analysed. These include models of the space-beam type as well as the shallow shell type. In the case of space-beams, the problem of nonlinear dynamic response, when the beam undergoes large overall rigid as well as elastic motion, is discussed. The beam is assumed to undergo large rotations as well as stretches. A simple finite element algorithm to predict the response is presented. When a shallow-shell type continuum model is used, a field-boundary element approach based on nonlinear integral equations is presented as a means to create a reduced-order dynamic model of the semi-discrete type. A simple algorithm to control the response predicted by these nonlinear semi-discrete equations is discussed.

In the second part of the article, detailed models of the lattice-type space-structures are discussed. Each member of the structural lattice is assumed to be either a "truss member", or as a "frame member". The "truss member" is assumed to carry only an axial load, and has three displacement degrees of freedom at each node. The "frame member" is assumed to carry an axial force, transverse shear forces, bending moments, and a twisting moment; and is assumed to have three displacement and three rotational degrees of freedom at each node. Explicit expressions for the tangent stiffness matrices of both "truss" type and "frame" type members, which undergo arbitrarily large displacements, arbitrarily large overall rigid rotations, and moderate local (relative) rotations, are derived. In all cases, each member (truss or frame type) is modelled by a single finite element, in the entire range of large deformations. Several examples are presented to illustrate the efficiency and on-board computational feasibility of these reduced-order models for lattice structures. In each instance, remarks on needs for future research are made.

1 Dynamics of Continuum Models of Highly Flexible Space-Structures Undergoing Large Deformations

In this section we deal with strategies for reduced-order structural dynamic modeling of beam and shell-type space structures which undergo large deformations. The space-structures are assumed to be represented by equivalent elastic continuua [see Noor and Mikulas, 1987].

The continuum model for a 3-d space-curved beam, that is employed here, is one wherein the effects of stretching, bending, torsion, and transverse shear deformations are accounted for. However, the cross-sectional warping is ignored. The case of conservative force loading, which may also lead to configuration-dependent moments on the beam, is treated. The beam is assumed to undergo arbitrarily large rotations and stretches. Using the three parameters associated with a conformal rotation vector representation of finite rotations, a well-defined Hamiltonian functional is established for the dynamic problem of the beam undergoing large rotational motion. The present approach leads to a symmetric tangent stiffness matrix at all times. In the present total Lagrangean description of motion, the mass-matrix of a finite element of the beam depends linearly on the linear acceleration, but nonlinearly on the rotational parameter and the attendant angular accelerations; the stiffness matrix depends nonlinearly on the deformation; and an "apparent" damping matrix depends nonlinearly on the rotations and the attendant velocities. A Newmark time-integration scheme is used to integrate the semi-discrete equations of motion. Examples of transient dynamic response of highly flexible beam-like structures in free-flight are presented to illustrate the presented methodologies.

Earlier notable contributions to 3-d beam theories undergoing finite deformations are due to Reissner (1973, 1981) and Simo and Vu-Quoc (1986). In these references, the existence of prescribed moments has been postulated a priori. The present consistent total Lagrangean approach, leading to a symmetric tangent stiffness, even when distributed external moments (which are configuration-dependent) are present, is due to Iura and Atluri (1986, 1987).

Let Y^m be the convected orthogonal curvilinear coordinate system. The coordinates Y^α are taken in the cross-section of the beam, while the coordinate Y^3 is taken along the beam axis, as shown in Fig. 1. The unit base vectors associated wlth the coordinates Y^m are denoted by $\underline{E}_m$. The well-known Frenet-Serret formulae lead to the following relations:

$$\underline{E}_{m,3} = \underline{K} \times \mathrm{E}_m, \qquad \underline{K} = K_m \underline{E}_m \tag{1}$$

where $(\ )_{,3} = d(\ )/dL$ where L is the arc length parameter of the line of origin of the coordinate system Y^m in the reference configuration; K_α are the components of initial curvature, and K_3 is the initial twist.

The undeformed base vectors at an arbitrary material point are given by

$$\underline{A}_\alpha = \underline{E}_\alpha, \quad \underline{A}_3 = -Y^2 K_3 \underline{E}_1 + Y^1 K_3 \underline{E}_2 + g_o \underline{E}_3, \quad g_o = 1 - Y^1 K_2 + Y^2 K_1 \tag{2}$$

Let $\overset{\circ}{\underline{e}}_3$ be the unit vector tangential to the deformed beam axis. Without loss of any generality, the base vectors $\underline{e}_\alpha$ and $\underline{e}_3$ are assumed to be the maps of the base vectors $\underline{E}_\alpha$ and $\underline{E}_3$ after a purely rigid rotation, denoted by the tensor $\mathbf{R}$, alone. In general, because of the transverse shear deformation, $\underline{e}_3 \neq \overset{\circ}{\underline{e}}_3$. The relationship between the unit orthogonal vectors $\underline{e}_m$ and $\underline{E}_m$ is written as

$$\underline{e}_m = \mathbf{R} \cdot \underline{E}_m, \qquad \mathbf{R} = R_{ij} \underline{E}_i \underline{E}_j \tag{3}$$

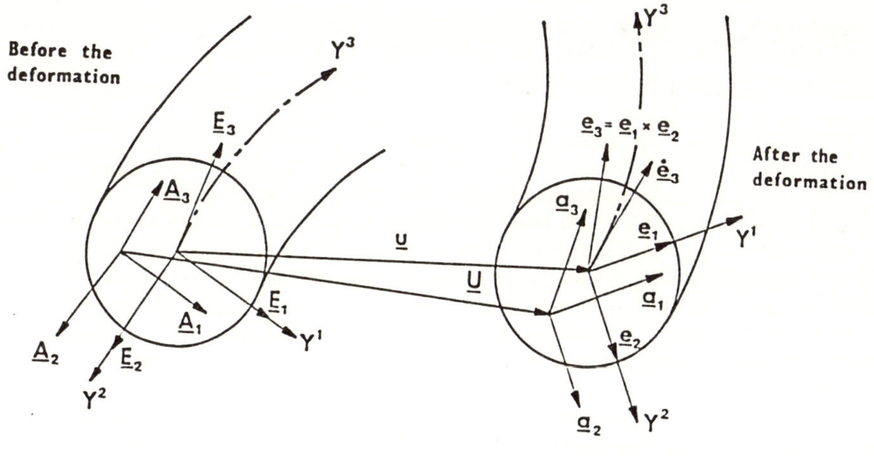

Fig. 1 Kinematic scheme for finitely stretched and rotated 3-D space-curved beam

From the definition of covariant base vectors, the vector $\overset{\circ}{\underline{e}}_3$ takes the natural form as:

$$\overset{\circ}{\underline{e}}_3 = (\delta_3^m + u^m|_3)\underline{E}_m/g, \qquad g = \sqrt{(u^1|_3)^2 + (u^2|_3)^2 + (1 + u^3|_3)^2} \tag{4}$$

where $\underline{u}(= u^m\underline{E}_m)$ is the displacement vector at the beam axis, δ_n^m the Kronecker delta, and $(\ )|_3$ the covariant differentiation by using the metric tensor $E_{ij} = \underline{E}_i \cdot \underline{E}_j$. The angles of shear deformations, denoted by β_α, are defined by $\sin\beta_\alpha = \underline{e}_\alpha \cdot \overset{\circ}{\underline{e}}_3$.

According to the hypotheses used, the displacement vector at an arbitrary material point is expressed as $\underline{U} = \underline{u} + Y^\alpha(\underline{e}_\alpha - \underline{E}_\alpha)$. The covariant base vectors at an arbitrary material point after the deformation are given by

$$\underline{a}_\alpha = \underline{e}_\alpha, \qquad \underline{a}_3 = (g\sin\beta_1 - Y^2k_3)\underline{e}_1 + (g\sin\beta_2 + Y^1k_3)\underline{e}_2 + (g\beta_3 - Y^1k_2 + k^2k_1)\underline{e}_3 \tag{5}$$

where

$$\beta_3 = \sqrt{1 - \sin^2\beta_1 - \sin^2\beta_2}, \qquad k_i = \frac{1}{2}\varepsilon_{ijk}[(\mathbf{R}\cdot\underline{E}_j)_{,3}]\cdot[\mathbf{R}\cdot\underline{E}_k] \tag{6}$$

in which ε_{ijk} is the permutation symbol. The vector $\underline{k}$, defined by $\underline{k} = k_m\underline{e}_m$, satisfies the following differential relation: $\underline{e}_{m,3} = \underline{k} \times \underline{e}_m$.

The Equations of Motion

With the help of the Green strain tensor $\boldsymbol{\varepsilon}(\varepsilon_{ij}\underline{A}_i\underline{A}_j)$ and the second Piola-Kirchhoff stress tensor $\mathbf{S}_1(S_1^{ij}\underline{A}_i\underline{A}_j)$, the internal virtual work is written as

$$IVW = \int S_1^{ij}\delta\varepsilon_{ij}dV, \qquad dV = g_o dY^1dY^2dL \tag{7}$$

The stress resultants and moments are defined, following Atluri (1984), as:

$$\underline{T} = \int g_o\underline{A}^3\cdot(\mathbf{S}_1\cdot\mathbf{F}^T)dA, \qquad \underline{M} = \int Y^\alpha\underline{e}_\alpha \times [g_o\underline{A}^3\cdot(\mathbf{S}_1\cdot\mathbf{F}^T)]dA \tag{8}$$

where $\mathbf{F}$ is the deformation gradient tensor, $(\ )^T$ a transpose and $dA = dY^1dY^2$. By using the component representation, we obtain the stress resultants and moments in the form

$$\underline{T} = T^j\underline{e}_j, \qquad \underline{M} = M^j\underline{e}_j, \qquad T^j = \int t^{3\bar{j}}g_odA,$$
$$M^1 = \int t^{3\bar{3}}Y^2g_odA\,, \quad M^2 = -\int t^{3\bar{3}}Y^1g_odA\,, \quad M^3 = \int (t^{3\bar{2}}Y^1 - t^{3\bar{1}}Y^2)g_odA \tag{9}$$

where $t^{m\bar{n}} = S_1^{mj}\underline{a}_j\cdot\underline{e}_n$. The $\overline{(\ )}$ in the contravariant tensor is used to emphasize that these are not components in convected coordinates Y^m.

As a rotational variation, we introduce, at first, a tensor $(\delta\mathbf{R}\cdot\mathbf{R}^T)$ introduced by Atluri (1984). Since $\mathbf{R}\cdot\mathbf{R}^T = \mathbf{I}$ where $\mathbf{I}$ is an identity tensor, $\delta\mathbf{R}\cdot\mathbf{R}^T$ is a skewsymmetric tensor. There exists, therefore, a vector $\delta\underline{\phi}$ satisfying $\delta\mathbf{R}\cdot\mathbf{R}^T = \delta\underline{\phi}\times\mathbf{I}$. It follows from Eqs. (7) to (9) that the IVW is rewritten, after some manipulations, as

$$IVW = -\int [\underline{T}_{,3}\cdot\delta\underline{u} + \{\underline{M}_{,3} + (\underline{x} + \underline{u})_{,3}\times\underline{T}\}\cdot\delta\underline{\phi}]dL + \int_{S_u+S_\sigma}\underline{T}\cdot\delta\underline{u} + \underline{M}\cdot\delta\underline{\phi}]_{L=o}^{L=l} \tag{10}$$

where $\underline{x}$ is the undeformed position vector of a point at the beam axis and l the length of the beam axis before the deformation; S_u and S_σ are parts of boundary on which geometrical and mechanical boundary conditions are prescribed respectively.

Let $\underline{P}_b(= P_b^j\underline{E}_j)$ be the vector of body force defined per unit volume of the undeformed beam, $\underline{P}_c(= P_c^j\underline{E}_j)$ the vector of distributed surface traction defined per unit area of the undeformed cylindrical surface of the beam, denoted as S_c; and $\underline{P}_e(=P_e^jE_j)$ the vector of distributed surface tractions at the end cross sections denoted as S_e. Then the external virtual work is written as

$$
\begin{aligned}
EVW &= \int \underline{P}_b \cdot \delta\underline{U} dV + \int \underline{P}_c \cdot \delta\underline{U} dS_c + \int_{S_\sigma}\int \underline{P}_e \cdot \delta\underline{U} dS_e]_{L=o}^{L=l} \\
&= \int [\underline{q} \cdot \delta\underline{u} + \underline{m} \cdot \delta\underline{\phi}] dL + \int_{S_\sigma} \overline{\underline{q}} \cdot \delta\underline{u} + \overline{\underline{m}} \cdot \delta\underline{\phi}]_{L=o}^{L=l}
\end{aligned}
\tag{11}
$$

where

$$
\begin{aligned}
&\underline{q} = q^j\underline{E}_j, \qquad \overline{\underline{q}} = \overline{q}^j\underline{E}_j, \qquad \underline{m} = m_{\alpha j}\underline{e}_\alpha \times \underline{E}_j, \\
&\overline{\underline{m}} = \overline{m}_{\alpha j}\underline{e}_\alpha \times \underline{E}_j, \qquad q^j = \int P_b^j g_o dA + \int p_c^j dS^*, \\
&\overline{q}^j = \textstyle\int P_e^j dS_e, \qquad m_{\alpha j} = \int Y^\alpha P_b^j g_o dA + \int Y^\alpha P_c^j dS^*, \\
&\overline{m}_{\alpha j} = \textstyle\int Y^\alpha P_e^j dS_e, \qquad dS_c = dS^* dL
\end{aligned}
\tag{12}
$$

The kinetic energy of the beam is written as

$$
T = \frac{1}{2}\int \rho \dot{\underline{U}} \cdot \dot{\underline{U}} dV \tag{13}
$$

where ρ is the density in the reference state. The principle of virtual work for the elastodynamic problem is represented as

$$
\int_{t_1}^{t_2} [\delta T - IVW + EVW] dt = 0 \tag{14}
$$

Using the conventional conditon that the variations of displacements at $t = t_1$ and $t = t_2$ are equal to zero, we obtain the LMB and the AMB conditions, expressed as:

$$
\begin{aligned}
&\underline{T}_{,3} + \underline{q} = \dot{\underline{L}}_t && (\text{ for arbitrary } \delta\underline{u}), \\
&\underline{M}_{,3} + (\underline{x} + \underline{u})_{,3} \times \underline{T} + \underline{m} = \dot{\underline{H}}_t && (\text{ for arbitrary } \delta\underline{\phi}),
\end{aligned}
\tag{15}
$$

where

$$
\begin{aligned}
&\underline{L}_t = A_\rho \dot{\underline{u}} + J_\alpha \dot{\underline{e}}_\alpha, && \underline{H}_t = J_\alpha \underline{e}_\alpha \times \dot{\underline{u}} + \mathbf{I}_\rho \cdot \underline{W} \\
&\mathbf{I}_\rho = J_{\alpha\beta}(\underline{e}_\alpha \cdot \underline{e}_\beta)\mathbf{I} - J_{\alpha\beta}\underline{e}_\alpha\underline{e}_\beta, && \underline{W} \times \mathbf{I} = \dot{\mathbf{R}} \cdot \mathbf{R}^T, \\
&A_\rho = \textstyle\int \rho g_o dA, \qquad J_\alpha = \int \rho Y^\alpha g_o dA, && J_{\alpha\beta} = \int \rho Y^\alpha Y^\beta g_o dA
\end{aligned}
\tag{16}
$$

The associated boundary conditions are written as

$$
\underline{T} = \overline{\underline{q}}, \qquad \underline{M} = \overline{\underline{m}} \quad \text{on } S_\sigma, \quad \underline{u} = \overline{\underline{u}}, \qquad \underline{\phi} = \overline{\underline{\phi}} \quad \text{on } S_u \tag{17}
$$

where $\overline{\underline{u}}$ and $\overline{\underline{\phi}}$ denote the prescribed value on S_u. It should be emphasized that the external moment vectors $\underline{m}$ and $\overline{\underline{m}}$, which are generated by the conservative forces, are configuration dependent, as shown in Eqs. (12c,d).

In the above development of equations of motion, the vector $\underline{\phi}$ is used as rotational variables. Simo and Vu-Quoc (1986) have pointed out, using the $\delta\underline{\phi}$ as a variation of rotational variables, that a well-defined functional exists at only an equilibrium configuration provided that no distributed external moments exist. Since the external moment vector $\underline{m}$, defined by Eq. (12c), is configuration dependent, the variation of inner product $(\underline{m} \cdot \underline{\phi})$ does not yield $\underline{m} \cdot \delta\underline{\phi}$. Therefore, the EVW, especially of the moments, does not, on first sight, appear to correspond to the first variation of an external energy functional.

In order to express $(\underline{m} \cdot \delta\underline{\phi})$ of Eq. (11b) as the first variation of an energy functional, we adopt a strategy wherein $(\underline{m} \cdot \delta\underline{\phi})$ can be expressed in terms of components of $\underline{m}$ and $\delta\underline{\phi}$ in the undeformed basis. It follows from Eqs. (3b) and (12c) and the definition of $\delta\underline{\phi}$ that

$$
\underline{m} \cdot \delta\underline{\phi} = m_{\alpha j}(\underline{e}_\alpha \times \underline{\mathrm{E}}_\mathrm{j}) \cdot \delta\underline{\phi} = \mathrm{m}_{\alpha\mathrm{j}}(\underline{\mathrm{E}}_\mathrm{j} \cdot \delta\mathbf{R} \cdot \underline{\mathrm{E}}_\alpha) = \delta(\mathrm{m}_{\alpha\mathrm{j}}\mathrm{R}_{\mathrm{j}\alpha}) \tag{18}
$$

Note that the Lagrangian components R_{ij} of $\mathbf{R}$ are expressed in terms of three arbitrary parameters α^j, such that ($\delta\mathbf{R} = R_{jk;i}\underline{E}_j\underline{E}_k\delta\alpha^i$), where ( ); denotes the differentiation with respect to α^i. Equation (18) indicates the possiblility of constructing a well-defined functional for the present problem with the use of α^i.

To show the equivalence between the AMB condition for $\delta\alpha^i$ and that for $\delta\underline{\phi}$ we consider, at first, the tensor equation of AMB condition for $\delta\underline{\phi}$. The inner product between the AMB condition and the variation $\delta\underline{\phi}$ is expressed as

$$\{\underline{M}_{,3} + (\underline{x} + \underline{u})_{,3} \times \underline{T} + \underline{m} - \dot{\underline{H}}_t\} \cdot \delta\underline{\phi} = \mathbf{C} : (\delta\mathbf{R} \cdot \mathbf{R}^T), \tag{19}$$

where

$$\mathbf{C} = Q^1\underline{e}_3\underline{e}_2 + Q^2\underline{e}_1\underline{e}_3 + Q^3\underline{e}_2\underline{e}_1 + m_{\alpha j}\underline{E}_j\underline{e}_\alpha - J_\alpha\underline{\ddot{u}}\underline{e}_\alpha - J_{\alpha\beta}\underline{\ddot{e}}_\alpha\underline{e}_\beta,$$
$$Q^j\underline{e}_j = \underline{M}_{,3} + (\underline{x} + \underline{u})_{,3} \times \underline{T}. \tag{20}$$

When $\mathbf{R}$ is expressed in terms of α^i, Eq. (19) is rewritten as

$$\mathbf{C} : (\delta\mathbf{R} \cdot \mathbf{R}^T) = \mathbf{C} : (\mathbf{R}_{;i} \cdot \mathbf{R}^T)\delta\alpha^i. \tag{21}$$

Since $\delta\mathbf{R} \cdot \mathbf{R}^T$ is a skewsymmetric tensor, the AMB condition for $\delta\underline{\phi}$ is represented from Eq. (19) as $\mathbf{C} = \mathbf{C}^T$. The AMB condition for $\delta\alpha^i$ is expressed from Eq. (21) as $\mathbf{C} : (\mathbf{R}_{;j} \cdot \mathbf{R}^T) = 0$. Since $\mathbf{R}_{;i} : \mathbf{R}^T$ is a skewsymmetric tensor, the AMB condition for $\delta\alpha^i$ is shown to be equivalent to that for $\delta\underline{\phi}$.

To complete a beam theory, we consider a stress-strain relationships. Equations (9) indicate that the use of the stress tensor $t^{m\bar{n}}$ yields the compact definition for the stress resultants and moments. Therefore, we use the stress tensor $t^{i\overline{m}}$ and the conjugate strain tensors $\Upsilon_{m\bar{n}}$ to construct the constitutive equation. The conjugate strain tensors $\Upsilon_{m\bar{n}}$ are defined as

$$\Upsilon_{m\bar{n}} = \underline{a}_m \cdot \underline{e}_n - \underline{A}_m \cdot \underline{E}_n. \tag{22}$$

For one-dimensional beams, we assume the following constitutive equations:

$$t^{3\bar{\alpha}} = G\Upsilon_{3\bar{\alpha}}\,, \qquad t^{3\bar{3}} = E\Upsilon_{3\bar{3}} \tag{23}$$

where G is the shearing modulus and E the Young modulus. Substituting Eqs. (23) into Eqs.(9) and using Eqs. (5) and (22) lead to

$$T^1 = GA_oh_1 - GI_1\tilde{k}_3, \qquad T^2 = GA_oh_2 + GI_2\tilde{k}_3,$$
$$T^3 = EAh_3 + EI\tilde{k}_1 - EI_2\tilde{k}_2, \qquad M^1 = EI_{11}\tilde{k}_1 + EI_1h_3 - EI_{12}\tilde{k}_2,$$
$$M^2 = EI_{22}\tilde{k}_2 - EI_2h_3 - EI_{12}\tilde{k}_1, \qquad M^3 = GJ\tilde{k}_3 - GI_1h_1 + GI_2h_2 \tag{24}$$

where

$$\tilde{k}_i = k_i - K_i, \qquad h_\alpha = g\sin\beta_\alpha, \qquad h_3 = g\beta_3 - 1, \quad A = \int g_o dA,$$
$$A_o = k_oA, \quad I_\alpha = \int Y \overset{\alpha}{g}_o\, dA, \quad I_{12} = \int Y^1Y^2g_odA,$$
$$I_{11} = \int (Y^2)^2g_odA, \quad I_{22} = \int (Y^1)^2g_odA, \quad J = \int [(Y^1)^2 + (Y^2)^2]g_odA \tag{25}$$

The factor k_o is the shear-correction factor [6].

The strain energy function W_s per unit length is expressed as

$$W_s = \frac{1}{2}GA_o(h_1)^2 + \frac{1}{2}GA_o(h_2)^2 + \frac{1}{2}EA(h_3)^2 + \frac{1}{2}EI_{11}(\tilde{k}_1)^2 + \frac{1}{2}EI_{22}(\tilde{k}_2)^2$$
$$+ \frac{1}{2}GJ(\tilde{k}_3)^2 + EI_1h_3\tilde{k}_1 - EI_2h_3\tilde{k}_2 - EI_{12}\tilde{k}_1\tilde{k}_2 - GI_1h_1\tilde{k}_3 + GI_2h_2\tilde{k}_3\,. \tag{26}$$

It should be noted that the well-known and commonly used expression, given by Eq. (26), for strain energy function is derived from the strain-stress relationships given by Eqs. (23).

As a basis of a numerical method, a variational principle often plays an important role. For many reasons, especially its flexibility in application, the mixed variational formulations are receiving a wide attention [Atluri, Gallagher, and Zienkiewicz (1983)]. On the basis of the governing equations given above, we will derive the functional for general mixed variational principle for elastostatic beams.

As first shown by Fraeijs de Veubeke (1972), and later generalized by Atluri and Murakawa (1977), a general mixed principle, for a 3-dimensional elastic material, and involving the first Piola-Kirchhoff stress tensor $\mathbf{t}_1$, the right stretch tensor $\mathbf{U}$, the finite rotation tensor $\mathbf{R}$ and the displacement vector $\underline{v}$ as variables, can be stated as the stationary condition of the functional F_1.

$$F_1(\mathbf{t}_1,\mathbf{U},\mathbf{R},\underline{v}) = \int_{V_o}[W_o(\mathbf{U})+\mathbf{t}_1^T : \{(\mathbf{I}+\underline{\nabla}_o\underline{v})^T-\mathbf{R}\cdot\mathbf{U}\}-\rho_o\underline{\bar{b}}\cdot\underline{v}]dV_o-\int_{S_\sigma}\underline{\bar{t}}\cdot\underline{v}\,ds-\int_{S_u}\underline{t}\cdot(\underline{v}-\underline{\bar{v}})ds \quad (27)$$

where W_o is a strain energy function, ρ_o the mass density in the undeformed state, $\underline{b}$ the body force vector per unit mass, $\underline{\bar{t}}$ the traction on the boundary per unit undeformed area and $\underline{\nabla}_o$ the gradient operator in the undeformed state.

The functional F_1 for a finitely deformed shell has been derived by Atluri (1984). Based on the resulting modified functional, some numerical results have been obtained by Punch and Atluri (1985). However, to the best of the authors' knowledge, no studies exist on the functional F_1 for a finitely deformed beam.

Employing α^i as rotational variables and after some manipulations, the functional for a finitely deformed beam is expressed as

$$\begin{aligned} G_1(\underline{T},\underline{M},\underline{h},\underline{\tilde{k}},\underline{u},\alpha^i,L_j^\dagger) = [W_s(\underline{h},\underline{\tilde{k}}) + \underline{T}\cdot\{(\underline{x}+\underline{u})_{,3} - \mathbf{R}\cdot(\underline{h}+\underline{E}_3)\} + \underline{M}\cdot\{\underline{l}_3 - \mathbf{R}\cdot\underline{\tilde{k}}\} \\ -\underline{q}\cdot\underline{u} - m_{\alpha j}R_{j\alpha}]dL - \lfloor_u \underline{T}\cdot(\underline{u}-\underline{\bar{u}}) + L_j^\dagger(\alpha^j - \alpha^{-j})]|_{L=o}^{L=1} - \lfloor_\sigma \underline{\bar{q}}\cdot\underline{u} + \bar{m}_{\alpha j}R_{j\alpha}|_{L=o}^{L=1} \end{aligned} \quad (28)$$

where $\underline{l}_3$ is the vector satisfying $\mathbf{R}_{,3}\cdot\mathbf{R}^T = \underline{l}_3 \times \mathbf{I}$, $L_j^\dagger$ a Lagrangean multiplier. The stationary condition, $\delta G_1 = 0$, yields the constitutive equations, the compatibility equations, the LMB and the AMB conditions, the mechanical and geometrical boundary conditions, and the physical meaning of the Lagrangean multiplier. For incremental functionals, see Iura and Atluri (1986).

When a potential energy π_p is obtained, Hamilton's principle for elastodynamic problems is expressed as

$$\delta\int_{t_3}^{t_2}[T-\pi_p]dt = 0 \quad (29)$$

where the subsidiary conditions are the geometrical boundary conditions and the conventional conditions that the variations of displacements at $t = t_1$ and $t = t_2$ are equal to zero. For the present problem, we obtain the π_p from Eq. (28) neglecting the compatibility conditions and the geometrical boundary conditions.

In the paper, to avoid using the four rotational parameters, the rotational variables α^i are defined as the Lagrangean components of the conformal rotation vector $\underline{\theta}^*$ in the following way:

$$\underline{\theta}^* = 4\tan\frac{\omega}{4}\underline{e} = \alpha^i\underline{E}_i \quad (30)$$

where $\underline{e}$ is a unit vector satisfying $\mathbf{R}\cdot\underline{e} = \underline{e}$ and ω a magnitude of rotation about the axis of rotation defined by $\underline{e}$. Because of singularity, the Rodrigues vector, defined by $\underline{\theta} = 2\tan\omega/2\ \underline{e}$, is valid only in the range of $|\omega| < \pi$. As shown in Eq. (30), however, the conformal rotation vector is valid even at $|\omega| = \pi$. Therefore, with the simple manipulation, the finite rotations are described in terms of only three rotation parameters.

The finite element formulation is used to derive the semidiscrete equations of motion. The displacement and the rotational components are interpolated by:

$$u^i = u^i_\beta N^\beta \quad , \qquad \alpha^i = \alpha^i_\beta N^\beta \tag{31}$$

where u^i_β and α^i_β denote the nodal displacement and rotational components, respectively, and N^β the shape functions defined by $N^1 = 1 - L/l_e$ and $N^2 = L/l_e$ where l_e is the element length. For later convenience, the following notations are introduced: $\underset{\sim}{d} = \{u^i_\beta\}$ and $\underset{\sim}{r} = \{\alpha^i_\beta\}$.

Following a standard finite element discretization, we obtain the following semidiscrete equations of motion:

$$\underset{\sim}{M}(\ddot{\underset{\sim}{d}}, \ddot{\underset{\sim}{r}}, \underset{\sim}{r}) + \underset{\sim}{C}(\dot{\underset{\sim}{r}}, \underset{\sim}{r}) + \underset{\sim}{K}(\underset{\sim}{d}, \underset{\sim}{r}) = \underset{\sim}{f}(\underset{\sim}{r}) \tag{32}$$

where $\underset{\sim}{M}$ depends linearly on $\ddot{\underset{\sim}{d}}$ but nonlinearly on $\ddot{\underset{\sim}{r}}$ and $\underset{\sim}{r}$, the $\underset{\sim}{C}$, $\underset{\sim}{K}$ and $\underset{\sim}{f}$ depend nonlinearly on their variables. Note that the vector $\underset{\sim}{C}$ is derived not from the damping effects but from the nonlinear effects of finite rotations, and that no simplification is made in this formulation in the sense that Coriolis and centrifugal effects as well as the inertia effects due to rotation are accounted for. The resulting semidiscrete equations are integrated by the Newmark algorithm. Consistent linearization procedures are employed to obtain linearized forms of the balance equations. A full Newton-Raphson method is used in the present calculations.

To illustrate the validity of the present theory in simulating large rotations, we analyze the case of a highly flexible right-angle beam in free flight, as shown in Fig. 2. Although we have used only three rotational parameters per node, the large deformations with finite rotations can be simulated without singularities. Other numerical examples and more detailed development of transient dynamic analysis of highly-flexible space-curved beams undergoing finite rotations and stretches may be found in Iura and Atluri (1987).

We now turn to the analysis of shell-type continuum models of space-structures such as antennae. Here the problem is one of creating a reliable reduced-order structural model. In the traditional finite element method, for thin shells; (i) either a fourth-order theory with C^1 continuous trial and test functions; or (ii) a continuum based shell theory with C^o continuous trial and test functions along with selective reduced integration schemes to alleviate shear/membrane locking, are used. In the present approach, a field-boundary-element method, based on unsymmetric variational statements and Petrov-Galerkin techniques, is developed for shallow shells undergoing large deformation (moderate rotation) dynamics. It is seen that the number of degrees of freedom in the current approach is reduced from that in the popular Galerkin finite-element-approach.

Consider a shallow shell of an isotropic elastic material with the mid-surface being designated by $z = z(x_\alpha)$. When finite deformations (only moderate rotations) are considered, the momentum balance laws may be written as [Zhang and Atluri, 1987]:

$$\text{(inplane)}: \qquad N_{\alpha\beta,\beta} + b_\alpha - \rho\ddot{u}_\alpha = 0 \qquad (\alpha = 1,2) \tag{33}$$

$$\text{(transverse)}: \quad D\nabla^4 w + \tfrac{N_{\alpha\beta}}{R_{\alpha\beta}} - (b_3 - \rho\ddot{w}) = f_3 + (N_{\alpha\beta} w_{,\beta})_{,\alpha} \tag{34}$$

where: $N_{\alpha\beta}$ are membrane forces; $(\ )_{,\beta} = \partial(\ )/\partial x_\beta$; w is the transverse displacement of the midsurface of the shell; $b_i (i = 1,2,3)$ are body forces; f_3 is the load normal to the shell midsurface; D is the bending rigidity; ∇^4 is the bi-harmonic operator in x_α; ρ is the mass-density; $(\ddot{\ }) = \partial(\ )/\partial t^2$; and

$$R_{\alpha\beta} = 1/(z_{,\alpha\beta}) \tag{35}$$

The non-linear stress-displacement relations are:

$$N_{11} = C(\varepsilon_{11} + \varepsilon_{22}), \quad N_{22} = C(\varepsilon_{22} + \nu\varepsilon_{11}), \quad N_{12} = C(1-\nu)\varepsilon_{12} \tag{36}$$

where C is the inplane rigidity, ν the Poisson ratio and

$$\varepsilon_{\alpha\beta} = (\frac{1}{2})[u_{\alpha,\beta} + u_{\beta,\alpha} + \frac{2w}{R_{\alpha\beta}} + w_{,\alpha} w_{,\beta}] \tag{37}$$

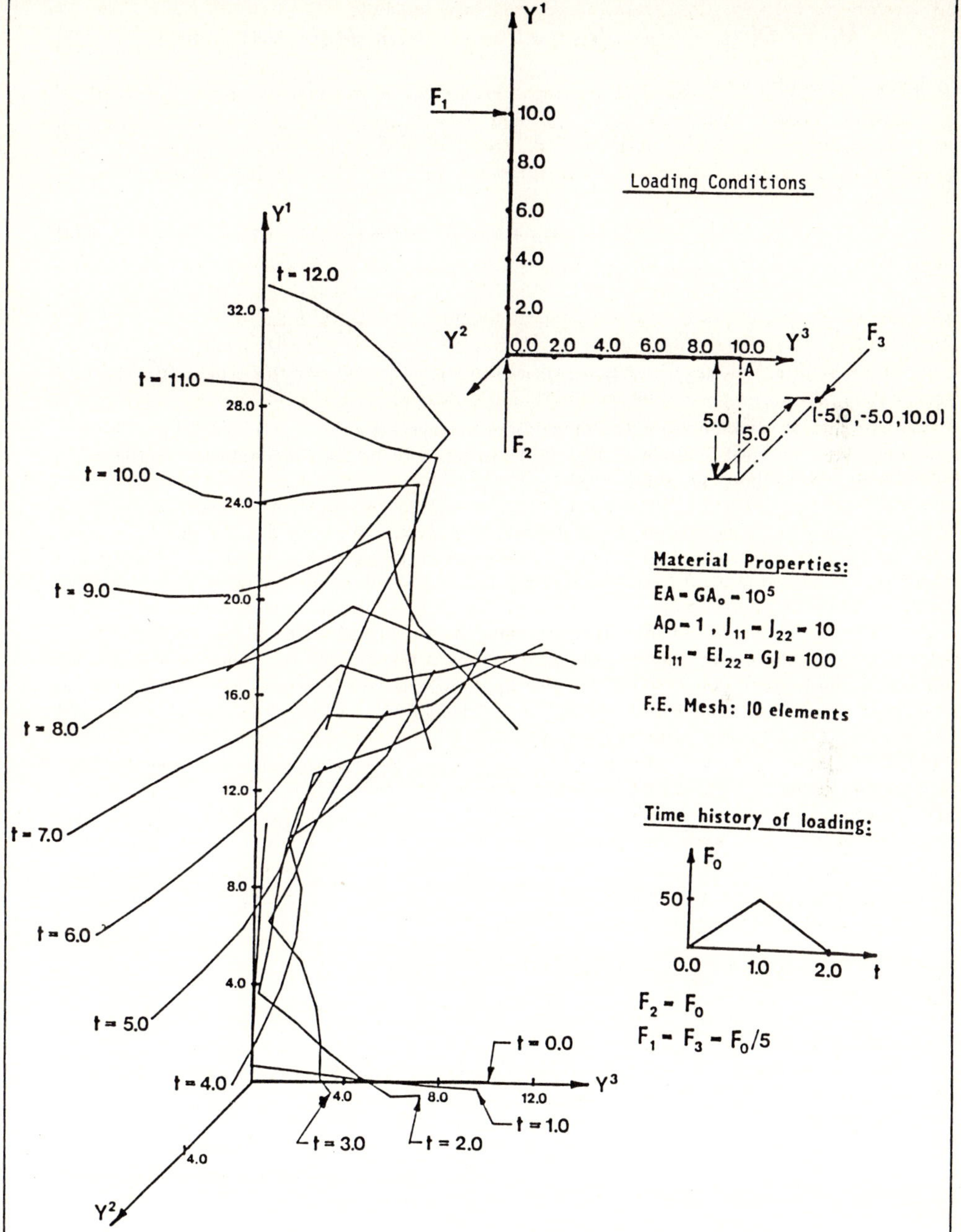

Fig. 2 Flexible right angle beam in free flight. Time step $\Delta t = 0.1$.

and

$$M_{11} = -D(w_{,11} + \nu w_{,22}); \quad M_{22} = -D(w_{,22} + \nu w_{,11}); \quad M_{12} = -D(1-\nu)w_{,12} \tag{38}$$

Using (37) in (36) we write:

$$N_{\alpha\beta} = N'_{\alpha\beta} + C\kappa_{\alpha\beta}w + N^n_{\alpha\beta} \tag{39}$$

where

$$N'_{11} = C(u_{1,1} + \nu u_{2,2}); \quad N'_{22} = C(u_{2,2} + \nu u_{1,1}); \quad N'_{12} = (\frac{1}{2})C(1-\nu)(u_{1,2} + u_{2,1}) \tag{40}$$

$$\kappa_{11} = \frac{1}{R_{11}} + \frac{\nu}{R_{22}}; \qquad \kappa_{22} = \frac{1}{R_{22}} + \frac{\nu}{R_{11}}; \qquad \kappa_{12} = \frac{1-\nu}{R_{12}};$$
$$N^n_{11} = (\frac{c}{2})[(w_{,1})^2 + (w_{,2})^2]; \quad N^n_{22} = (\frac{c}{2})[(w_{,2})^2 + \nu(w_{,1})^2]; \quad N^n_{12} = (\frac{c}{2})(1-\nu)(w_{,1})(w_{,2}) \tag{41}$$

We now define weak-forms of the momentum-balance laws, Eq. (33) and (34). The test functions u^*_β in the weak-solution of Eq. (33) are taken to be the "fundamental solutions" in infinite space of the linear equation:

$$[N'_{\alpha\beta}(u^*_\sigma)]_{,\beta} + \delta(x_u - \xi_u)\delta_{\alpha\theta}e_\theta = 0 \tag{42}$$

where $\delta(x_\mu - \xi_\mu)$ is a Dirac delta function at $x_\mu = \xi_\mu$; $\delta_{\alpha\theta}$ is the Kronecker delta, and e_θ denotes the direction of the applied point load (in the x_θ direction). This "fundamental solution" will be denoted as $u^*_{\theta(\sigma)}(x_\mu, \xi_\mu)$ ie. , the displacement along the x_σ direction, in a plane infinite body, at the location x_μ, due to a unit load in the x_σ direction at the location ξ_μ. Likewise, $P^*_{(\theta)\alpha}(x_\mu, \xi_\mu)$ will denote the traction along the x_α direction on an oriented surface at x_μ (with a unit normal η_σ) due to a unit load along x_θ at the location ξ_μ. Thus,

$$u^*_{(\theta)\alpha}(x_\mu, \xi_\mu) = \frac{1}{8\pi G}[(\nu - 3)(\ln R)\delta_{\theta\alpha} + (1+\nu)R_{,\theta}R_{,\alpha}]$$
$$P^*_{(\theta)\alpha}(x_\mu, \xi_\mu) = \frac{-h}{4\pi R}\{\frac{\partial R}{\partial n_\sigma}[(1-\nu)\delta_{\theta\alpha} + 2(1+\nu)R_{,\theta}R_{,\alpha}] - (1-\nu)(n_\theta R_{,\alpha} - n_\alpha R_{,\theta})\} \tag{43}$$

where $R = |\xi_\mu - x_\mu|$ and $G = E/2(1+\nu)$. With the test functions being chosen as in Eq. (42), the weak-form of Eq. (33) may be shown [See Zhong and Atluri] to yield:

$$\begin{aligned}\gamma u_\theta(\xi_\mu) = & \int_\Omega [b_\alpha(x_\mu) - \rho\ddot{u}_\alpha(x_u)]u^*_{(\theta)\alpha}(\xi_\mu, x_\mu)d\Omega \\ & + \int_\Gamma [P_\alpha(x_\mu)u^*_{(\theta)\alpha}(x_\mu, \xi_\mu) - u_\alpha(x_\mu)p^*_{(\theta)\alpha}(x_\mu, \xi_\mu)]d\Gamma \\ & - \int_\Omega [c\kappa_{\alpha\beta}w(x_\mu)u^*_{(\theta)\alpha,\beta}(x_\mu, \xi_\mu) + N^n_{\alpha\beta}(x_\mu)u^*_{(\theta)\alpha,\beta}(x_\mu, \xi_\mu)]d\Omega\end{aligned} \tag{44}$$

It can be shown that γ in Eq. (44) is unity when ξ_μ is in the interior of Ω, and is equal to (1/2) when ξ_μ falls on the "smooth" boundary Γ. In Eq. (44), P_α are inplane tractions at Γ.

The test function w^* in the weak-form of Eq. (34) is chosen to be the "fundamental solution", in infinite space, of the equation:

$$D\nabla^4 w^* = \delta(x_\mu - \xi_\mu) \tag{45}$$

i. e.

$$w^*(x_\mu, \xi_\mu) = \frac{1}{8\pi}R^2 \ln R; \quad R = |\xi_\mu - x_\mu| \tag{46}$$

Using w^* as a test function in the weak-form of Eq. (34), one can show [Zhang and Atluri, 1987]:

$$\begin{aligned}\gamma_w D w(\xi_\mu) &= \int_\Gamma [V_n(x_\mu) w^*(x_\mu,\xi_\mu) - M_n(x_\mu)\Psi_n^*(x_\mu,\xi_\mu) + \Psi_\eta(x_\mu) M_n^*(x_\mu,\xi_\mu) \\ &\quad - w(x_\mu) V_n^*(x_\mu,\xi_\mu)] d\Omega - \int_\Gamma C\kappa_{\alpha\beta}\eta_\beta u_\alpha(x_\mu) w^*(x_\mu,\xi_\mu) d\Gamma \\ &\quad + \int_\Omega \{C[\kappa_{\alpha\beta} w^*(x_\mu,\xi_\mu)]_{,\beta} u_\alpha(x_\mu) - C\frac{\kappa_{\alpha\beta}}{R_{\alpha\beta}} w(x_\mu) w^*(x_\mu,\xi_\mu) \\ &\quad - \frac{N^n_{\alpha\beta}}{R_{\alpha\beta}}(x_\mu) w^*(x_\mu,\xi_\mu) + [b_3 - \rho\ddot{w} + f_3 + (N_{\alpha\beta} w_{,\beta})_{,\alpha}](x_\mu) w^*(x_\mu,\xi_\mu)\} d\Omega \\ &\quad + \sum_{i=1}^{K} [< M_t > w^* - < M_t^* > w]_{(i)} \end{aligned} \tag{47}$$

where V_n, M_n, and Ψ_n are trial solutions (part of which is specified) for shear, moment, and rotation respectively at Γ; V_n^*, M_n^* and Ψ_n^* are the corresponding shear, moment, and rotation due to the test solution w^*; $<(\ )>$ denotes a jump in the quantity $(\ )$ at a corner on Γ in the direction of the increasing arc length as Γ.

Since $\frac{\partial w}{\partial n}$ is an independent variable at Γ, an independent integral relation for $\frac{\partial w}{\partial n}$ is derived. Towards this purpose, consider a second fundamental solution,

$$w_\nu^* = (1/2\pi) R(\ln R)\cos\phi \tag{48}$$

The integral relation for $\frac{\partial w}{\partial n}$ may be derived [see Zhang and Atluri, 1987] as:

$$\begin{aligned}\gamma_\nu D\frac{\partial w}{\partial n} &= \int_\Gamma [V_\mu(x_\mu) w_\nu^*(x_\mu,\xi_\mu) - M_n(x_\mu)\Psi_\nu(x_\mu,\xi_\mu) - < w(x_\mu) - w(\xi_\mu) > V_\nu^*(x_\mu,\xi_\mu) \\ &\quad - C\kappa_{\alpha\beta}\eta_\beta u_\alpha w_\nu^*(x_\mu,\xi_\mu)] d\Gamma - \int_\Omega \{C(\kappa_{\alpha\beta} w_\nu^*(x_\mu,\xi_\mu))_{,\beta} u_\alpha(x_\mu) - C\frac{\kappa_{\alpha\beta}}{R_{\alpha\beta}} w(x_\mu) w_\nu^*(x_\mu,\xi_\mu) \\ &\quad - \frac{N^n_{\alpha\beta}}{R_{\alpha\beta}}(x_\mu) w_\nu^*(x_\mu,\xi_\mu) + [b_3 - \rho\ddot{w} + f_3 + (N_{\alpha\beta} w_{,\beta})_{,\alpha}](x_\mu) w_\nu^*(x_\mu,\xi_\mu)\} d\Omega \\ &\quad + \sum_{i=1}^{K} [< M_t > w_\nu^* - < M_{\nu t}^* > w]_{(i)} \end{aligned} \tag{49}$$

An examination of Eqs. (44), (47), and (49) reveals that:

(1) For given body forces b_α, the integral relations for u_θ (Eq. 44) involve the trial functions u_θ only at the boundary Γ. On the other hand, due to the curvature induced coupling of the trial functions u and w, the integral relation for u_θ contains a domain-integral (over Ω), involving the trial solution w. If the inplane inertia ($\ddot{u}_\alpha$) is included, the integral equation for u_θ involves the integral of $\ddot{u}_\alpha$ over Ω as well.

(2) Due to the curvature induced coupling of u_θ and w, the integral relations for w and $\frac{\partial w}{\partial n}$ contain the domain-integrals involving u_α and w.

(3) In the nonlinear problem, the nonlinear terms $N^n_{\alpha\beta}$ and $(N_{\alpha\beta} w_{,\beta})_{,\alpha}$, involving both the trial solutions u_α and w, enter the domain integrals.

(4) For reasons (1) to (3) above, unlike the case of classical homogeneous isotropic elasto-statics [Atluri and Grannell (1978)] wherein a descritization of the relevant integral equations requires the use of basis functions for displacements at the boundary alone, the present nonlinear shallow-shell formulation requires the assumption of basis functions for trial solutions u_α and w, at the boundary Γ as well as in the interior Ω.

(5) In order to descritize the integral equations (44), (47) and (49), one needs to assume only very simple trial functions u_α, w, and $\frac{\partial w}{\partial n}$. For instance, the domain Ω may be descritized into a number of finite elements, and the boundary Γ into a number of boundary elements. It is

seen that w and u_α need only be piecewise differentiable and need not even be C^o continuous at element boundaries in Ω.

(6) At each point on the boundary, two of the in-plane variables M_α ($\alpha = 1,2$) and P_α ($\alpha = 1,2$) are specified and the other two are unknown. Likewise, two of the out-of-plane variables V_n, M_n, Ψ_n, and w are specified, and the other two are unknown. At each point in the interior, the three displacements, u_α and w, are unknown. Thus, from Eqs. (44), (47), and (49), one obtains exactly as many equations as the number of unknowns, so that the problem is well-posed.

The above approach based on integral equations has been used in the control of linear or nonlinear dynamic response of continuum plate models of large-space-structures in O'Donoghue and Atluri (1986, 1987), and Atluri, Zhang, and O'Donoghue (1986). It has also been applied in creating reduced-order-dynamic-models for shallow shells undergoing small as well as large deformations in Zhang and Atluri (1986a, 1986b, and 1987).

To illustrate the advantangeous factors of the present field/ boundary element method, we present here the results of the analysis of linear vibration and transient response of a simply supported shallow spherical shell. In these computations, the basis for the trial functions are assumed as follows: (i) over each boundary element at Γ, the variables u_i, w, $w_{,n}$, P_i, M_n, and V_n are interpolated linearly, (ii) over each domain element, u_i and w are interpolated bilinearly. Only a quadrant of the shell is modeled, to consider the doubly-symmetric modes of vibration. The domain is descritized into interior elements such that the nodes are equidistant from each other in the radial as well as angular coordinates. The first mesh consists of nodes at $\Delta R = R$; $\Delta\theta = \pi/2$; the second one with $\Delta R = R/2$ and $\Delta\theta = \pi/4$; the third with $\Delta R = R/3$ and $\Delta\theta = \pi/6$; and the fourth with $\Delta R = R/4$ and $\Delta\theta = \pi/8$; with this last mesh being illustrated in Fig. (3). Note

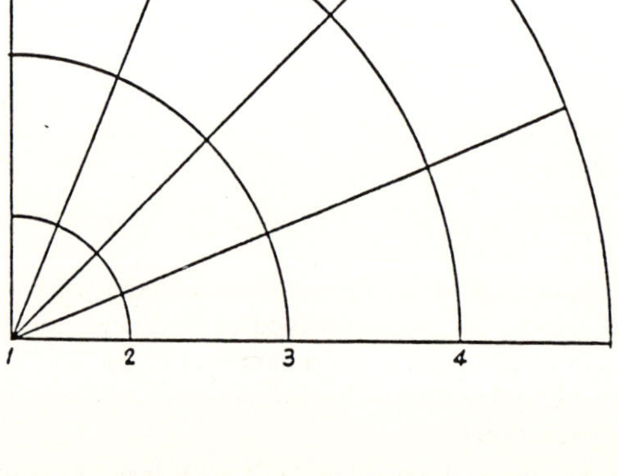

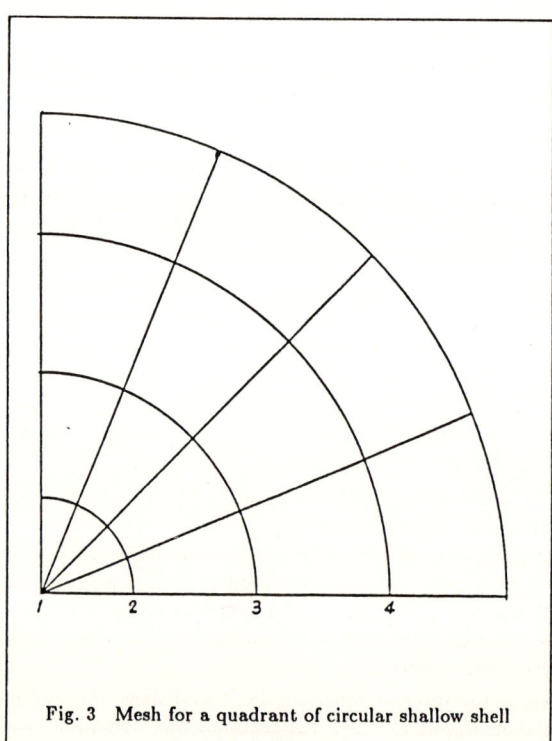

Fig. 3 Mesh for a quadrant of circular shallow shell

that in the present field/boundary element approach, the total number of degrees of freedom are: 11 (for mesh 1); 24 (mesh 2); 53 (mesh 3); 81 (mesh 4). The computed eigenvalues for various values of (r_o/R) [where r_o is the base-radius of the shallow shell, and R is the principal radius of curvature of the sphere] and $E = 1000$, $h = 1$, and $\nu = 0.25$ are shown in Table 1.

Table 1

r_o/R		.05	.083	.111	.143	.2	.333	.5
Reissner[8] Mode 1		3.1988	3.2344	3.2774	3.3409	3.4926	4.0181	4.9839
Mode 1	Mesh 2	3.2394	3.2825	3.3395	3.4229	3.6180	4.2546	5.2856
	Mesh 3	3.2009	3.2466	3.3016	3.3823	3.5726	4.2023	5.2338
Mode 2	Mesh 2	7.0871	7.1094	7.1364	7.1764	7.2729	7.6126	8.2359
	Mesh 3	6.8866	6.9095	6.9373	6.9785	7.0776	7.4364	8.0641
Mode 3	Mesh 2	7.087	7.1094	7.1364	7.1764	7.2729	7.6126	8.2359
	Mesh 3	6.8866	6.9095	6.9373	6.9785	7.0776	7.4264	8.0641
Mode 4	Mesh 2	12.1855	12.1985	12.2142	12.2377	12.2945	12.4984	12.8875
	Mesh 3	11.5442	11.5579	11.5745	11.5993	11.6592	11.8741	12.2829
Mode 5	Mesh 2	12.1855	12.1985	12.2142	12.2377	12.2945	12.4984	12.8875
	Mesh 3	11.5442	11.5579	11.5745	11.5993	11.6592	11.8741	12.2829
Mode 6	Mesh 2	14.0712	14.0825	14.0961	14.1165	14.1659	14.3436	14.6845
	Mesh 3	13.5024	13.5144	13.5289	13.5505	13.6029	13.7913	14.1514

Table 1 indicates that even mesh 2 with only 24 degrees of freedom gives the eigenvalues for the first 6 (doubly symmetric) modes with acceptable accuracy.

The above results indicate that the present reduced-order structural modeling strategy for shallow-shell type antenna structure is of considerable benefit in implementing algorithms for control, wherein the primary limiting factor, in a computational sense, is the size of the system of equations in the Riccati equation governing the feed-back gain matrix.

Using the mesh with 24 degrees of freedom, results for undamped nonlinear, axisymmetric, transient response are obtained for two initial conditions: (i) at $t = 0$, the crown of the shallow spherical shell has an outward normal velocity of $v_o = 15.0$, which represents an outward point impulse applied at the crown; (ii) at $t = 0$, $v_o = 3\cos(r\pi/10)$ where r is the distance from the center of the base plate. The material density is taken as $\rho = 0.8$, and the time step is taken as $\Delta t = 0.002$ in the time integration of the semi-discrete equations. The results for the former case are shown in Fig. (4), while those for the latter are shown in Fig. (5).

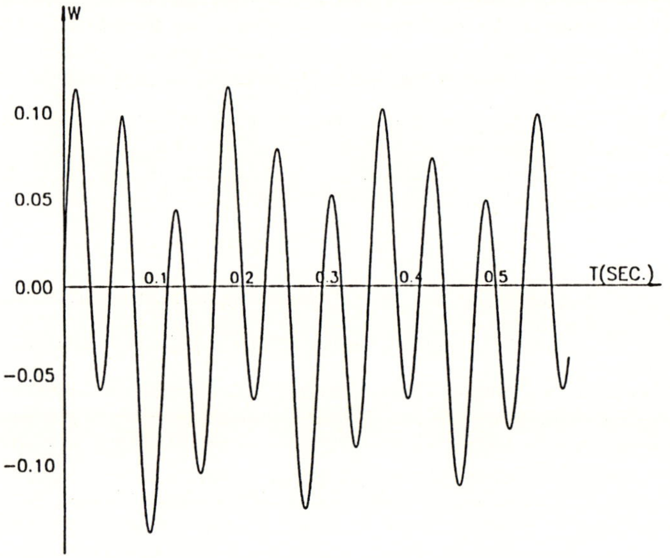

Fig. 4 Nonlinear transient response at crown of a shallow spherical shell due to a central impulse

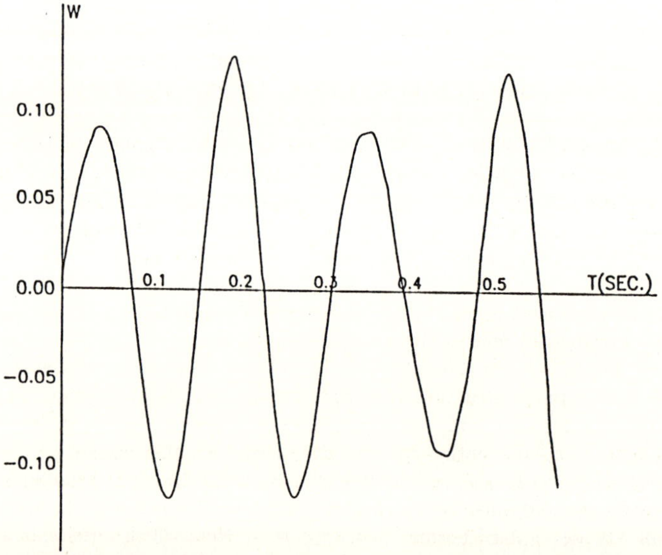

Fig. 5 Nonlinear transient response at crown of a shallow spherical shell due to initial velocity distribution $v_o = 3. * \cos(r\pi/10.)$

The system of matrix equations describing the nonlinear transient response of a shallow shell, derived from the present field/boundary element approach, are of the form [Zhang (1987). Zhang and Atluri (1987)]:

$$\mathbf{M}\Delta\ddot{\mathbf{W}} + \mathbf{C}\Delta\dot{\mathbf{W}} + \mathbf{K}\Delta\mathbf{W} = \Delta\mathbf{F} + \mathbf{b}\Delta\mathbf{f}_c \tag{50}$$

where **M**, **C**, and **K** are tangent inertia, damping, and stiffness matrices which, in general, depend on the prior state of deformation.

The control forces $\mathbf{f}_c$ must be designed so as to damp out the response of the system for any disturbance of the type:

$$\mathbf{W}(t=0) = \mathbf{W}_o; \qquad \text{and } \dot{\mathbf{W}}(t=0) = \dot{\mathbf{W}}_o \tag{51}$$

Towards a simple solution for the nonlinear control problem, we consider first the problem of control of the linear system:

$$\mathbf{M}\ddot{\mathbf{W}} + \mathbf{C}\dot{\mathbf{W}} + \mathbf{K}^L\mathbf{W}_s = \mathbf{b}\mathbf{f}_c \tag{52}$$

where **M**, **C**, and $\mathbf{K}^L$ are the mass, damping, and stiffness matrices, respectively, of the linear system, and are thus assumed to be independent of deformation. Eqn. (52) may be recast in the state variable form, as:

$$\dot{\mathbf{S}} = \mathbf{A}\mathbf{S} + \mathbf{B}\mathbf{f}_c; \qquad \mathbf{S}(0) = \mathbf{S}_o \tag{53}$$

where

$$\mathbf{S}^b = \lfloor \mathbf{W}, \dot{\mathbf{W}} \rfloor \tag{54}$$

$$\mathbf{A} = \begin{bmatrix} 0 & \mathbf{I} \\ -\mathbf{M}^{-1}\mathbf{K}^L & -\mathbf{M}^{-1}\mathbf{C} \end{bmatrix} \tag{55}$$

$$\mathbf{B} = \begin{bmatrix} 0 \\ -\mathbf{M}^{-1}\mathbf{b} \end{bmatrix} \tag{56}$$

For a linear regulator problem, a typical quadratic performance index may be defined as:

$$J = \frac{1}{2}\underline{S}^t(t_f)\mathbf{P}\underline{S}(t_f) + \frac{1}{2}\int_{t_o}^{t_f}(\underline{S}^t\mathbf{Q}\underline{S} + \underline{f}_c^t\mathbf{R}\underline{f}_c)dt \tag{57}$$

where t_o is the initial time, t_f is the fixed final time, and the non-unique constant matrices Q and **R** are positive definite weighting matrices which determine the magnitudes of the control forces and the quantitative decay of the controlled responses. The minimization of J in (57) subject to the constraint equation (53), using the Hamiltonian-Jacobi-Bellman equation [Bryson and Ho, 1974], leads to the following relation for feed-back gain matrix **G**:

$$\underline{f}_G = -\mathbf{R}^{-1}\mathbf{B}^t\mathbf{G}\underline{S} \tag{58}$$

where **G** is the solution of the Riccati equation:

$$\dot{\mathbf{G}} = -\mathbf{G}\mathbf{A} - \mathbf{A}^t\mathbf{G} + \mathbf{G}\mathbf{B}\mathbf{R}^{-1}\mathbf{B}^t\mathbf{G} - \mathbf{Q}; \qquad \mathbf{G}(t_f) = \mathbf{Q} \tag{59}$$

where, the matrix **A**, for the linear system, is defined in Eq. (55). Once the Riccati equation (59) is solved for **G**, and (58) is used in (53), the thus modified Eq. (53) becomes a standard initial value problem for the controlled response.

An efficient algorithm, based on the Schur vector approach [Laub, 1979], has been devised for the solution of Eq. (59) for systems of up to 100 degrees of freedom, in O'Donoghue (1986) and Zhang (1987).

A simple technique for controlling the nonlinear motion of structural systems considered so far (i. e. systems wherein the source of nonlinearity is due to large deformations) is as follows. The feed-back gain matrix **G** is assumed to be determined from Eq. (59), using the parameters of the linear system [i. e. , the matrices **A** and **B** for the system, assuming that the system behaves linearly]. However, the feed-back control forces are determined by applying this gain-matrix G to the sensed actual i. e. nonlinear state $[\mathbf{W}, \dot{\mathbf{W}}]$ of the nonlinear system. Splitting the gain-matrix G, as determined for a psuedo-linear system, relating to the displacement and velocity vectors of the nodes, give the control forces on the nonlinear system as:

$$\Delta \underline{f}_c = -\mathbf{R}^{-1}\mathbf{B}^t[\mathrm{G}_1 \Delta \underline{W} + \mathrm{G}_2 \Delta \underline{\dot{W}}] \tag{60}$$

where $\Delta\underline{W}$, and $\Delta\underline{\dot{W}}$ refer to the components of the incremental state vector of the actual nonlinear system. The controlled nonlinear response is computed by using the following equation, which results from using Eq. (60) in Eq. (50):

$$\mathbf{M}\Delta\underline{\ddot{W}} + [\mathbf{C} + \mathbf{b}\mathbf{R}^{-1}\mathbf{B}^t\mathrm{G}_1]\Delta\underline{W} + [\mathbf{K} + \mathbf{b}\mathbf{R}^{-1}\mathbf{B}^t\mathrm{G}_2]\Delta\underline{W} = \Delta\underline{F} \tag{61}$$

where, it should be recalled that, **M**, **C**, and **K** are the tangent mass, damping, and stiffness matrices of the nonlinear system.

The asymptotic stability of such controlled nonlinear systems as described through Eq. (61) are discussed in O'Donoghue (1985) and Zhang (1987).

In the examples to follow, for simplicity, the final time t_f is taken to be infinity, and only the steady state Riccati equation is solved. The control actuators are assumed to be located at the interior nodal points. The requirement of controllability [O'Donoghue (1986) and Zhang (1987)] has been satisfied in choosing the number and locations of these controllers. In the present simply-supported shell problem, since the transverse displacement w is much bigger than the inplane displacements u_i, only w is taken as the response quantity to be controlled. The field boundary element mesh consists of nodes at $r = (r_o/4)$ and $\theta = \pi/8$ in the circular baseplane. In the nonlinear response control problem, the weighting matrix **R** is taken to be $\mathbf{R}$ = diagonal $[1, 1, ...1]$; and the matrix **Q** is taken to be: **Q** = diagonal $[0.2, 0.2, ...0.2]$; thus satisfying the asymptotic stability criteria [Zhang, 1987].

The initial disturbance is taken to be an upward point impulse at the crown of the shell, resulting in an initial upward crown velocity of $v_o = 15.0$. The controlled as well as uncontrolled nonlinear responses of the shell at the crown node, and the nodes corresponding to $(r_o/4)$; $(r_o/2)$, and $(3r_o/4)$, are shown respectively, in Figs. (6), (7), (8), and (9). The variations of the corresponding control-actuator forces are shown in Figs. (10), (11), (12), and (13), respectively.

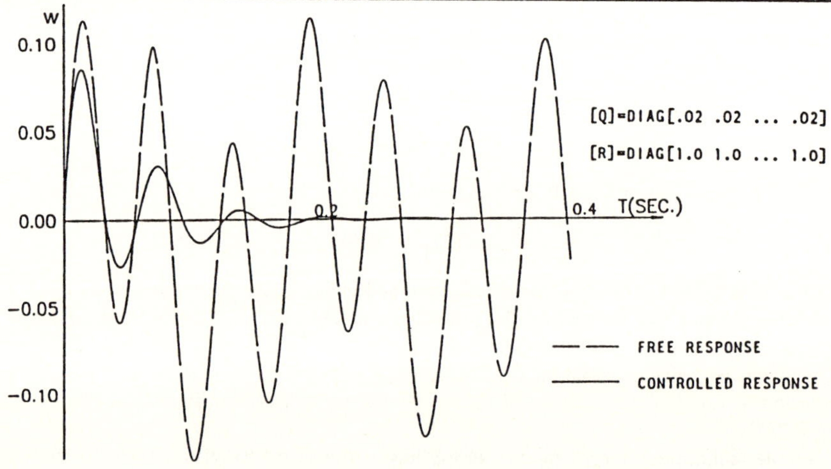

Fig. 6 Controlled nonlinear response at node 1

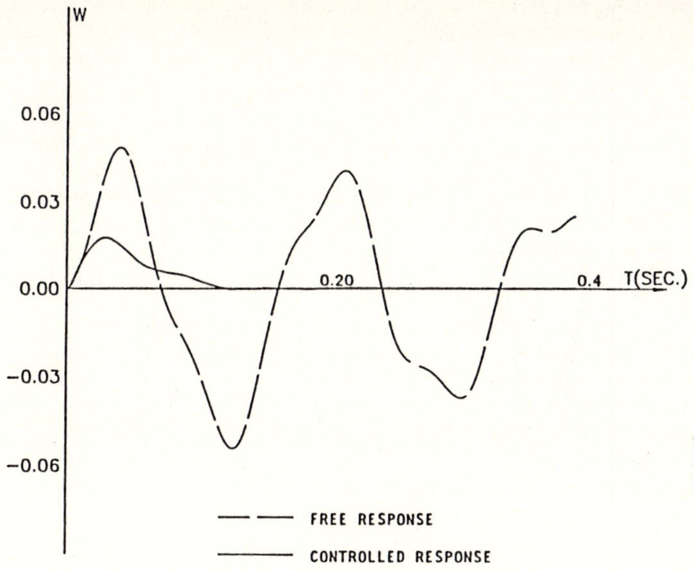

Fig. 7 Controlled nonlinear response at node 2

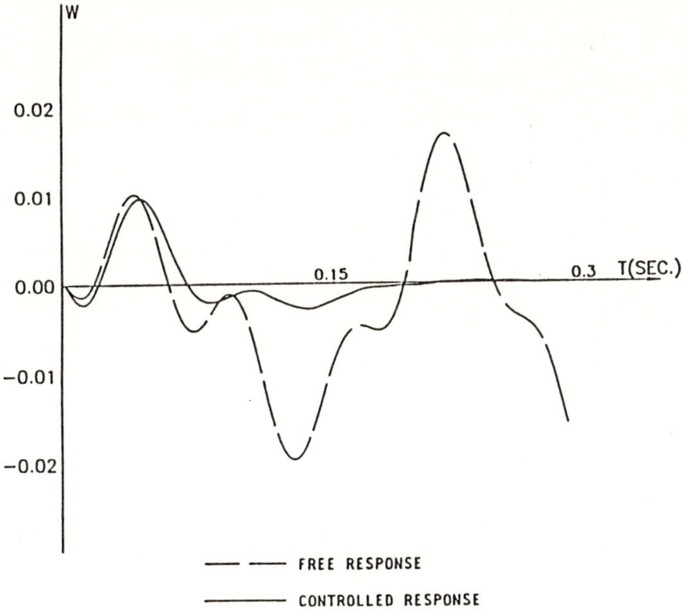

Fig. 8 Controlled nonlinear response at node 3

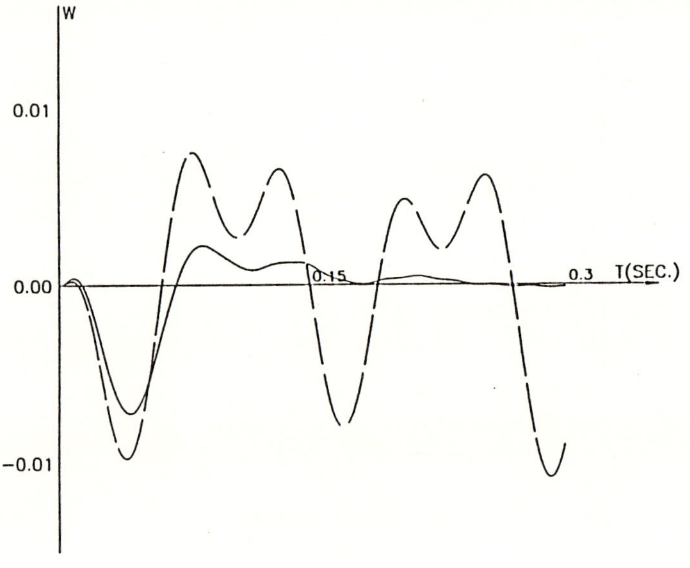

Fig. 9 Controlled nonlinear response at node 4

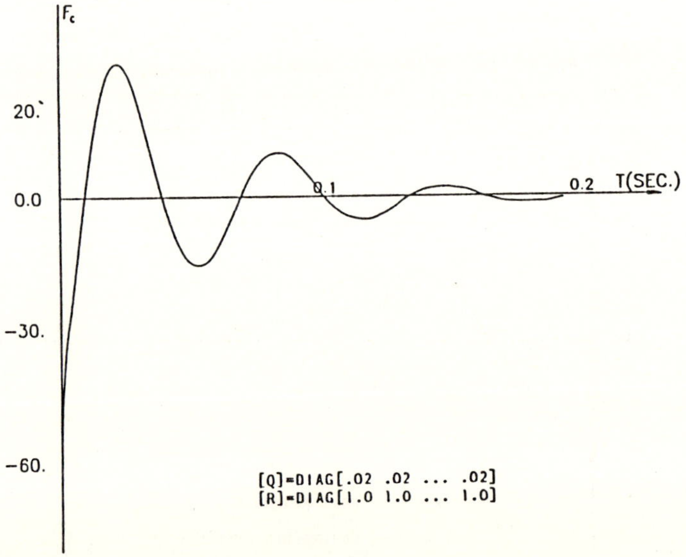

Fig. 10 Control force output by the actuator located at node 1

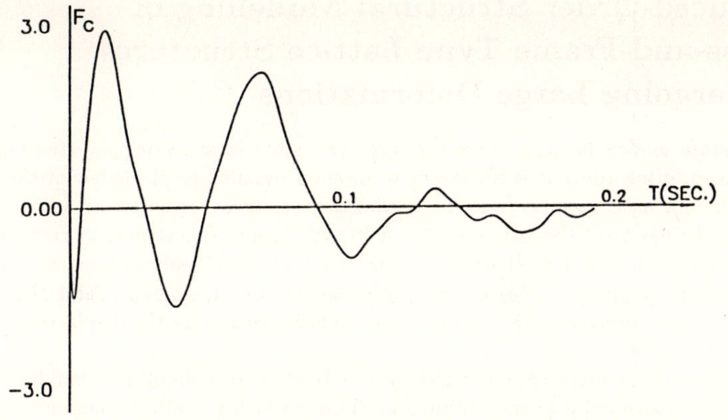

Fig. 11 Control force output by the actuator located at node 2

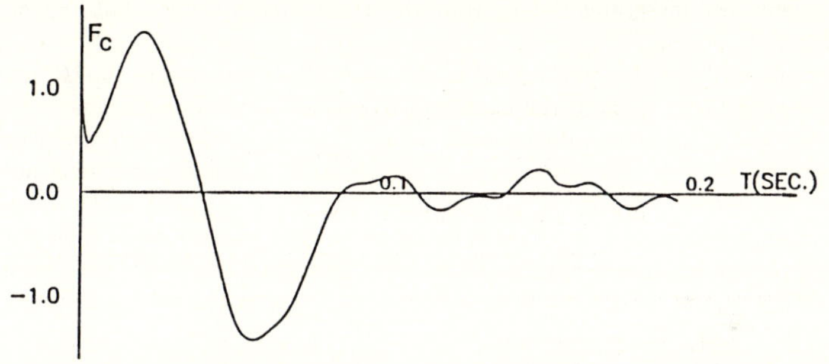

Fig. 12 Control force output by the actuator located at node 3

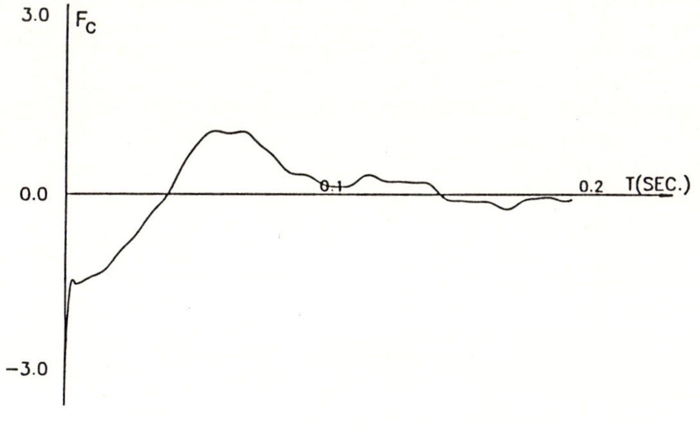

Fig. 13 Control force output by the actuator located at node 4

2 Reduced-Order Structural Modelling of Truss-and-Frame Type Lattice Structures Undergoing Large Deformations

While continuum models of space-structures are convenient in analyzing the standing-wave type global responses, often times it is necessary to develop detailed local models of the lattice type structure, wherein each member may be modeled as a 'truss' member or a 'frame' member, depending on the design of the lattice joints. A 'truss' member is assumed to carry only an axial force, and the kinematics of deformaion is characterized by the three displacements of each 'truss' joint. The 'frame' member carries bending moments, twisting moments, and lateral forces, in addition to the axial force; and the deformation is characterized by the displacements as well as rotations at each node.

In the nonlinear transient response analyses, it is usual to employ the semidiscrete method, with tangent inertia, damping, and stiffness matrices as in Eq. (50), for instance. In the usual finite element analyses, much of the effort is expended in evaluating these tangent matrices. For instance, the tangent stiffness matrix ${}^N\mathbf{K}$ (at the Nth increment of deformation) accounts for the effects of initial displacements and initial stresses. Usually, one assumes basis functions over each finite element, and integrates the appropriate strain energy terms (that depend nonlinearly on displacements) to obtain ${}^N\mathbf{K}$.

Recently, research had been performed to obtain explicit expressions for ${}^t\mathbf{K}$ for lattice type truss and frame structures, without the use of the asssumed element displacement basis functions and without the use of element integrations. In the case of a truss, each member is assumed to undergo arbitrarily large rigid rotations, moderate axial stretches, and local buckling [in its buckled state, the axial contraction of each member may be about 20% of the initial length]. In the case of a beam member, each member is assumed to undergo arbitrarily large rigid rotations, moderate relative rotations, and have nonlinear bending-stretching coupling. Furthermore, for the range of deformations considered, each member of the space-truss or space-frame was sought to be modeled by a single finite element.

Such explicit tangent stiffness matrices for the space-truss members were derived by Kondoh and Atluri [1985a, b], and Tanaka, Kondoh, and Atluri (1985). As for frame members, Kondoh and Atluri (1985b) first presented a displacement-based approach for plane frames, wherein the exact solution for the displacement of each member, modeled as a beam-column, was employed. Later, Kondoh, Tanaka, and Atluri (1985, 1986) extended the displacement approach to the three-dimensional case, using the concept of semi-tangential rotations. However, these displacement approaches did not consider the effect of plastic deformations, nor of nonconservative loading. Later work by Kondoh and Atluri (1987) showed that a complementary energy approach, based on assumed stress, is a much simpler alternative to derive explicit expressions for tangent stiffness of plane frames undergoing large elasto-plastic deformations. This was later extended to the case of space-frames by Shi and Atluri (1987).

Consider an initially straight slender truss member spanning two nodes $[x_{1i}$, and $x_{2i}]$ in space (where 1 and 2 denote the nodes; and $i = 1, 2$, and 3 denote the Cartesian directions). The initial length of the member is $\ell[\ell^2 = (x_{2i}-x_{1i})(x_{2i}-x_{1i}) = (\underline{x}_2-\underline{x}_1)\cdot(\underline{x}_2-\underline{x}_1)]$, and it is considered to have a uniform cross- section. Let $\underline{u}_1(u_{1i})$ and $\underline{u}_2(u_{2i})$ be the arbitrary displacements at the cetroidal axis of the member at nodes 1 and 2 respectively. From the polar-decomposition theorem, the total axial stretch of the member is given by:

$$\begin{aligned} \delta = \ell^* - \ell &= \{[(\underline{x}_2 + \underline{u}_2) - (\underline{x}_1 + \underline{u}_1)] \cdot [(\underline{x}_2 + \underline{u}_2) - (\underline{x}_1 + \underline{u}_1)]\}^{\frac{1}{2}} - \ell \\ &= \{(D\underline{x} + D\underline{u}) \cdot (D\underline{x} + D\underline{u})\}^{\frac{1}{2}} - \{D\underline{x} \cdot D\underline{x}\}^{\frac{1}{2}} \\ &= \{(Dx_i + Du_i)(Dx_i + Du_i)\}^{\frac{1}{2}} - \{Dx_i Dx_i\}^{\frac{1}{2}} \end{aligned} \tag{62}$$

where $D\underline{x} = \underline{x}_2 - \underline{x}_1$; and $D\underline{u} = \underline{u}_2 - \underline{u}_1$.

By taking the Taylor Series expansions of (62), the relation between the incremental axial stretch $\Delta\delta$, and the incremental nodal displacements $\Delta(Du_i)[= \Delta u_{2i} - \Delta u_{1i}]$ may be derived as:

$$\Delta\delta = \frac{\partial\delta}{\partial(Du_i)}\Delta(Du_i) + \frac{\partial^2\delta}{\partial(Du_i)\partial(Du_j)}\ \Delta(Du_i) \quad \Delta(Du_j)\ +\ \text{higher} \tag{63}$$

The truss member is assumed to transmit only an axial force N while the material is assumed to remain elastic, the truss member may undergo local buckling. Since large deformations and local buckling are considered, N is in general a nonlinear function of δ. The tangent spring-constant of the truss member is assumed to be k i. e.

$$k = \partial N/\partial\delta \tag{64}$$

or, in a linearized form,

$$\Delta N = k \cdot \Delta\delta \tag{65}$$

When the truss member remains straight and unbuckled, from linear elasticity and has:

$$k = EA/\ell; \qquad N < N_{cr} \tag{66}$$

where N_{cr} is the Euler buckling load of the truss-member treated as a simply supported beam-column.

On the other hand, when $N \geq N_{cr}$, the truss member undergoes local-buckling, and the well-known elastica solution [Timoshenko and Gere 1961] leads to:

$$N = N_{cr} \cdot \frac{1}{[1 + (\hat{\delta}/4\ell)]^2} \tag{67}$$

where $N_{cr} = (-\pi^2 EI/\ell^2)$ i. e. , N_{cr} is compressive and critical. In Eq. (67) $\hat{\delta}$ is the axial length after the buckling of the member.

For small values of $(\hat{\delta}/\ell)$, Eq. (67) may be approximated as:

$$N = N_{cr}[1 - \frac{1}{2}(\frac{\hat{\delta}}{\ell})] \tag{68}$$

Eq. (68) is valid for $(\hat{\delta}/\ell)$ in the range of 0 to 0.2 [See Tanaka, Kondoh, and Atluri 1985]. This range of values of $(\hat{\delta}/\ell)$ is adequate for considerations of realistic large deformations of a truss-member as a part of an overall large space structure.

Thus, in the post-buckling-range, the axial stiffness of a truss member is given from Eq. (68) by:

$$k = \frac{\partial N}{\partial\hat{\delta}} = -\frac{1}{2\ell}N_{cr} = \frac{\pi^2 EI}{2\ell^3} \qquad N \geq N_{cr} \tag{69}$$

Thus;

$$k = \frac{EA}{\ell}, \quad N < N_{cr}; \qquad k = \frac{\pi^2 EI}{\ell^3}, \qquad N \geq N_{cr} \tag{70}$$

which indicates a substantial loss of stiffness of a truss member due to lacal buckling.

If N is the current axial force (either before or after local buckling) in a member that undergoes an incremental deformation resulting in an incremental axial stretch $\Delta\delta$, the incremental strain-energy in the member is given by:

$$\Delta U = N\Delta\delta + \frac{1}{2}k(\Delta\delta)^2 \tag{71}$$

where k is given by Eqs. (70) depending on whether $N < N_{cr}$ or $N \geq N_{cr}$. Using (63) in (71), one has:

$$\Delta U = N\{\frac{\partial \delta}{\partial(Du_i)}\Delta(Du_i) + \frac{\partial^2 \delta}{\partial(Du_i)\partial(Du_j)}\Delta(Du_i)\Delta(Du_j)\}$$
$$+\frac{1}{2}k\{\frac{\partial \delta}{\partial(Du_i)}\frac{\partial \delta}{\partial(Du_j)}\Delta(Du_i)\Delta(Du_j)\} + h \cdot o \cdot t \tag{72}$$

where $\Delta(Du_i) = \Delta(u_{2i} - u_{1i}) = \Delta u_{2i} - \Delta u_{1i}$. The coefficients of the (6 x 6) tangent stiffness matrix ${}^t\mathbf{K}$ of the 2 noded truss member, with incremental nodal displacements $\Delta u_{\alpha i}$ ($\alpha = 1$, 2 nodes; and $i = 1$, 2, 3 Cartesian directions), are derived from evaluating the second-partial-derivatives as below:

$$\frac{\partial^2 \Delta U}{\partial \Delta u_{\alpha i} \partial \Delta u_{\beta j}} \qquad \begin{pmatrix} \alpha, \beta = 1,2 \\ i,j = 1,2,3 \end{pmatrix} \tag{73}$$

The vector of residual nodal forces on the element, at the current stage of incremental deformation, is given by:

$$\underline{R} = \frac{\partial \Delta U}{\partial \Delta u_{\alpha i}} \qquad \begin{pmatrix} \alpha = 1,2 \\ i = 1,2,3 \end{pmatrix} \tag{74}$$

When the tangent stiffness matrix of the element is derived in an explicit fashion as in Eq. (73), the tangent stiffness of the lattice structure as a whole is also obtained in an explicit form, through the usual element-assembly process.

In the present work, to determine the quasi-static finite deformations of a lattice structure, an "arc-length" method [Crisfield, 1983; Riks 1979; and Kondoh and Atluri 1985a,b] is used.

An example of a beam-like space-truss (the PACOSS truss), subjected to axial and bending loads is shown in Fig. (14). The structure is that of a twelve-bay truss whose member properties are indicated in Fig. (14). In order to trigger the coupling between the axial and transverse displacements, which is characteristic of the buckling mode, in the case when only an axial load is applied, a "load-imperfection" equal to $(P/1000)$ is added in the transverse direction at one of the end-nodes, as shown in Fig. (14). For this predominately axial-load case, Fig. (15) shows the relation between the magnitudes of the axial load and that of the transverse displacement at the loaded end for two scenarios: (i) when local (member) buckling is assumed to be suppressed through an <u>active control</u> mechanism (such as piezo-electric ceramics bonded to the member along segments of its length) and thus each member is assumed to remain straight and stable; and (ii) when each member is allowed to undergo local buckling. Fig. (15) clearly demonstrates the advantageous effects of controlling the local buckling deformations of individual members and forcing them to remain straight and stable.

Fig. (16) shows the case when the PACOSS truss is subjected primarily to bending loads. This figure shows the relation between the magnitudes of transverse (bending) loads and transverse displacement, respectively, once again for two scenarios: (i) when local member buckling is suppressed, and (ii) when member buckling is allowed. Fig. (16) shows that a nearly <u>five-field</u> increase the load-carrying capacity of the truss results when active control mechanisms are employed to keep each individual truss member from buckling.

It should be remarked that in Figs. (15) and (16), the letters A, B, C etc. , indicate the stages at which the respective members, whose numbers are identified in Figs. (15) and (16), respectively, undergo local buckling.

We now consider the case of lattice structures whose joints are designed such that each of the members may be modelled as an initially straight beam. Let x_i and $\underline{e}_i$ represent a fixed set of Cartesian coordinates and unit bases respectively [See Fig. (17)]. Let x'_i and $\underline{e}'_i$ be the local

MATERIAL PROPERTY

MEMBER	EA LB	EI LB IN^2
LONGERONS	7.08×10^6	2.16×10^6
DIAGONAL, BATTENS	2.70×10^5	6.43×10^4
SHORT LONGERONS	1.65×10^8	2.20×10^8
HORIZONTAL DIAGONAL	1.37×10^8	1.52×10^8

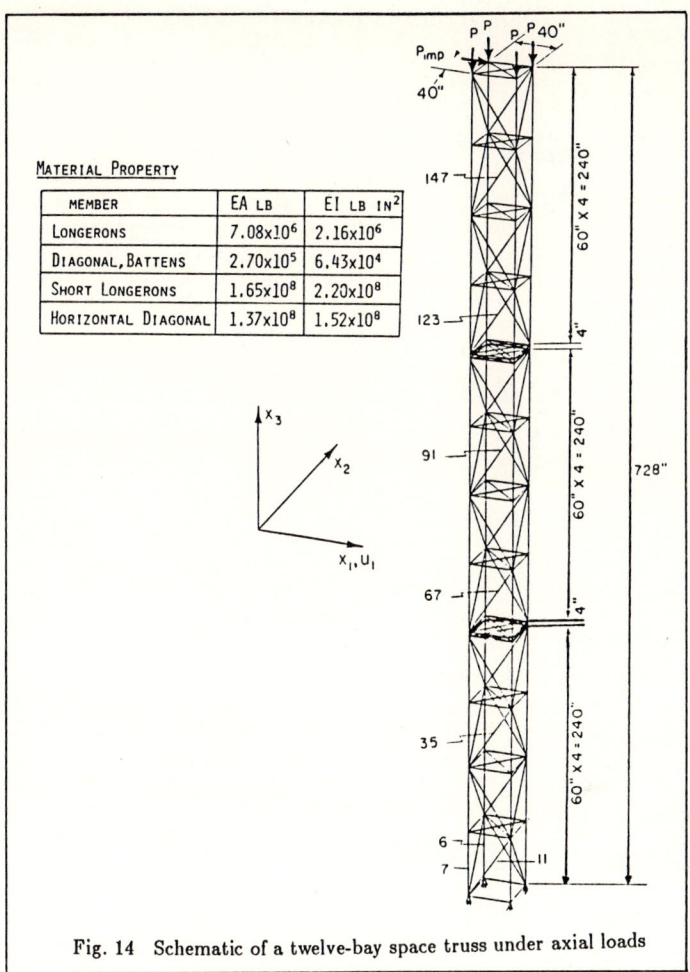

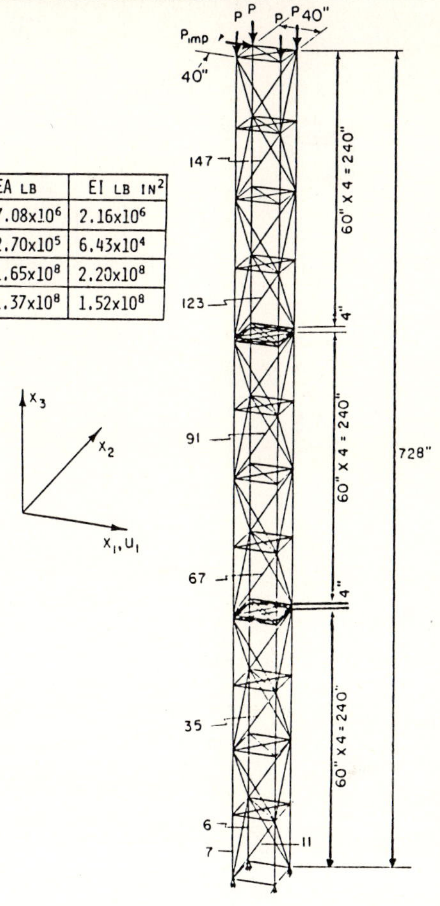

Fig. 14 Schematic of a twelve-bay space truss under axial loads

coordinates and bases along the arbitrarily oriented undeformed member. Let $x_{\alpha i}$ be the global coordinates ($i = 1; 2, 3$) of the node α ($\alpha = 1, 2$) of the beam in space. The base vector $\underline{e}_i'$ are chosen such that:

$$\underline{e}_i' = (Dx_i)\underline{e}_i/\ell; \qquad \ell^2 = Dx_i \cdot Dx_i$$
$$\underline{e}_2' = (\underline{e}_3 \times \underline{e}_1')/|\underline{e}_3 \times \underline{e}_1'| \qquad \underline{e}_3' = \underline{e}_1' \times \underline{e}_2' \tag{75}$$

where $Dx_i = (x_{2i} - x_{1i})$, and subscripts 1 and 2 denote the nodes. Let $\hat{x}_i$ and $\hat{\underline{e}}_i$ be the local coordinates and basis as shown in Fig. (17), where $\hat{\underline{e}}_1$ is along the straight line joining nodes 1 and 2 after deformation. Another basis system is $\underline{e}_i^*$ which is locally tangential and normal to the deformed centroidal axis of the beam.

Let $u_{\alpha i}$ ($\alpha = 1, 2$; $i = 1, 2, 3$) be the displacements of nodes 1 and 2 in the global Cartesian system. Defining the quantities:

$$Du_i = u_{2i} - u_{1i}; \qquad \text{and} \quad D\hat{x}_i = Dx_i + Du_i \tag{76}$$

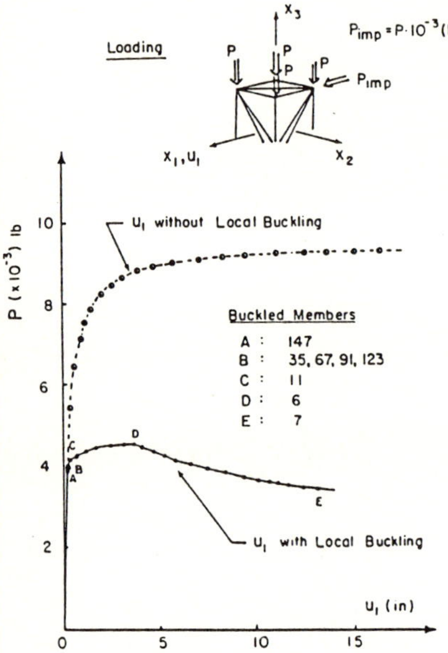

Fig. 15 Deflections at free end under axial loads with and without the influence of local buckling of truss members

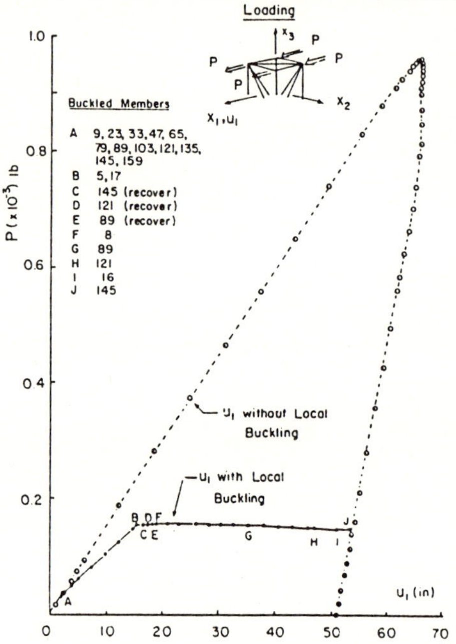

Fig. 16 Deflection at free end under bending loads with and without the influence of local buckling of truss members

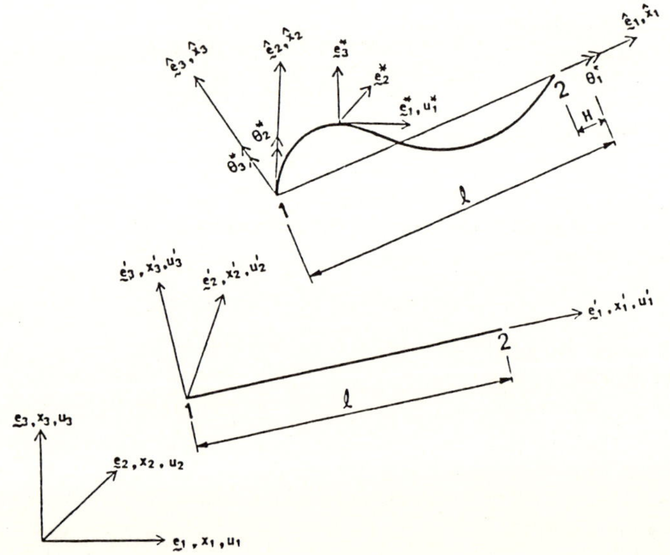

Fig. 17 Nomenclature for kinematics of deformation of a space framed member

we have:

$$\hat{e}_1 = (D\hat{x}_i)\underline{e}_i/L$$
$$\hat{e}_2 = (\underline{e}_3 x \hat{\underline{e}}_1)/|\underline{e}_3 x \hat{\underline{e}}_1 \qquad \hat{e}_3 = \hat{e}_1 x \hat{\underline{e}}_2 \tag{77}$$

where $L^2 = D\hat{x}_i \cdot D\hat{x}_i$; or Eq. (77) may be written as:

$$\begin{Bmatrix} \hat{\underline{e}}_1 \\ \hat{\underline{e}}_2 \\ \hat{\underline{e}}_3 \end{Bmatrix} = \begin{bmatrix} D\hat{x}_1/L & D\hat{x}_2/L & D\hat{x}_3/L \\ -D\hat{x}_2/S & D\hat{x}_1/S & 0 \\ -D\hat{x}_1 D\hat{x}_3/LS & -D\hat{x}_2 D\hat{x}_3/LS & S/L \end{bmatrix} \begin{Bmatrix} \underline{e}_1 \\ \underline{e}_2 \\ \underline{e}_3 \end{Bmatrix} \tag{78}$$

where $S^2 = (D\hat{x}_1)^2 + (D\hat{x}_2)^2$ or

$$\hat{\underline{e}}_i = R_{ik}\underline{e}_k \tag{79}$$

Equation (75) may likewise be written as:

$$\underline{e}_i' =^o R_{ik}\underline{e}_k \tag{80}$$

Thus, $^o\mathbf{R}$ characterizes the orientation of the undeformed beam in a fixed Cartesian system $\underline{e}_i$; and $\mathbf{R} \cdot {}^o\mathbf{R}^t$ characterizes the arbitrary rigid rotation of the deformed beam relative to its undeformed configuration. Note, however, $\mathbf{R} \cdot {}^o\mathbf{R}^t$ locates only the line joining the two nodes in the deformed configuration of the beam.

The relative rotations θ_i^* of the deformed centroidal axis of the beam, with respect to the line joining the nodes 1 and 2 in the deformed configuration, are assumed to be small. Under these assumptions, the change in the length of the beam is given by:

$$\begin{aligned} H &= L - \ell = \{(Dx_i + Du_i) \cdot (Dx_i + Du_i)\}^{\frac{1}{2}} - \ell \\ &= \{D\hat{x}_i \cdot D\hat{x}_i\}^{\frac{1}{2}} - \ell \end{aligned} \tag{81}$$

In view of (81), the finite rotation matrix $\mathbf{R}$ of Eq. (78) may be written as:

$$\mathbf{R} = \begin{Bmatrix} \partial H/\partial(Du_1) & \partial H/\partial(Du_2) & \partial H/\partial(Du_2) \\ -\partial S/\partial(Du_2) & \partial S/\partial(Du_1) & 0 \\ -[\partial H/\partial(Du_3)] \text{ x } [\partial S/\partial(Du_1)] & -[\partial H/\partial(Du_3)] \text{ x } [\partial S/\partial(Du_2)] & S/L \end{Bmatrix} \tag{82}$$

When the element is parallel to the x_3 axis, $s^2 = (Dx_1)^2 + (Dx_2)^2 = 0$ and hence Eqs. (78) are not valid. In this special case, we may take:

$$\underline{e}_1' = \underline{e}_3; \quad \underline{e}_2' = \underline{e}_2; \quad \underline{e}_3' = -\underline{e}_1.$$

The displacements of the nodes 1 and 2, in the basis vector system $\underline{e}_i'$ are taken to $u_{\alpha i}'$ ($\alpha =$ 1, 2; $i =$ 1, 2, 3). It is seen that

$$u_{\alpha i}' = (^oR)_{ik} u_{\alpha k} \qquad (\alpha = 1, 2; k = 1, 2, 3) \tag{83}$$

where $^o\mathbf{R}$ characterizes the initial configuration of the beam, as in Eq. (80).

As explained in detail in Kondoh and Atluri (1987), the consistent forms of linear and angular momentum balance relations for the Jaumann Stress resultants and stress couples, defined in Atluri (1983), may be simply written as:

$$\frac{\partial N}{\partial \hat{x}_i} + \hat{q}_1 = 0 \tag{84}$$

$$\frac{\partial \hat{M}_1}{\partial \hat{x}_1} = 0; \qquad \frac{\partial^2 \hat{M}_2}{\partial \hat{x}_1^2} + \frac{\partial}{\partial \hat{x}_1}(N\theta_2^*) + \hat{q}_3 = 0 \tag{85}$$

$$\frac{\partial^2 M_3}{\partial \hat{x}_1^2} + \frac{\partial}{\partial \hat{x}_1}(N\theta_3^\star) - \hat{q}_2 = 0 \tag{86}$$

where $\hat{q}_i$ are the distributed loads along the $\hat{e}_i$ directions per unit length of the undeformed element.

The load acting on the member may be considered to be both "dead-type" and "follower-type" [i. e. always along the $\hat{e}_i$ axis]. Thus,

$$\hat{q}_i = R_{ij}q_{cj} + q_{ni} \tag{87}$$

where 'c' denotes conservative and 'n' denotes non-conservative. We assume trial functions for N and $\hat{M}_1$, which satisfy Eqs. (84) and (85a) identically; but trial functions for $\hat{M}_2$ and $\hat{M}_3$ satisfy only the linear parts of Eqs. (85b) and (86) respectively.

Thus, the chosen trial stress fields in the member are:

$$\begin{aligned} N = \quad & n + R_{1j}\{-\int_0^{\hat{x}_1} q_{cj}d\hat{x}_1 + \frac{1}{\ell}\int_0^\ell[\int_0^{\hat{x}_1} q_{cj}d\hat{x}]d\hat{x}_1\} \\ & -\int_0^{\hat{x}_1} q_{n1}d\hat{x}_1 + \frac{1}{\ell}\int_0^\ell(\int_0^{\hat{x}_1} q_{n1}d\hat{x}_1)d\hat{x}_1 \end{aligned} \tag{88}$$

$$\hat{M}_1 = m_1 \tag{89}$$

$$\hat{M}_2 = (1 - \frac{\hat{x}_1}{\ell})m_{12} + \frac{\hat{x}_1}{\ell}m_{22} + R_{3j}M_{pcj} + M_{pn2} \tag{90}$$

$$\hat{M}_3 = (1 - \frac{\hat{x}_1}{\ell})m_{13} + \frac{\hat{x}_1}{\ell}m_{23} + R_{2j}M_{pcj} + M_{pn3} \tag{91}$$

where

$$M_{pcj} = -\int_0^{\hat{x}_1}[\int_0^{\hat{x}_1} q_{cj}d\hat{x}_1]d\hat{x}_1 + \frac{\hat{x}_1}{\ell}\int_0^\ell[\int_0^{\hat{x}_1} q_{cj}d\hat{x}_1]d\hat{x}_1 \tag{92}$$

and

$$M_{pni} = -\int_0^{\hat{x}_1}[\int_0^{\hat{x}_1} q_{ni}d\hat{x}_1]d\hat{x}_1 + \frac{\hat{x}_1}{\ell}\int_0^\ell[\int_0^{\hat{x}_1} q_{ni}d\hat{x}_1]d\hat{x}_1 \tag{93}$$

The corresponding test functions (or variations in N and $\hat{M}_i$) are taken to be:

$$\begin{aligned} &\delta N = \nu \\ &\delta\hat{M}_1 = \mu_1 \\ &\delta\hat{M}_i = \mu_i = \mu_{1i}(1 - \frac{\hat{x}_1}{\ell}) + \mu_{2i}(\frac{\hat{x}_1}{\ell}) \quad i = 2,3 \end{aligned} \tag{94}$$

Assuming that the material is linear elastic, we define a complementary energy density W_c such that:

$$\partial W_c/\partial N = h; \qquad \frac{\partial W_c}{\partial \hat{M}_i} = \kappa_i \tag{95}$$

where h is the axial strain, and κ_i the curvature strains.

Rather than considering the point-wise compatibility conditions, we now write the weak-form of the compatibility conditions for the beam as a whole. These are:

$$\begin{aligned} &\int_0^\ell \frac{\partial W_c}{\partial N}\nu d\hat{x}_1 = \int_0^\ell u_{1,1}^\star \nu d\hat{x} = \nu H = \nu[(D\hat{x}_i \cdot D\hat{x}_i)^{\frac{1}{2}} - \ell] \\ &\int_0^\ell \frac{\partial W_c}{\partial \hat{M}_1}\mu_1 d\hat{x}_1 = \int_0^\ell \theta_{1,1}^\star \mu_1 d\hat{x}_1 = (\theta_{21}^\star - \theta_{11}^\star)\mu_1 \\ &\int_0^\ell \frac{\partial W_c}{\partial \hat{M}_2}\mu_2 d\hat{x}_1 = -\int_0^\ell \theta_{2,1}^\star \mu_2 d\hat{x}_1 = -\int_0^\ell (\theta_2^\star\mu_2)_{,1}d\hat{x}_1 + \int_0^\ell \theta_2^\star \frac{\partial \mu_2}{\partial \hat{x}_1}d\hat{x}_1 = -\theta_{22}^\star\mu_{22} + \theta_{12}^\star\mu_{12} \end{aligned} \tag{96}$$

Since $\int_0^\ell \theta^* d\hat{x}_1 = 0$.

Likewise,

$$\int_0^\ell \frac{\partial W_c}{\partial \hat{M}_3} \mu_3 d\hat{x}_1 = -\theta^*_{23}\mu_{23} + \theta^*_{13} M_{13} \tag{97}$$

where $\theta^*_{\alpha i}$ ($\alpha = 1, 2$; $i = 1, 2, 3$) is the ith relative rotation (assumed small) at the α node.

Note that the balance relations (84 and 85a) are satisfied identically. The weak forms of Eqs. (85b) and (86) may be written for each member, as:

$$\int_0^\ell \left(\frac{\partial \hat{M}_2}{\partial \hat{x}_1} - \hat{Q}_3 + N\theta^*_2\right)\beta^*_2 d\hat{x}_1 = 0 \tag{98}$$

and

$$\int_0^\ell \left(\frac{\partial \hat{M}_3}{\partial \hat{x}_1} + \hat{Q}_2 + N\theta^*_3\right)\beta^*_3 d\hat{x}_1 = 0 \tag{99}$$

where

$$\hat{Q}_2 = \frac{-\partial \hat{M}_3}{\partial \hat{x}_1} = \frac{-m_{23} + m_{13}}{\ell} + M_{p3,1} \tag{100}$$

and

$$\hat{Q}_3 = \frac{\partial \hat{M}_2}{\partial \hat{x}_1} = \frac{m_{22} - m_{12}}{\ell} + M_{p2,1} \tag{101}$$

Finally, when 2 or more beam-members are joined at a lattice-joint in the space structure, the joint equilibrium must be satisfied. The forces scting on any joint are: (i) the given external forces and moments, if any, at the joint; and (ii) the internal forces and moments in all members meeting at the joint.

Let $\underline{Q}^e$ be the generalized internal nodal force vector:

$$\underline{Q}^e = \{N_1; \hat{Q}_{12}; \hat{Q}_{13}; \hat{M}_1; \hat{M}_{12}; \hat{M}_{13}; N_2; \hat{Q}_{22}; \hat{Q}_{23}; \hat{M}_i; \hat{M}_{2i}; \hat{M}_{23}\} \tag{102}$$

where N_α, $\hat{Q}_{\alpha 2}$ etc. are nodal forces along $\hat{e}_i$ axes. Let the external nodal forces (and moments) be specified along (and about) the global $\underline{e}_i$ axes. Let

$$\underline{f}^e = \lfloor F_{1i}; \overline{M}_{1i}; \overline{F}_{2i}; \overline{M}_{2i} \rfloor \qquad i = 1,2,3 \tag{103}$$

Let a 12 x 12 transformation matrix $\mathbf{R}^*$ be defined such that

$$\mathbf{R}^* = \begin{bmatrix} -\mathbf{R}^t & 0 & 0 & 0 \\ 0 & -\mathbf{R}^t & 0 & 0 \\ 0 & 0 & \mathbf{R}^t & 0 \\ 0 & 0 & 0 & \mathbf{R}^t \end{bmatrix} \tag{104}$$

where $\mathbf{R}$ is the 3 x 3 rotation matrix defined in Eq. (82).

Then, the joint equilibrium may be written as:

$$\sum_{\text{elem}} \mathbf{R}^* \underline{Q}^e - \underline{\overline{f}} = 0 \tag{105}$$

which, in weak form, may be written as:

$$\sum_{\text{elem}} \underline{v}^t \mathbf{R}^* \underline{Q}^e - \underline{v}^t \cdot \underline{\overline{f}} = 0 \tag{106}$$

where $\underline{v}$ is a (12 x 1) vector of trial functions, which may interpreted as the virtual displacements (along $\underline{e}_i$) and virtual rotations (around $\underline{e}_i$) at the two nodes of each member. Thus,

$$\lfloor \underline{v} \rfloor = \lfloor \delta u_{11}; \delta u_{12}; \delta u_{13}; \delta\theta_{11}; \delta\theta_{12}; \delta\theta_{13}; \delta u_{21}; \delta u_{22}; \delta u_{23}; \delta\theta_{21}; \delta\theta_{22}; \delta\theta_{23} \rfloor \tag{107}$$

when $\delta u_{\alpha i}$ ($\alpha = 1, 2$; $i = 1, 2, 3$) denote the ith component of virtual displacement at αth node of the member.

Since the relative rotations $\theta^*_{\alpha i}$ are assumed to be small, one may define a generalized member deformation vector:

$$\underline{D}^t = \lfloor H; (\theta^*_{21} - \theta^*_{11}); (\theta^*_{12} - \theta^*_{22}); (\theta^*_{13} - \theta^*_{23}) \rfloor \tag{108}$$

Let $\underline{\sigma}$ denote the generalized member-internal-force vector which satisfies only the homogeneous linear part of the equilibrium equations (84-86), i. e. ,

$$\underline{\sigma} = \lfloor n; m_1; m_{12}; m_{22}; m_{13}; m_{23} \rfloor \tag{109}$$

and

$$\lfloor \delta\underline{\sigma} \rfloor = \lfloor \nu; \mu_1; \mu_{12}; \mu_{22}; \mu_{13}; \mu_{23} \rfloor \tag{110}$$

From Eqs. (88-91), we see that the homogeneous parts of the trial functions for generalized member forces are:

$$\begin{bmatrix} n \\ M_1 \\ M_2 \\ M_3 \end{bmatrix} = \begin{bmatrix} 1 & 0 & 0 & 0 & 0 & 0 \\ 0 & 1 & 0 & 0 & 0 & 0 \\ 0 & 0 & (1 - \hat{x}_1/\ell) & (\hat{x}_1/\ell) & 0 & 0 \\ 0 & 0 & 0 & 0 & (1 - \hat{x}_1/\ell) & (\hat{x}_1/\ell) \end{bmatrix} \begin{Bmatrix} n \\ m_1 \\ m_{12} \\ m_{22} \\ m_{13} \\ m_{23} \end{Bmatrix} \tag{111}$$

$$= \mathbf{F} \cdot \underline{\sigma} \tag{112}$$

and

$$\begin{Bmatrix} \nu \\ \mu_1 \\ \mu_2 \\ \mu_3 \end{Bmatrix} = \mathbf{F} \cdot \delta\underline{\sigma} \tag{113}$$

Finally, we define a vector:

$$\underline{W}^t = \left\{ \frac{\partial W_c}{\partial N}; \frac{\partial W_c}{\partial \hat{M}_1}; \frac{\partial W_c}{\partial \hat{M}_2}; \frac{\partial W_c}{\partial \hat{M}_3} \right\} \tag{114}$$

Using the notations in Eqs. (108-114), the combined weak forms of member compatibility equations (96 and 97); the member equilibrium equations (98 and 99); and the joint equilibrium equations (106); may be written as:

$$\sum_{\text{members}} \{ -\int_0^\ell \underline{W}^t \cdot \mathbf{F} \cdot \delta\underline{\sigma} d\hat{x}_1 + \mathbf{D}^t \cdot \delta\underline{\sigma} + \underline{v}^t \cdot \mathbf{R}^* \cdot \underline{Q}^c - \underline{v}^t \cdot \underline{\bar{f}} + \int_0^\ell N\theta^*_2\beta^*_2 d\hat{x}_1 + \int_0^\ell N\theta^*_3\beta^*_3 d\hat{x}_1 \} = 0 \tag{115}$$

Thus, the only integrals to be evaluated over the element are those involving the integrand $\underline{W}^t \cdot \mathbf{F}$. Since the integrands are simple, they can be trivally evaluated in closed form. On the other hand, omitting the terms $N\theta^*_2\beta^*_2$ and $N\theta^*_3\beta^*_3$ from Eq. (115) will lead to some errors in the tangent stiffness matrix, but it is entirely permissible in the context of the iterative arc-length method used in solving the present finite-deformation problem, as discussed in Kondoh and Atluri (1987). The details of the algebraic derivation of the explicit expression for the tangent stiffness matrix of the beam undergoing arbitrarily large over-all rotations as a part of a deforming latttice-structure, may be found in Kondoh and Atluri (1987) and Shi and Atluri (1987).

When plastic hinges form at any location in the member, only the member compatibility conditions, Eqs. (96 and 97), are changed accordingly. Even when such plastic hinges form in a member, the tangent stiffness matrix of the member may be evaluated explicitly [see Kondoh and Atluri (1987) and Shi and Atluri (1987) for further details].

Here we present the example of a space-frame as shown in Fig. (18). The loading consists of a simple point load at the crown. Here, each member is modelled by a single element with a (12 x 12) stiffness matrix. The computed force-displacement relations are shown in Fig. (19). The two sets of results correspond to those obtained from: (i) the stress-based weak formulation of Eq. (115) and (ii) the displacement-based formulation presented in Kondoh, Tanaka, and Atluri (1986).

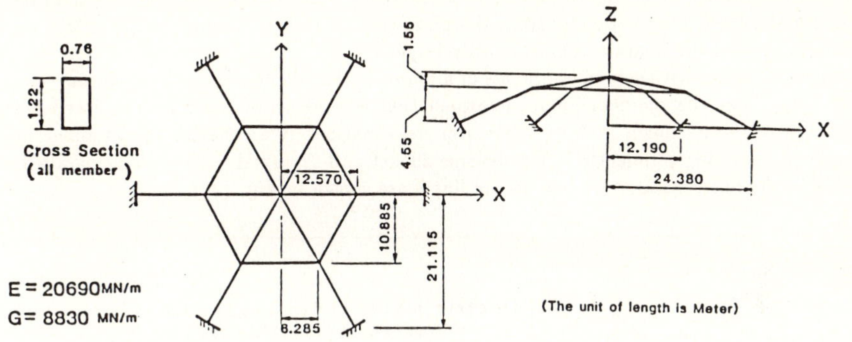

Fig. 18 Framed dome (the unit of length is meter)

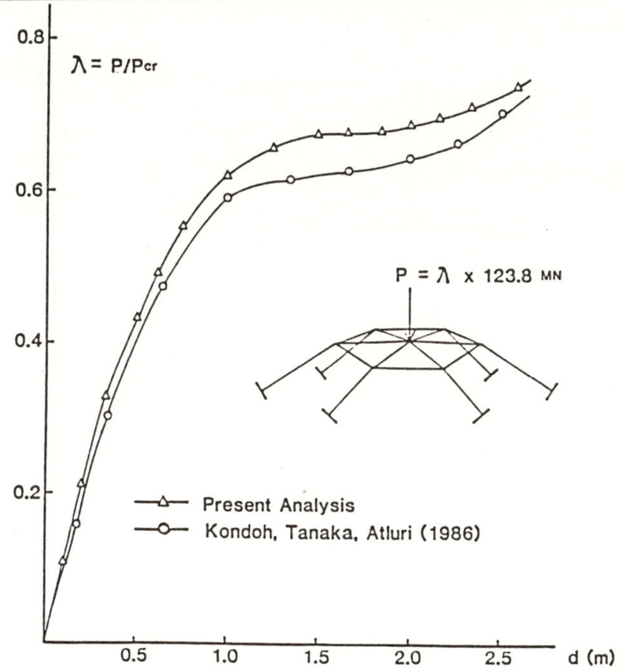

Fig. 19 Load-displacement curve when a concentrated load alone acts at the crown point

As for active control of deformations of a truss or a beam member of a lattice structure, the concept of piezo-electric layers bonded to the surfaces of the members, in segments along the length of a member, appears promising. These actuators transfer shear forces to the underlying structural surfaces depending on the magnitude of the excitation voltage applied. Crawley and de Luis (1984) presented a static model for the shear transfer from a peizo-liner to a beam member which was assumed to be subjected to a pure bending moment, when the deformation of the beam is infinitesmal. Im and Atluri (1987) presented a more complete solution which considers axial forces, tarnsvese shear forces, as well as bending moments in the structural member itself, in solving for the shear stress transfer from the piezo-liner. Further, Im and Atluri (1987) consider the deformation of the beam to be arbitrarily large.

Consider a beam with several segments of a piezo-liner at its bottom and top surface as shown in Fig. (20). Let the member be so deformed that a segment of the member, lined with the piezo-actuator, is as shown in Fig. (21) after deformation. We consider the member internal forces to M (bending moment); V (transverse shear) and H (axial frame). The detailed stress patterns in the beam-column as well as the liners are shown in Fig. (22).

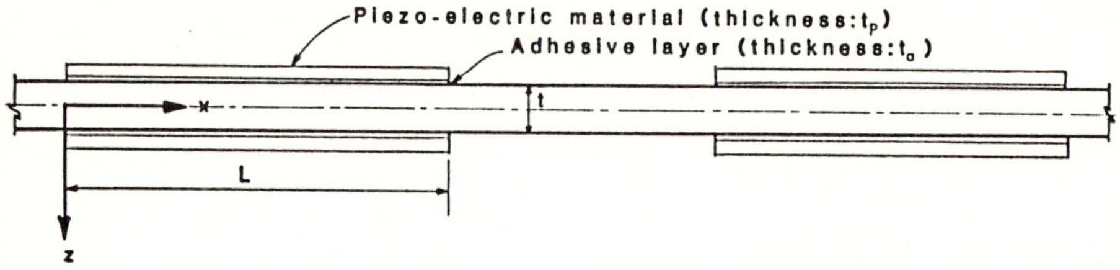

Fig. 20 Beam-column lined with a piezo-electric material

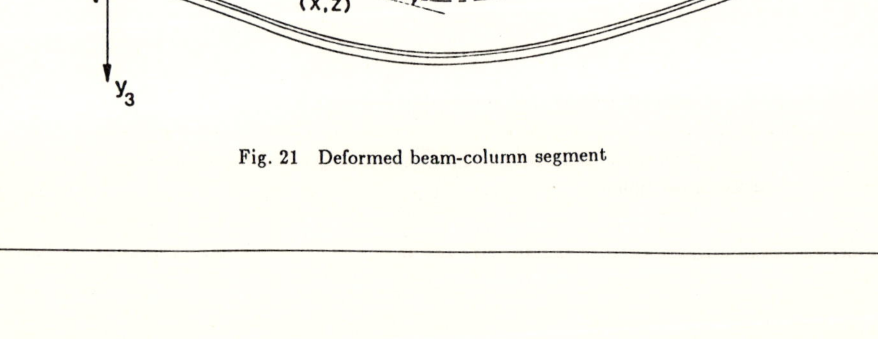

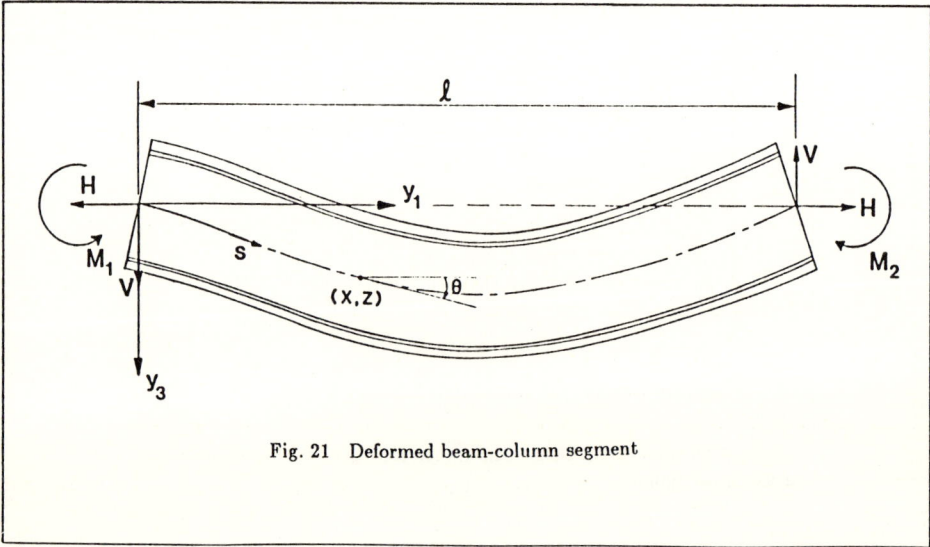

Fig. 21 Deformed beam-column segment

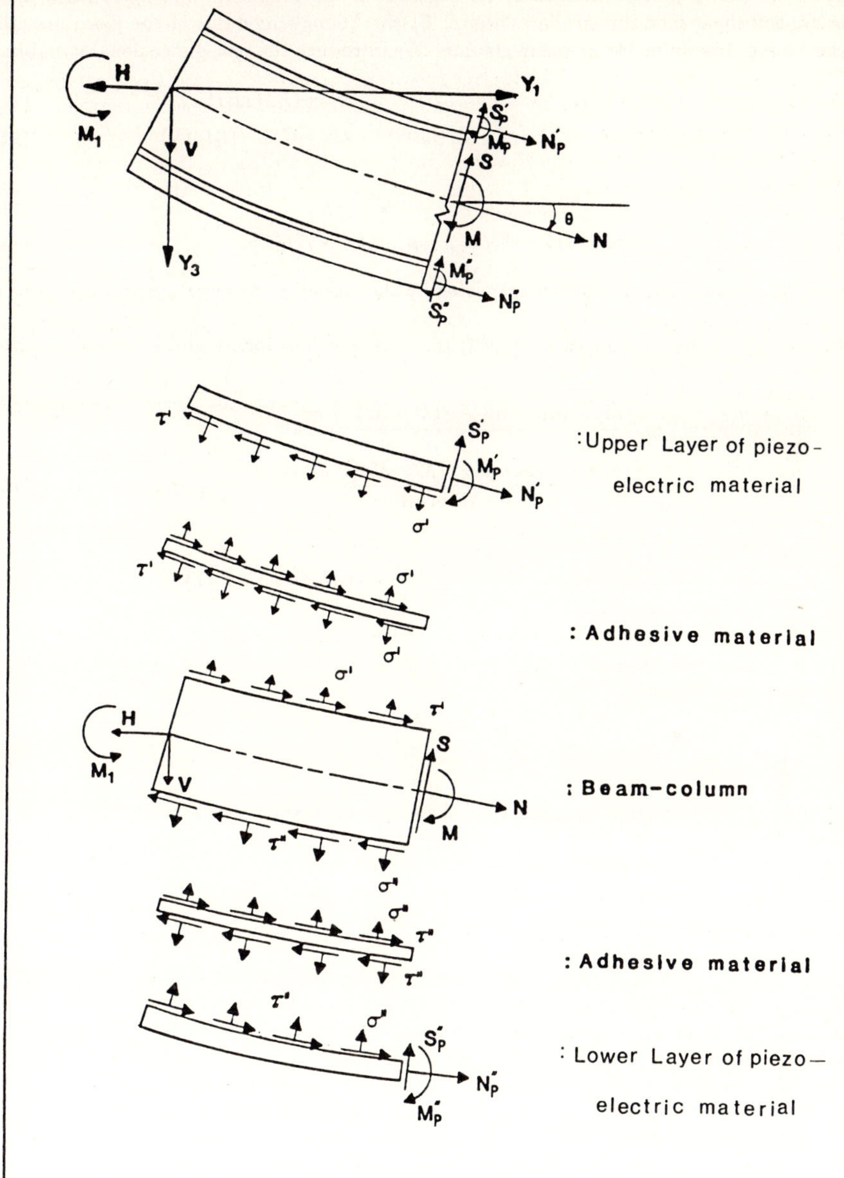

Fig. 22 Free body diagram of each layer

Let t_a be the thickness of the adhesive, t_p be the thickness of the piezo-electric liner; t the thickness of the beam; L is the length of the segment of the beam-column lined with a piezo-actuator; G_a the shear modulus of the adhesive; E_p the Young's modulus of the piezo material; and E the Young's modulus of the beam column. We introduce the non-dimensional variables:

$$\bar{x} = x/L; \qquad \xi = L/t; \qquad m = (LM/EI) \tag{116}$$

$$\alpha = (6h + h_p)^{\frac{1}{2}}; \quad h = G_a L^2/(t_a t E); \quad h_p = G_a L^2/(t_a t_p E_p) \tag{117}$$

$$\eta = t/t_a; \qquad \overline{G} = G_a/E_p \tag{118}$$

$$p^+ = (\tau' + \tau'')/E_p; \quad p^- = (\tau' - \tau'')/E_p \tag{119}$$

where τ' and τ'' are the shear stresses transmitted by the 'upper' and 'lower' piezo-liners [see Fig. (22)].

It has been shown by Im and Atluri (1987) that the solutions for p^+ and p^- may be written as:

$$p^+ = \eta\overline{G}[\frac{m_2 - m_1}{\alpha^2} - \frac{m_1 - \xi(\Lambda' - \Lambda'')}{\alpha}\sin h\alpha\bar{x} - \frac{\{m_2 - m_1\cos h\alpha - \xi(\Lambda' - \Lambda'')(1 - \cos h\alpha)\}}{\alpha \sin h\alpha}\cos h\alpha\bar{x}] \tag{120}$$

and

$$p^- = \frac{\xi\eta\overline{G}(\Lambda' + \Lambda'' - \frac{2N}{Et})}{\beta}\{\frac{1 - \cos h\beta}{\sin h\beta}\cos h\beta\bar{x} + \sin h\beta\bar{x}\} \tag{121}$$

where $\beta = (2h + h_p)^{\frac{1}{2}}$, m_1 and m_2 are the non-dimensional bending moments in the beam at the end points of the segment lined with a piezo-actuator; and Λ' and Λ'' are the mechanical strains induced by the piezo-liners:

$$\Lambda' = \frac{CV'}{tp}; \qquad \Lambda'' = \frac{CV''}{tp} \tag{122}$$

where V' and V'' are excitation voltages applied to the upper and lower actuators, and c is the piezo-electric constant.

For example, we consider the case when $\Lambda' = -\Lambda''$, and when the axial forces in the member are zero. Then $p^- = 0$ and $\tau'/Ep = \tau''/Ep = p^+/2$. We assume the values:

$$\xi = 10; \quad \eta = 10; \quad G = 1/63; \quad h = 57$$

$$h_p = 423; \quad \Lambda' = -\Lambda'' = 10^{-3}, \text{ and } m_1 = 10^{-3} \tag{123}$$

which corresponds roughly to that of an aluminum column, epoxy adhesive, and ceramic piezo-actuator.

We first plot the shear stress distribution as in Fig. (23) when there is no axial force in the beam-column. The present results agree with those in Crawley and de Luis (1984) when the transverse shear in the beam is taken to be zero ($V = 0$). On the other hand, when $(VL/M_1) =$ 0.8, their result is not different from the result for zero shear force over the range $0 \leq (x/\ell) \leq 0.8$; and such is not the case in the present results. It is also seen that the transverse shear force in the beam contributes a significant change in the distribution of the shear stress exerted by the actuator; and that the degrees of the localization of the transmitted shear stress at the two ends of the actuator segment may be very different depending upon the magnitudes of the transverse shear force in the beam.

Because of the assumption that the piezo-actuator segment is short, the flexural deformation of the actuator segment is decoupled with its axial deformation, and the effect of the axial force

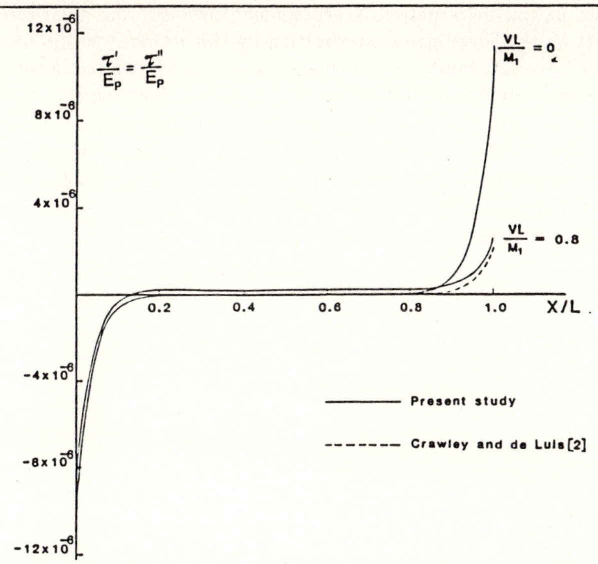

Fig. 23 Effect of shear force in the beam upon the shear stress transmitted to the beam by the actuator when there is no axial force in the beam; Crawley and de Luis' result agrees with the result of the present study when there is no shear force in the beam. However, for $VL/M_1 = 0.8$, their result is not distinct from the result for zero shear force over the range $0 \leq X/L < 0.8$ under the current scale. Such is not the case with the present results.

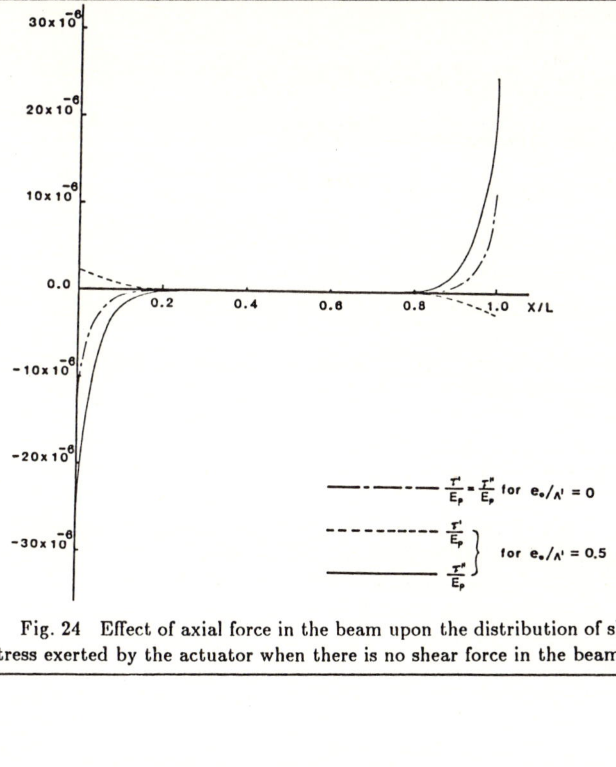

Fig. 24 Effect of axial force in the beam upon the distribution of shear stress exerted by the actuator when there is no shear force in the beam.

in the beam upon its flxural deformation is negligible. However, the axial force in the beam has a significant effect on the shear-stress transmitted by the actuator to the beam-column. For numerical illustration, the distribution of the shear stress exerted by the actuator is plotted in Fig. (24) for a case when there is no transverse shear force, but nonzero axial force in the beam. It is seen that, for this case, the shear stress exerted by the upper actuator has a totally different distribution as compared to the shear stress exerted by the lower actuator. As another example, we consider the case when the transverse shear force as well as the axial force in the beam are nonzero. As seen from Fig. (25), for this case, the distribution of the transmitted shear stress to the beam may be more complex as compared to the two earlier cases.

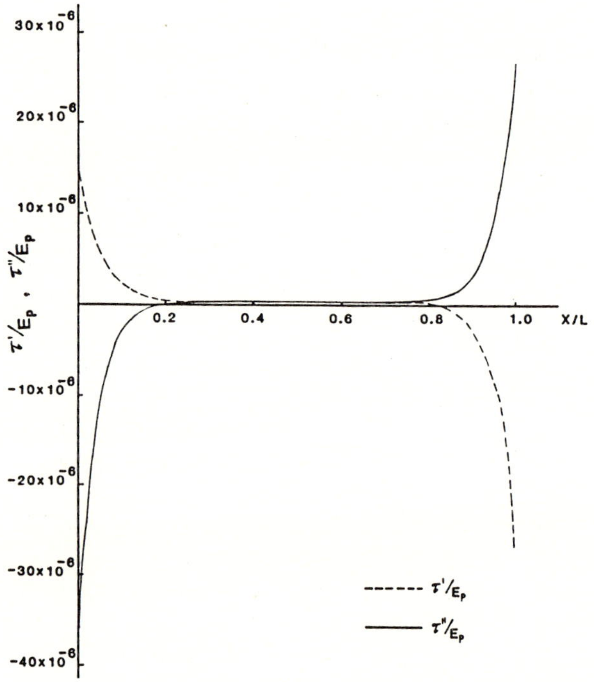

Fig. 25 Distribution of shear stress exerted by the actuator for the case of nonzero shear and axial forces in the beam: $\frac{VL}{M_1} = 1.0$, $e_o/\Lambda' = 1.0$

Finally it is recalled that in the present study, only the rotation of one end of the beam-column, relative to the other, is assumed to be small, because the beam-column segment (with the attached actuator) is assumed to be small. Further, we imposed an approximate rigid rotation, which can be finite, to bring the deformed beam-column segment to the configuration in Fig. (21), so that the line connecting the two-nodes is horizonal. Therefore, the present results are applicable to the case of slender flexible structures undergoing large deformations and rotations. When these controlling shear stresses are included in the special space-truss and space-frame analysis methods described earlier in the paper [Kondoh and Atluri (1987); Shi and Atluri (1987), Tanaka, Kondoh, and Atluri (1985)], the controlled dynamic transient responses may be determined.

Acknowledgements

This work was supported by the U. S. Air Force Office of Scientific Research, and the office of SDIO/IST. The thoughtful encouragement of Dr. A. K. Amos is sincerely appreciated. It is a pleasure to thank Ms. Deanna Winkler for her assistance in preparing this article.

References

Atluri, S. N., and Murakawa, H. (1977) "On Hybrid Finite Element Models in Nonlinear Solid Mechanics" in Finite Elements in Nonlinear Mechanics (Eds. P. G. Bergan, et al) Tapir Press, Norway, Vol. 1, pp. 3-41.

Atluri, S. N., and Grannel, J. J. (1978): Boundary Element Methods (BEM) and Combination BEM-FEM, Report GIT-ESM-SNA-78-16, Georgia Tech, 78 pp.

Atluri, S. N., Gallagher, R. H., and Zienkiewicz, O. C. (Eds.) (1983) Hybrid and Mixed Finite Element Methods, J. Wiley & Sons.

Atluri, S. N. (1984): "Alternate Stress and Conjugate Strain Measures, and Mixed Variational Formulations Involving Rigid Rotations, for Computational Analyses of Finitely Deformed Solids, with Application to Plate and Shell Theory" Computers & Structures, Vol. 18, No. 1, pp. 93-116.

Atluri, S. N., Zhang, J-D., and O'Donoghue, P. E., (1986) "Analysis and Control of Finite Deformations of Plates and Shells: Formulations and Interior/Boundary Element Algorithms" in Finite Element Methods For Plate & Shell Structures, Vol. 2 Formulations & Algorithms (Eds. T. J. R. Hughes & E. Hinton), Pineridge Press, pp. 127-153.

Bryson, A. E., and Ho, Y. (1975) Applied Optimal Control, Hemisphere Publishing Co. Washington.

Crawley, E. F., and de Luis, J. (1984) "Use of Piezo-Ceramics as Distributed Acuators in Large Space Structures" Proc. 22nd AIAA Structures, Structural Dynamics, and Materials Conference, Lake Tahoe, Nevada.

Crisfield, M. A. (1981) "A Fast Incremental/Iterative Solution Procedure That Handles Snap-Through", Computers & Structures, Vol. 13, pp. 55-62.

Fraeijis de Veubeke, B. (1972): "A New Variational Principle for Finite Elastic Displacements" Int. J. Engrg. Science, Vol. 10, pp. 745-763.

Im, S., and Atluri, S. N. (1987) "Force Transfer To a Beam-Column by a Piezo-Actuator in Conjunction with Nonlinear Control of Slender Structures", Computers & Structures (In Press).

Iura, M., and Atluri, S. N. (1986) "On a Consistent Theory, and Variational Formulation of Finitely Stretched and Rotated 3-D Space Curved Beams" Computational Mechanics (In Press).

Iura, M., and Atluri, S. N. (1987): "Dynamic Analysis of Finitely Stretched and Rotated 3-D Space-Curved Beams" Computers & Structures (In Press).

Kondoh, K., Tanaka, K., and Atluri, S. N. (1985) "A Method for Simplified Nonlinear Analysis of Large Space-Tresses and Frames, Using Explicitly Derived Tangent Stiffnesses, and Accounting for Local Buckling" U. S. Air Force Wright Aero Labs, AFFDL-TR-85-3079, 171 pp.

Kondoh, K., and Atluri, S. N. (1985a) "Influence of Local Buckling on Global Instability: Simplified, Large Deformation, Post-Buckling Analysis of Plane Trusses" Computers & Structures, Vol. 21, No. 4, pp. 613-627.

Kondoh, K., and Atluri, S. N. (1985b) "A Simplified Finite Element Method for Large Deformation, Post-Buckling Analyses of Large Frame Structures, Using Explicitly Derived Tangent Stiffness Matrices" Int. Jnl. for Num. Meth. in Engg. , Vol. 23, pp. 69-90.

Kondoh, K., Tanaka, K., and Atluri, S. N. (1986) "An Explicit Expression for Tangent Stiffnesses of a Finitely Deformed 3-D Beam and its Use in the Analysis of Space Frames" Computers & Structures, Vol. 24, No. 2, pp. 253-272.

Kondoh, K., and Atluri, S. N. (1987) "Large-Deformation, Elasto-Plastic Analysis of Frames Under Non-Conservative Loading, Using Explicitly Derived Tangent Stiffness Based on Assumed Stresses" Computational Mechanics, Vol. 2, No. 1, pp. 1-25.

Laub, A. J. (1979) "A Shur Method for Solving Algebraic Riccati Equations" IEEE Transactions on Automatic Control, Vol. AC-24, pp. 913-921.

Noor, A. K., and Mikulas, M. M. (1987) "Continuum Modeling of Large Lattice Structures-Status and Projections" in Large Space Structures: Dynamics & Control (Eds: S. N. Atluri and A. K. Amos), Springer-Verlag (this volume).

O'Donoghue, P. E. (1985) "Boundary Integral Equation Approach to Nonlinear Response Control of Large Space Structures: Alternating Technique Applied to Multiple Flaws in Three-Dimensional Bodies" Ph. D. Thesis, Georgia Tech, 232 pages.

O'Donoghue, P. E., and Atluri, S. N. (1986) "Control of Dynamic Response of a Continuum Model of a Large Space Structure" Computers & Structures, Vol. 23, pp. 199-211.

O'Donoghue, P. E., and Atluri, S. N. (1987) "Field/Boundary Element Approach to the Large Deflection of Thin Plates" Computers & Structures, Vol. 27, No. 3, pp. 427-435.

Punch, E. F., and Atluri, S. N. (1985) "Large Displacement Analysis of Plates by a Stress-Based Finite Element Approach" Computers & Structures, Vol. 24, No. 1, pp. 107-117.

Reissner, E. (1973) "On a One-Dimensional Large-Displacement Finite-Beam Theory" Studies in Applied Math. , Vol. 52, pp. 87-95.

Reissner, E. (1981) "On Finite Deformations of Space-Curved Beams" J. Appl. Math. Physics (ZAMP), Vol. 32, pp. 734-744.

Riks, E. (1972) "The Application of Newton's Method to the Problem of Elastic Stability" J. Applied Mech. , Vol. 39, pp. 1060-1066.

Shi, G. Y., and Atluri, S. N. (1987) "Elasto-Plastic Large Deformation Analysis of Space-Frames: A Plastic-Hinge and Stress-Based Explicit Derivation of Tangent Stiffness" Int. Jrl. for Num. Meth. in Engg. (In Press).

Simo, J. C., and Vu-Quoc, L. (1986) "A Three-Dimensional Finite-Strain Rod Model. Part II: Computational Aspects" Comp. Meth. Appl. Mech. & Engg. , Vol. 58, pp. 79-116.

Tanaka, K., Kondoh, K., and Atluri, S. N. (1985) "Instability Analysis of Space Trusses Using Exact Tangent Stiffness Matrices" Finite Elements in Analysis & Design, Vol. 1, pp. 291-311.

Timoshenko, S. P., and Gere, J. M. (1961) Theory of Elastic Stability 2nd ed. , McGraw-Hill, N. Y., pp. 76-82.

Zhang, J-D., and Atluri, S. N. (1986a) "A Boundary/Interior Element Method for Quasi-static and Transient Response Analyses of Shallow Shells" Computers & Structures, Vol. 24, pp. 213-224.

Zhang, J-D., and Atluri, S. N. (1986b) "Nonlinear Quasi-static and Transient Response Analyses of Shallow Shells: Formulations and Interior/Boundary Element Algorithms" in Boundary Elements (Ed: Q. Du) Pergamon Press, Oxford, pp. 87-110.

Zhang, J-D. (1987) "Nonlinear Dynamic Analysis and Optimal Control of Shallow Shells by the Field-Boundary-Element Approach" Ph. D. Thesis, Georgia Tech, 190 pages.

Zhang, J-D., and Atluri, S. N. (1987) "Post-Buckling Analysis of Shallow Shells by the Field-Boundary-Element Method" Int. Jrl. for Num. Meth. in Engg. (In Press).

Modal Cost Analysis for Simple Continua

A. Hu*, R.E. Skelton** and T.Y. Yang***

School of Aeronautics and Astronautics
Purdue University
West Lafayette, Indiana 47907, U.S.A.

ABSTRACT

The most popular finite element codes are based upon appealing theories of convergence of modal frequencies. For example the popularity of cubic elements for beam-like structures is due to the rapid convergence of modal frequencies and stiffness properties. However, for those problems in which the primary consideration is the accuracy of response of the structure at specified locations it is more important to obtain accuracy in the modal costs than in the modal frequencies. The modal cost represents the contribution of a mode in the norm of the response vector. This paper provides a complete modal cost analysis for simple continua such as beam-like structures. Upper bounds are developed for mode truncation errors in the model reduction process and modal cost analysis dictates which modes to retain in order to reduce the model for control design purposes.

1.0 INTRODUCTION

This paper presents a complete modal analysis for simple continua such as beam-like structures. Convergence properties of the modal cost are developed with finite element models of flexible structures and the efficiency of these elements is discussed when the model is to be used in control design. We wish to show that in this case the finite element decisions may be quite different than those the structural analyst might make using the standard modal frequency convergence criterion.

We shall illustrate concepts using the simplest possible example, a simply supported beam and show that even this simple structure lacks a complete theory of finite element modeling for control design. Although the analysis is performed here mainly for this simple example, the conclusions can be shown also true for other space structures with multi-input multi-output [15]. Much has been written about the control of flexible structures, and the beam example has been used to illustrate the stability problems present with feedback control [1-3]. However, there is yet to emerge a theory on how to modify the original finite element model to improve these stability properties. Instead, attention in the past has been focused on the model reduction problem after a finite element model is given. [4-8].

---

* Graduate Research Assistant, School of Aeronautics and Astronautics.

** Professor, School of Aeronautics and Astronautics

*** Professor of Aeronautics and Astronautics and Dean of Engineering Schools.

Springer Series in Computational Mechanics
S. N. Atluri, A. K. Amos (Eds.)
Large Space Structures: Dynamics and Control

Traditionally, a finite element model is generated using such model error criteria as accuracy of a) stress and strain [9], b) kinetic and potential energies [10], c) frequencies [11], and d) momentum [5]. These criteria are all important in the design of the structure where the structural integrity is of paramount importance. However, when applied to modeling for control design, the weakness of these criteria is that the time varying nature of the applied load is ignored. Indeed, the control forces cannot be known *a priori*. Typically, some assumption about the force is required to proceed. But if one assumes the locations of the forces then he precludes any subsequent optimization of the actuator location. If one assumes the magnitude and time history of the forces then he precludes any subsequent optimization of the control forces. In other words, one cannot optimize the control prior to the development of mathematical model of the structure, and yet one cannot develop the most appropriate mathematical model prior to specific knowledge of the control forces. This leads to

THE MODEL-CONTROL INSEPARABILITY PRINCIPLE [17]

The modeling and control problems are not separable and are necessarily iterative.

The significance of this principle is that one can *never* provide a reduced order model of the structure which promises satisfactory performance after the control law is designed (over the whole range of control law possibilities), no matter how closely the frequencies (or other quantities such as stress, strain, kinetic energy, potential energy, momentum coefficients, etc.) are matched between the physical structure and the approximate model. The second point to be made about the principle is that there currently exists no systematic means to iterate between the modeling and control problems such that improvement is guaranteed on subsequent iterations.

The purpose of this study is to better understand the interaction between finite element modeling decisions and control system design, by using the control objective function in the modal reduction decisions. This leads to modal cost analysis [4,18,19] as a tool for determining sufficient accuracy in models.

The guidelines for finite element modeling for *structural* analysis are well-developed and sophisticated [9-11]. Among the principal conclusions are these (as applied, where appropriate, to the simply-supported Euler-Bernoulli beam):

(a) Uniform beam elements using cubic polynomial shape functions gives no greater than 11% error in the first half frequencies for beam-like structures [12].

(b) The convergence of the first half frequencies improves monotonically with increasing number of finite elements [12].

The importance of a mode in the *control* problem may be measured by the contribution a mode makes in the scalar control objective function. This contribution is a scalar called the "modal cost". This suggests that control objectives will affect modeling decisions and that modal cost analysis may help provide a unified criteria for the selection of finite element models for control design. In [13] and [14] it was shown that: 1) the modal costs are not ordered by frequency; 2) modal cost errors may exceed (by large margins) the errors in modal frequency; 3) convergence of the modal costs is *not monotonic* with the number of uniform cubic beam elements.

A quintic beam element is shown to be more efficient than the cubic beam elements for the control design of beam-like structures. This is consistent with the observation that the elements (increasing the order of shape functions) is more efficient than refining the mesh (decreasing the size of the element) [21-23].

This paper is organized as follows. Section 2.0 briefly reviews the necessary modal cost analysis. Sections 3.0 concerns the convergence properties for simple continua such as beam-like structures and presents some useful theorems. In Section 4.0 open-loop modal cost analysis is applied to finite element models of a simply supported beam. The finite element models are obtained using both cubic and quintic beam elements. Section 5.0 uses the finite element models derived from the cubic and quintic beam elements for the control design and evaluates the closed loop performance of the resultant "optimal" regulators.

2.0 MODAL COST ANALYSIS

Any partial differential equation of the form

$$\rho(r)\ddot{\mu}(r,t) + \tilde{K}\mu(r,t) = f(r,t) \tag{1}$$

can be viewed as arising from an elastic system with kinetic energy

$$T = \frac{1}{2}\int_E \dot{\mu}^T(r,t)\rho(r)\dot{\mu}(r,t)dr \tag{2}$$

where integration is performed over the entire structure, E, and potential energy

$$U = \frac{1}{2}\int_E \mu^T(r,t)\tilde{K}\mu(r,t)dr \tag{3}$$

where $\tilde{K}$ is a self-adjoint operator defined to include the boundary conditions. Using the Ritz approximation

$$\mu(r,t) = \Psi(r)q(t)\ , \quad \Psi(r) \triangleq [\psi_1, \psi_2, ..., \psi_N] \tag{4}$$

the partial differential equation (1) may be modeled by a finite number of ordinary differential equations

$$M\ddot{q} + Kq = Bu\ , \quad y = Pq + R\dot{q} \tag{5}$$

where $q^T \triangleq [q_1(t), q_2(t), \cdots, q_N(t)]$ and y is the output of the system, and

$$M = \int_E \Psi^T(r)\rho(r)\Psi(r)dr \tag{6a}$$

$$K = \int_E \Psi^T(r)\tilde{K}\Psi(r)dr \tag{6b}$$

$$Bu = \int_E \Psi^T(r)f(r,t)dr \tag{6c}$$

and ψ_i, i = 1, ... is a complete set of basis functions. (If the basis functions ψ_i are not *complete,* then the models (1) and (5) are not equivalent and (5) is an approximation of (1)).

Now if the applied control force u_k (or torque u_j) is applied at discrete location r_k (or r_j), then

$$B_{(kth\ col.)} = [\psi^T(r_k)) \quad (\text{or } B_{(jth\ col.)} = [\frac{d}{dr}\psi^T(r_j)]) \tag{6d}$$

If the l-th (or m-th) output is rectilinear (or rotational) displacement at r_l (or r_n), then

$$P_{(lth\ row)} = [\psi^T(r_l)] \quad (\text{or } P_{(mth\ row)} = [\frac{d}{dr}\psi^T(r_m)]) \tag{6e}$$

If the l-th (or m-th) output is rectilinear (or rotational) displacement rate at r_l (or r_m), then

$$R_{(lth\ row)} = [\psi^T(r_l)]\ (\text{or } R(mth\ row) = [\frac{d}{dr}\ \psi^T(r_m)]) \tag{6f}$$

Assume that the control objective is to regulate (that is, to keep small) a quadratic function V_y of the selected output variables, $y^T = (y_1, \ldots, y_k)$,

$$V_y = E_\infty y^T Q y,\ E_\infty \triangleq \lim_{t\to\infty} E \tag{7a}$$

where E is the expectation operator which is needed because the electromechanical actuators usually have additive white noise. Some trade-off is required with the amount of control used, as measured by $V_u = E_\infty u^T R u$. Hence, the optimal control is determined to minimize $V_y + \rho V_u$ where the best value of ρ is judged empirically by evaluating the controller in the presence of a more complete evaluation model of the system. This step always yields V_y versus V_u plots of the form in Fig. 1. That is, the difference between the optimal control theoretical results (solid curve in Fig. 1)

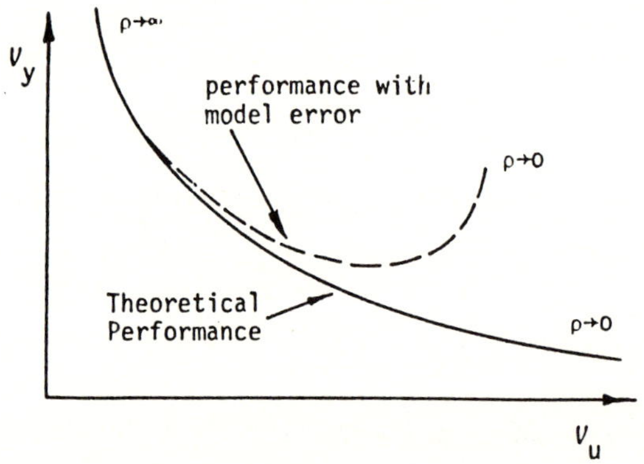

Fig. 1 Theoretical and Realizable Performance

and the actual response characteristics (dotted curve in Fig. 1) is due to modeling errors. These modeling errors always exist and can be decomposed into four categories [16]: (1) model error due to incorrect model order, (2) model error due to neglected disturbances, (3) model error due to incorrect parameters (coefficients in the differential equations) and (4) model error due to neglected nonlinearities. Any one or any combination of these types of errors has the effect described by Fig. 1. By focusing upon revisions of the finite element model we seek to reduce the effect of errors in model parameters and model order.

Since the actuator produces the desired control u(t) as well as the undesired noise w(t), Bu is replaced by $B(u+w)$ where w(t) represents a zero-mean white noise with intensity W and uncorrelated with the initial state x(0). That is,

$$E w(t) = 0 \tag{7b}$$

$$E w(t) w^T(\tau) = W\delta(t-\tau) \tag{7c}$$

$$E w(t) x^T(0) = 0 \tag{7d}$$

where $\delta(t)$ is the Dirac delta and x is the state vector defined below. In the first of our studies we shall set $u = 0$ in order to study the open-loop situation - the behavior of the structure without feedback control.

The elastic system represented by eqn (5) always admits the state representation

$$\dot{x} = Ax + B(u+w) \; ; \quad x \epsilon R^{2N}, \; u, w \epsilon R^m$$

$$y = Cx \qquad ; \quad y \epsilon R^k \tag{8a}$$

where

$$A = \text{block diag } (S_1, S_2, ..., S_N), \; S_i = \begin{bmatrix} 0 & 1 \\ -\omega_i^2 & -2\varsigma_i\omega_i \end{bmatrix} \tag{8b}$$

$$B^T = [0 \; b_1^T \; 0 \; b_2^T \; \cdots \; 0 \; b_N^T] \; , \; b_i^T = B(\text{ith column}) \tag{8c}$$

$$C = [p_1 \; r_1 \; p_2 \; r_2 \; \cdots \; p_N \; r_N], \; p_i = P(\text{ith row}), \; r_i = R(\text{ith row}) \tag{8d}$$

The state vector may be expressed as $x^T \triangleq [x_1 \; x_2 \; \cdots \; x_N]$ where $x_i^T \triangleq [\eta_i \dot{\eta}_i]$, η_i is the ith modal displacement, and $\dot{\eta}_i$ is the corresponding rate. Also the modal damping coefficients ς_i have been assumed.

Assume $u = 0$. Then the quadratic cost function associated with output y and white noise excitation w, $V(y,w) \triangleq V_y$, may be expressed in terms of the contribution V_i of the ith mode. Thus,

$$V_y \triangleq V(y,w) = \sum_{i=1}^{N} V_i(y,w), \text{ or } V_y = \sum_{i=1}^{N} V_i \tag{8e}$$

where the modal cost V_i may be calculated by [18]

$$V_i = \text{tr}[C^TQCX]_{ii} \; ; \; i = 1, 2, ..., N \; , \; 0 = XA^T + AX + BWB^T \tag{8f}$$

It has been shown [3, 18, 19] that

$$V_i = \frac{1}{4\varsigma_i\omega_i^3} \{p_i^T Q p_i + r_i^T Q r_i \omega_i^2\} \; (b_i^T W b_i) + 0(\varsigma_i) \; , \tag{9a}$$

where $0(\varsigma_i)$ is a term of order ς_i. We are interested in *lightly* damped space structures, where $\varsigma_i \simeq 0.005$. Hence we ignore the $0(\varsigma_i)$ term in (9a). It is shown in [3,18] that the first bracketed terms in (9a) is zero if and only if mode i is unobservable in the output $y(t_i)$. The second bracketed term in (9a) is zero if and only if mode i is uncontrollable from w(t). The first bracketed term is also similar to the mode "participation factor" in structural dynamics, but the importance of a mode *cannot* be determined solely from this term due to the dependence of this term on the scaling of the right eigenvectors. The second bracketed term depends upon the scaling of left eigenvectors (and hence cannot be used alone to rank modes). The *product* of the two terms remains *invariant* with such scaling. Hence the modal cost analysis is much more useful than mode participation factors.

The modal cost analysis (9a) is for a multi-input multi-output (MIMO) system. It is trivial to show that if Q and W are diagonal, then (9a) reduces to (let p_{i_α} denote the αth element of the vector p_i),

$$V_i = \sum_{\alpha,=1}^{k} \sum_{\beta=1}^{m} V_i(y_\alpha, w_\beta) \ , \ V_i(y_\alpha, w_\beta) \triangleq \frac{(p_{i_\alpha}^2 + r_{i_\alpha}^2 \omega_{i_\beta}^2)}{4\varsigma_i \omega_i^3} b_{i_\beta}^2 W_{\beta\beta} Q_{\alpha\alpha} \tag{9b}$$

where $V(y_\alpha, w_\beta)$ is the modal cost associated with a scalar output, y_α and a scalar input w_β. Hence, by summing over the number of inputs and outputs the modal costs associated with all the possible scalar input-output combinations yields the total modal cost for the MIMO case.

We shall discuss only the scalar input-output extensively for a variety of input-output combinations, since eqn. (9b) shows exactly how to handle the MIMO cases in terms of SISO results. For beam-like structures the modal cost expressions (9a) may now be explicitly written in terms of frequencies ω_i, mode shapes ψ_i, and mode slopes ϕ_i for the specific scalar input, scalar output cases of eqn. (10a-10h), where $V_i(y,w)$ denotes the ith modal cost when the scalar output is y measured at r_o and the scalar white noise input is w applied at r_c (y will be chosen as rectilinear or angular displacement or displacement rate (four possibilities, μ, θ, $\dot{\mu}$, $\dot{\theta}$, respectively) and w will be chosen as force f or torque T respectively). Note that θ is defined as $\theta \triangleq \frac{\partial \mu}{\partial r}(r_o, t)$. Also zero initial conditions and unity weighting are assumed, that is, $q(o) = \dot{q}(o) = 0$, and $Q = W = 1$.

$$V_i(\mu,f) = \psi_i^2(r_o)\psi_i^2(r_c)/4\varsigma_i\omega_i^3 \tag{10a}$$

$$V_i(\mu,T) = \psi_i^2(r_o)\phi_i^2(r_c)/4\varsigma_i\omega_i^3, \quad \phi_i \triangleq \frac{\partial}{\partial r}\psi_i(r) \tag{10b}$$

$$V_i(\theta,f) = \phi_i^2(r_o)\psi_i^2(r_c)/4\varsigma_i\omega_i^3 \ , \tag{10c}$$

$$V_i(\theta,T) = \phi_i^2(r_o)\phi_i^2(r_c)/4\varsigma_i\omega_i^3 \ , \tag{10d}$$

$$V_i(\dot{\mu},f) = \psi_i^2(r_o)\psi_i^2(r_c)/4\varsigma_i\omega_i \ , \tag{10e}$$

$$V_i(\dot{\mu},T) = \psi_i^2(r_o)\phi_i^2(r_c)/4\varsigma_i\omega_i \ , \tag{10f}$$

$$V_i(\dot{\theta},f) = \phi_i^2(r_o)\psi_i^2(r_c)/4\varsigma_i\omega_i \ , \tag{10g}$$

$$V_i(\dot{\theta},T) = \phi_i^2(r_o)\phi_i^2(r_c)/4\varsigma_i\omega_i \ . \tag{10h}$$

In general, the expressions for ω_i, $\psi_i(r)$, and $\psi_i(r)$ in eqn (10) can be obtained for various kinds of simple continua from standard textbooks. For simply-supported, sliding-sliding (vertically), and pinned-sliding (vertically) beams, ω_i, $\psi_i(r)$ and $\phi_i(r)$ are listed as follows

Table 1 Expressions for ω_i, $\psi_i(r)$ and $\phi_i(r)$ of Euler-Bernoulli beams

	simply-supported	sliding-sliding	pinned-sliding
ω_i	$k_\omega i^2$	$k_\omega i^2$	$k_\omega(\frac{2i-1}{2})^2$
$\psi_i(r)$	$k_\psi \sin(\pi ri/L)$	$k_\psi \cos(\pi ri/L)$	$k_\psi \cos(\frac{2i-1}{2}\frac{\pi r}{L})$
$\phi_i(r)$	$k_\phi i\cos(\pi ri/L)$	$-k_\phi i\sin(\pi ri/L)$	$-k_\phi(\frac{2i-1}{2})\sin(\frac{2i-1}{2}\frac{\pi r}{L})$

$$k_\omega = \sqrt{\frac{EI}{\rho}}\left(\frac{\pi}{L}\right)^2, \; k_\psi = \sqrt{2/\rho L}, \; k_\phi = k_\psi \pi / L. \tag{11}$$

The upper bounds on modal costs on V_i are given by TABLE 2. It should be mentioned that physical actuator and sensor devices have only a finite bandwidth so that TABLE 2 only has significance for ω_i within the smallest bandwidth of the sensor or actuator devices, ω_{BW}. Then the highest structural mode for which TABLES 1 and 2 apply is $\omega_i \leq \omega_{BW}$, or in terms of mode number, $i^2 \leq \omega_{BW}/k_\omega$. It is clear from Table 2 that the infinite sum $\sum_{i=1}^{\infty} V_i$ exists for five of the eight cases (let $i \rightarrow \infty$ in eqn (10a-10e)). Ignoring the bandwidth limitation we will compute the infinite sums $V = \sum_{i=1}^{\infty} V_i$ in Section 3.0 for these cases.

TABLE 2: Upper Bounds of Modal Costs for Beam-like Structures (pinned-pinned, sliding-sliding, pinned-sliding)

	w = force = f	w = torque = T
$y = \mu(r_o,t)$	$\lvert V_i \rvert \leq \left(\frac{L^4}{\pi^6\sqrt{(EI)^3\rho}}\right)\frac{i^{-6}}{\varsigma_i}$	$\lvert V_i \rvert \leq \left(\frac{L^2}{\pi^4\sqrt{(EI)^3\rho}}\right)\frac{i^{-4}}{\varsigma_i}$
$y = \theta = \frac{\partial\mu}{\partial r}(r_o,t)$	$\lvert V_i \rvert \leq \left(\frac{L^2}{\pi^4\sqrt{(EI)^3\rho}}\right)\frac{i^{-4}}{\varsigma_i}$	$\lvert V_i \rvert \leq \left(\frac{1}{\pi^2\sqrt{(EI)^3\rho}}\right)\frac{i^{-2}}{\varsigma_i}$
$y = \dot{\mu}(r_o,t)$	$\lvert V_i \rvert \leq \left(\frac{1}{\pi^2\sqrt{EI\rho^3}}\right)\frac{i^{-2}}{\varsigma_i}$	$\lvert V_i \rvert \leq \frac{1}{L^2\sqrt{EI\rho^3}\varsigma_i}$
$y = \theta = \frac{\partial\dot{\mu}}{\partial r}(r_o,t)$	$\lvert V_i \rvert \leq \frac{1}{L^2\sqrt{EI\rho^3}\varsigma_i}$	$\lvert V_i \rvert \leq \frac{\pi^2}{L^4\sqrt{EI\rho^3}\varsigma_i} i^2$

3.0 CONVERGENCE OF THE RESPONSE MEASURE V_y

In this section, some analytical expressions for the convergence of the modal costs are developed. For simplicity, Theorems 1-4 are established only for the case of single input and single output. However, modal cost analysis is not limited to single inputs and outputs and hence Theorems 1-4 can be easily generalized to multi-input and multi-output case by use of eqn. (9b). Theorems 1-3 are established for the infinite sum of the modal cost (when it exists) for the pinned-pinned, sliding-sliding and pinned-sliding beams. Theorem 4 determines the number of modes retained so that the total cost error is less than a given small number. Moreover, it can be shown that similar theorems for the beams under the effect of axial forces or on an elastic foundation as well as for bars and shafts can be also developed, and may be found in [15].

Theorem 1: Let $V(y,w)$ *denote* V_y*, the value of eqn* (7a) *with the scalar output* y *and the scalar white noise input* w. *For the Euler-Bernoulli simply supported beam (Fig. 2) with parameters* $W = Q = 1$, $q(o) = \dot{q}(o) = 0$ *of the form of eqn* (11)*, the sum of the modal costs*

$$V(\mu,f) = \sum_{i=1}^{\infty} V_i = \frac{L^4}{4\varsigma\sqrt{\rho(EI)^3}} \{\frac{2r_o^2r_c^2}{3L^6} (r_o^2 + r_c^2)$$

$$+ \frac{2}{15L^5} [r_o^5 + r_c^5 - \frac{(r_o+r_c)^5+|r_o-r_c|^5}{2}] + \frac{2}{3L^4} r_o^2r_c^2\} \tag{12}$$

Or, for displacement outputs $y = \mu(r_o,t)$ *and torque inputs,* $w = T$,

$$V(\mu,T) = \frac{L^2}{4\varsigma\sqrt{\rho(EI)^3}} \{\frac{2r_o^2}{3L^4} (r_o^2+3r_c^2) - \frac{2}{3L^3} [r_o^3-r_c^3$$

$$+ \frac{(r_o+r_c)^3 + |r_o-r_c|^3}{2}] + \frac{2}{3} \frac{r_o^2}{L^2}\} \tag{13}$$

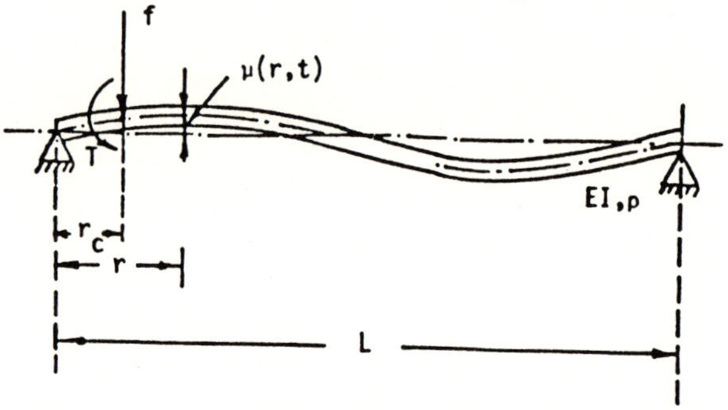

Fig. 2 Simply Supported Beam.

Or, for rotation outputs $y = \theta(r_o,t)$ *and torque inputs* $w = T$,

$$V(\theta,T) = \frac{1}{4\varsigma\sqrt{\rho(EI)^3}} \{\frac{2}{3} + \frac{2}{L^2} (r_o^2 + r_c^2) - \frac{1}{2L} [3(r_o+r_c)$$

$$+ |r_o-r_c|]\} \tag{14}$$

Theorem 2

Let $V(y,w)$ *denote the value of the modal cost with the scalar output* y *and the scalar white noise input* w*. For the sliding-sliding beam, the sum of the modal costs for displacement output* $y = \mu(r_o,t)$ *and force input* $w = f$ *converges to*

$$V(\mu,f) = \frac{L^4}{4\varsigma\sqrt{\rho(EI)^3}} \{\frac{2}{45L^6} [2r_o^6 + 15r_o^2r_c^2(r_o^2+r_c^2) + 2r_c^6]$$

$$- \frac{2}{15L^5} [r_o^5 + r_c^5 + \frac{(r_o+r_c)^5 + |r_o-r_c|^5}{2}]$$

$$+ \frac{2}{9L^4} [r_o^4 + 3r_o^2r_c^2 + r_c^4] - \frac{2}{45L^2} [r_o^2 + r_c^2] + \frac{4}{945}\} \tag{15}$$

Or, for displacement output $y = \mu(r_o,t)$ *and torque input* $w=T$

$$V(\mu,T) = \frac{L^2}{4\varsigma\sqrt{\rho(EI)^3}} \{\frac{2r_c^2}{3L^4}(r_c^2+3r_o^2) \frac{2}{3L^3}[r_c^3 - r_o^3 + \frac{(r_o+r_c)^3 + |r_o-r_c|^3}{2}]$$

$$+ \frac{2r_c^2}{3L^2}\} \tag{16}$$

Or, for rotation displacement output $y = \theta(r_o,t)$ *and torque input* $w=T$

$$V(\theta,T) = \frac{1}{8\varsigma\sqrt{\rho(EI)^3}} \{\frac{(r_o+r_c) - |r_o-r_c|}{L}\} \tag{17}$$

Theorem 3

Let $V(y,w)$ *denote the value of the modal cost with the scalar output* y *and the scalar white noise input* w. *For the pinned-sliding beams, the sum of the modal costs for displacement outputy*$=\mu(r_o,t)$ *and force input* $w=f$ *converges to*

$$V(\mu,f) = \frac{L^4}{\varsigma\sqrt{\rho(EI)^3}} \{-\frac{1}{30L^5}[r_o^5 + r_c^5 + \frac{(r_o+r_c)^5+|r_o-r_c|^5}{2}]$$

$$+ \frac{1}{6L^4}[r_o^4 + 3r_o^2r_c^2 + r_c^4] + \frac{1}{6L^2}[r_o^2 + r_c^2] + \frac{1}{15}\} \tag{18}$$

Or, for displacement output $y = \mu(r_o,t)$ *and torque input* $w=T$

$$V(\mu,T) = \frac{L^2}{\varsigma\sqrt{\rho(EI)^3}} \{\frac{1}{6L^3}[r_o^3 - r_c^3 - \frac{(r_o+r_c)^3 + |r_o-r_c|^3}{2}] + \frac{r_c^2}{2L^2}\} \tag{19}$$

Or, for rotation displacement output $y=\theta(r_o,t)$ *and torque input* $w=T$

$$V(\theta,T) = \frac{1}{8\varsigma\sqrt{\rho(EI)^3}} [\frac{(r_o+r_c) - |r_o-r_c|}{L}] \tag{20}$$

It should be noted that eqns (12-20) hold only when $r_o + r_c \leq L$, and that if $r_o + r_c > L$, r_o should be replaced by $L-r_o$, and r_c by $L-r_c$. We also point out that for all other choices of inputs and outputs, the costs $V(\theta,f)$, $V(\dot{\mu},f)$ have similar forms as above, and may be found in [15].

If the finite sum

$$\sum_{i=1}^{N} V_i/V$$

is sufficiently close to 1, then enough modes N have been retained to make the cost function "complete", i.e., $V - \sum_{i=1}^{N} V_i \leq \epsilon V$ for a sufficiently small $\epsilon > 0$. In order to achieve the most rapid convergence of this partial sum toward V, one must retain the N modes with the largest modal costs V_i. Thus, the relative importance of a mode is established by the modal cost ranking $V_1 \geq V_2 \geq V_3 \geq \ldots$ etc. The advantage of this method of mode selection is that the control objectives (V_y) influence the mode selection process.

The rapid convergence of the model error criterion is illustrated for a simply-supported beam with $V(y,w) = V(\mu,T)$ in Fig. 3, for $r_c = 0$ and a variety of r_o values. Thus, just 4 modes give a modal cost error e_V in eqn. (21) of less than 0.1%,

$$e_V = (1 - \sum_{i=1}^{N} V_i/V) \tag{21}$$

and the critical modes, listed in order of their modal cost, are modes (numbered by frequency) 1,2,4,5. Fig. 3 is drawn for a variety of r_o values to show that the choice of r_o does not alter the basic trends and conclusions drawn from modal cost analysis, although the exact answers depend upon r_o and should be computed for the specific r_o for the problem at hand.

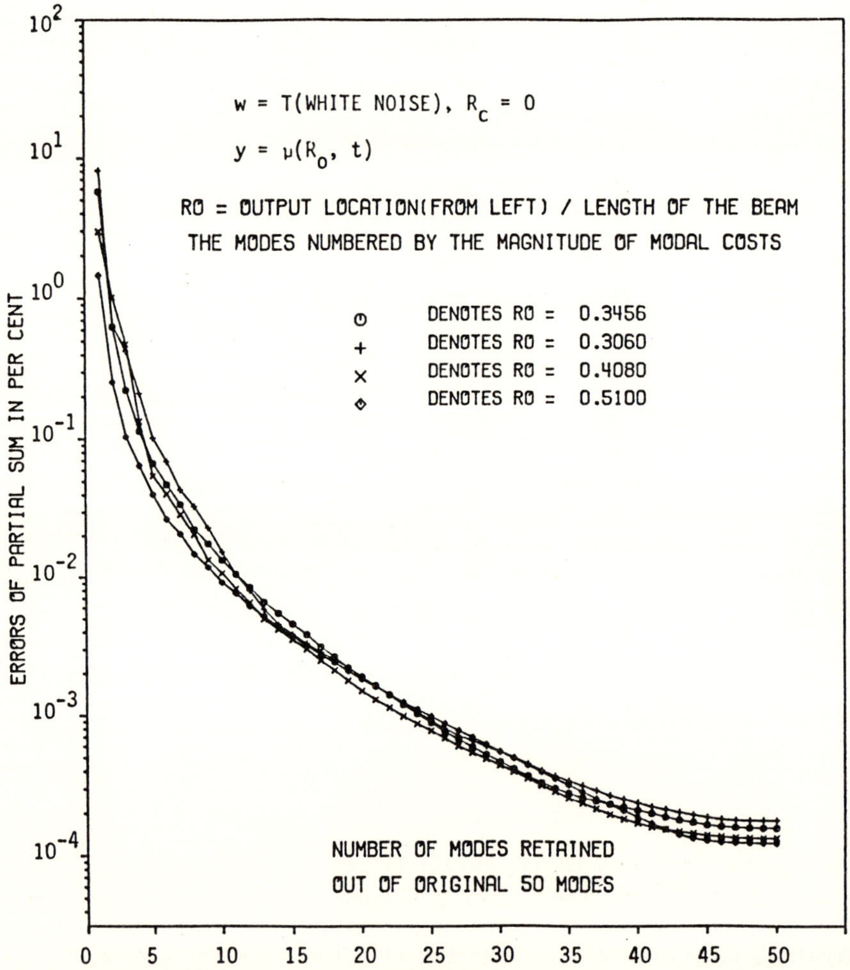

FIGURE 3

CONVERGENCE OF OPEN LOOP COST FUNCTION VS. MODES RETAINED

Theorem 4

Let N(y, w) *denote the number of modes retained for the simply supported Euler-Bernoulli beam with modal costs given by eqn.* (10a–10e), *and with parameters in eqn* (10–11). *The required number of modes retained such that* $(V - \sum_{i=1}^{N} V_i) \leq \epsilon V$, *for any specificed* ϵ *is for displacement outputs* $y = \mu(r_o,t)$ *and force inputs* $w = f$,

$$N(\mu,f) \geq \bar{N}(\mu,f) = \left(\frac{k_\psi^4}{20\varsigma k_\omega^3 \epsilon V}\right)^{\frac{1}{5}} = \left(\frac{L^4}{5\pi^6 V \varsigma\epsilon\sqrt{(EI)^3\rho}}\right)^{\frac{1}{5}} \; ; \tag{22a}$$

Or, for displacement outputs $y = \mu(r_o,t)$ *and torque inputs* $w = T$,

$$N(\mu,T) \geq \bar{N}(\mu,T) = \left(\frac{k_\psi^2 k_\phi^2}{12\varsigma k_\omega^3 \epsilon V}\right)^{\frac{1}{3}} = \left(\frac{L^2}{3\pi^4 V \varsigma\epsilon\sqrt{(EI)^3\rho}}\right)^{\frac{1}{3}} \; ; \tag{22b}$$

Or, for rotation outputs $y = \theta(r_o,t)$ *and force inputs* $w = f$,

$$N(\theta,f) \geq \bar{N}(\theta,f) = \left(\frac{k_\psi^2 k_\phi^2}{12\varsigma k_\omega^3 \epsilon V}\right)^{\frac{1}{3}} = \left(\frac{L^2}{3\pi^4 V \varsigma\epsilon\sqrt{(EI)^3\rho}}\right)^{\frac{1}{3}} \; ; \tag{22c}$$

Or, for rotation outputs $y = \theta(r_o,t)$ *and torque inputs* $w = T$,

$$N(\theta,T) \geq \bar{N}(\theta,T) = \frac{k_\phi^4}{4\varsigma k_\omega^3 \epsilon V} = \frac{1}{\pi^2 V \varsigma\epsilon\sqrt{(EI)^3\rho}} \; ; \tag{22d}$$

Or, for displacement rate output $y = \dot{\mu}(r_o,t)$ *and force input* $w = f$,

$$N(\dot{\mu},f) \geq \bar{N}(\dot{\mu},f) = \frac{k_\psi^4}{4\varsigma k_\omega \epsilon V} = \frac{1}{\pi^2 V \varsigma\epsilon\sqrt{EI\rho^3}} \, . \tag{22e}$$

Theorem 4 is useful in finding the number of modes to be retained so that the total cost error is less than a given small number. The proofs for Theorems 1-4 are omitted and can be found in [15]. Modal cost analysis reveals *which* modes to retain and Theorem 4 reveals *how many* modes to retain in a lightly damped structure.

Now suppose the system in eqn. (8) represents a two-input two-output system, that is, $u = (u_1,u_2)^T = (f,T)^T$ and $y = (y_1,y_2)^T = (\mu,\theta)^T$. If $W = Q = \begin{bmatrix} 1 & 0 \\ 0 & 1 \end{bmatrix}$, then the overall cost function V is

$$V = V(y_1,u_1) + V(y_1,u_2) + V(y_2,u_1) + V(y_2,u_2)$$

$$= V(\mu,f) + V(\mu,T) + V(\theta,f) + V(\theta,T) \tag{23}$$

When the linear system (8) represents a pinned-pinned beam (or sliding-sliding, pinned-sliding beams), then we can use Theorems 1-3 successively for the modal cost corresponding to each input-output combination in the right hand side of eqn. (23), and then sum them up to get the overall cost function V. Obviously this can be easily generalized to simple beams with multiple input-outputs.

Finally, let us use the simply supported beam with $EI = \rho$, $\rho = \frac{2}{L}$, $L = \pi$ (yielding $k_\psi = k_\phi = k_\omega = 1$) and $\varsigma = 0.005$ as an example. The input location is either $r_c = 0.5L$ or $r_c = 0$ for either force or torque actuator. The output location is always $r_o = 0.35L$ for μ, θ, and $\dot{\mu}$ outputs. The total modal costs corresponding to eqn (10a-10e) for the above input and output combinations can be computed as follows.

TABLE 3: Total modal cost corresponding to eqn (10a-10e)

	$r_c = 0.5L$	$r_c = 0$	$r_c = 0.5L$	$r_c = 0$	$r_c = 0.5L$
V_i	$V_i(\mu,f)$	$V_i(\mu,T)$	$V_i(\theta,f)$	$V_i(\theta,T)$	$V_i(\dot{\mu},f)$
$V = \sum_{i=1}^{\infty} V_i$	39.6984	42.0127	10.9585	26.1133	61.6850

If we require that $\epsilon = 0.01$, then the number of modes required, $\overline{N}$, can be found in Table 4.

TABLE 4: Number of Modes Required to Ensure Total Cost Error $\epsilon \leq 0.01$ (pinned-pinned beam, $EI = \rho$, $\rho = \frac{2}{L}$, $L = \pi$)

$\overline{N}(\mu,f)$	$\overline{N}(\mu,T)$	$\overline{N}(\theta,f)$	$\overline{N}(\theta,T)$	$\overline{N}(\dot{\mu},f)$
2	4	6	192	82

For example, when $w = f$, $r_c = 0.5L$, $y = \mu(r_o,t)$, $r_o = 0.35L$, the total modal cost $V = 39.6984$. In order to guarantee that the total cost error $(1 - \sum_{i=1}^{N} V_i/V) < 0.01$, it was found from Table 4 that at least two modes must be retained. In fact, since V is already known, the actual error of total cost retaining 2 modes can be easily computed to be 0.005% < 1%. It should be noted that the error of total cost retaining 1 mode in this case is 0.01% < 1%, which means that equation (22a) gives a conservative estimate for the number $\overline{N}$. It should also be noted that in the fourth column of Table 4, $\overline{N}(\theta,T) = 192$. This suggests that 192 modes are needed for accurate ($\epsilon < .01$) description of the angular position $\theta(t)$ of a point on a beam-like structure with torque actuators acting. In control design practice, it may not be practical to try to control so many modes. Hence one may have to make some compromise in *control performance* since retaining fewer than 192 modes will describe $|\theta(t)|^2$ much less accurately (than $\epsilon = .01$) and hence will make it more *difficult* to tightly control $\theta(t)$ by feedback control design based upon the reduced model.

4.0 OPEN LOOP FINITE ELEMENT MODAL COST ANALYSIS

In this section, open loop modal cost analysis is performed on the finite element models of a simply supported beam. However, in [15] it is shown that the results are valid for beams having other boundary conditions with multi-input and multi-output, since the modal cost formula takes very similar forms in the later cases as that of simply supported beam. The finite elment models for the beam are derived using either cubic or quintic beam elements. The results are compared with the exact modal costs computed in the previous section in order to evaluate the convergence of the modal costs when finite element methods are used. The beam used is shown in Fig. 2 where a white noise torque w is applied at $r_c = 0$ and the output is either linear displacement $y = \mu(r_o, t)$ or rotation rate $y = \dot{\theta}(r_o,t)$, $r_o = .35L$ where L is the length of the beam.

The displacement function for the cubic shape function is

$$\mu(r,t)=a_1(t) + a_2(t)r + a_3(t)r^2 + a_4(t)r^3 \tag{24}$$

and for the quintic shape function is

$$\mu(r,t) = c_1(t)+c_2(t)r+c_3(t)r^2+c_4(t)r^3+c_5(t)r^4+c_6(t)r^5 \tag{25}$$

where the a_i's and c_i's are generalized coordinates [20,21].

The degrees of freedom at a nodal point for a beam modelled with cubic shape functions are displacement (μ) and rotation ($\theta \triangleq \frac{\partial \mu}{\partial r}$). When using quintic shape functions there is an additional degree of freedom called curvature ($\kappa \triangleq \frac{\partial^2 \mu}{\partial r^2}$) at each nodal point [20]. It is this additional degree of freedom that greatly improves the crude curvature distribution of cubic beam elements and hence makes the quintic beam elements more resemble to system eigenfunctions.

Using standard finite element procedure, the second order form and the output equation corresponding to eqn. (8) can be easily obtained as follows [14,15].

$$\ddot{\eta}_i(t) + 2\varsigma_i\hat{\omega}_i\dot{\eta}_i(t) + \hat{\omega}_i^2\eta_i(t) = \hat{b}_i^T w(t), \quad i = 1, ..., N \tag{26}$$

$$y = \sum_{i=1}^{N} [\hat{p}_i\eta_i(t)+\hat{r}_i\dot{\eta}_i(t)] \tag{27}$$

where $\hat{\omega}_i$, $\hat{b}_i$, $\hat{p}_i$, and $\hat{r}_i$ are the approximate values of ω_i, b_i, p_i and r_i computed from finite element method. Also the modal costs formula in eqn (9a) takes the following form

$$\hat{V}_i = \frac{1}{4\varsigma_i\hat{\omega}_i^3} \{\hat{p}_i^T Q\hat{p}_i+\hat{r}_i^T Q\hat{r}_i\hat{\omega}_i^2\}(\hat{b}_i^T W\hat{b}_i) , \quad i = 1, ..., N \tag{28}$$

Note that the modal cost magnitudes are not necessarily ordered by frequency since $\hat{V}_i$ in eqn. (28) is a function not only of frequency but also of mode shape and mode slope [15].

The relative errors in frequency and modal cost e_{ω_i} and e_{v_i} are defined by

$$e_{\omega_i} \triangleq \left|\frac{\hat{\omega}_i-\omega_i}{\omega_i}\right| \times 100\% \tag{29}$$

$$e_{v_i} \triangleq \left|\frac{\hat{V}_i-V_i}{V_i}\right| \times 100\% \tag{30}$$

where V_i and $\hat{V}_i$ are computed using eqns. (9a) and (28) respectively.

Using the procedures in eqn (26-30) with both cubic and quintic beam elements we can compute the errors in natural frequencies and modal costs. Fig. 4 - 9 compare the *trends* of frequency and modal cost errors as the number of uniform finite elements increases. Fig. 10 and 11 are used to compare the numerical *efficiency* in computing above quantities based on 20 degrees of freedom finite element models.

Note that although Fig. 6-11 are plotted only for specific input and output locations, they do represent the general trends of our numerical simulation for a variety of input and output locations, which is briefly explained later in this section.

Fig. 4 shows that the frequency errors monotonically decrease as the number of finite elements increases for cubic beam elements. Note that the errors in the first half of modal frequencies are no greater than 11% for cubic beam elements. In contrast to Fig. 4, Fig. 6 shows that when using the cubic shape function the modal cost errors *do not monotonically* decrease with an increase in the number of finite elements for $y = \mu(r_o,t)$. From Fig. 6 note that the 5-element model predicts the modal cost of mode 2 ($e_{v_2} = 4\text{x}10^{-3}$) much more accu-

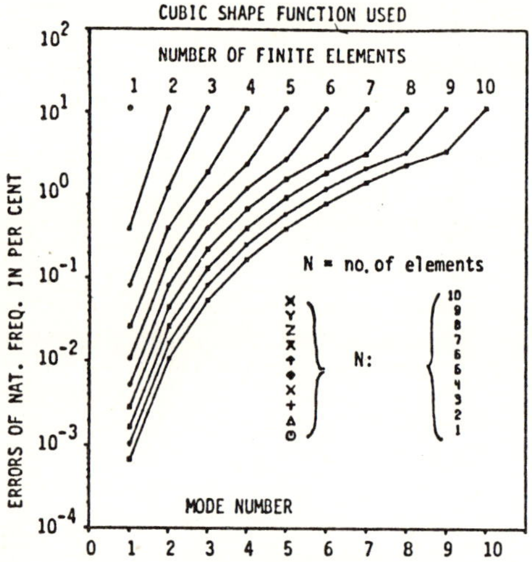

FIGURE 4 ERROR IN FREQUENCY VS. MODE NUMBER AS THE NUMBER OF ELEMENTS INCREASES

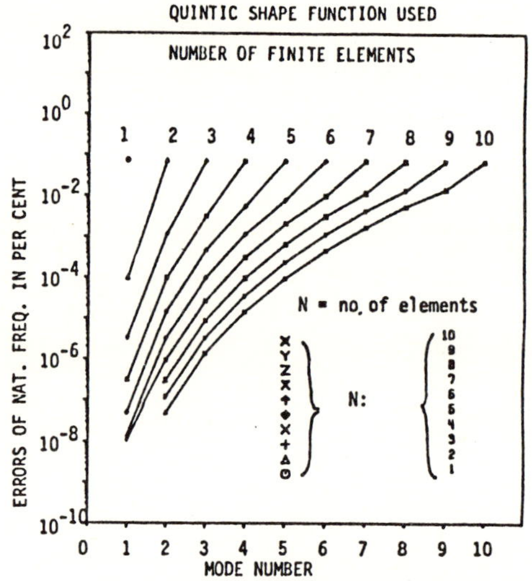

FIGURE 5 ERROR IN FREQUENCY VS. MODE NUMBER AS THE NUMBER OF ELEMENTS

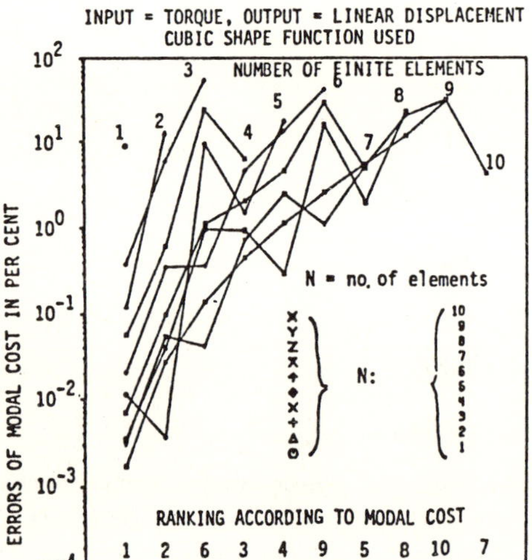

FIGURE 6 ERROR IN MODAL COST VS. MODE NUMBER AS THE NUMBER OF ELEMENTS INCREASES

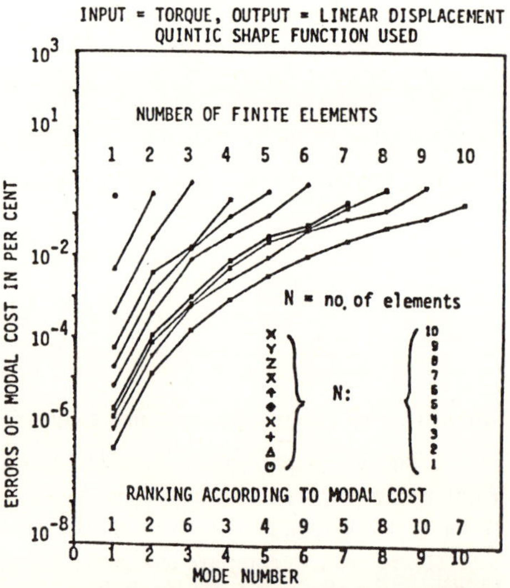

FIGURE 7 ERROR IN MODAL COST VS. MODE NUMBER AS THE NUMBER OF ELEMENTS INCREASES

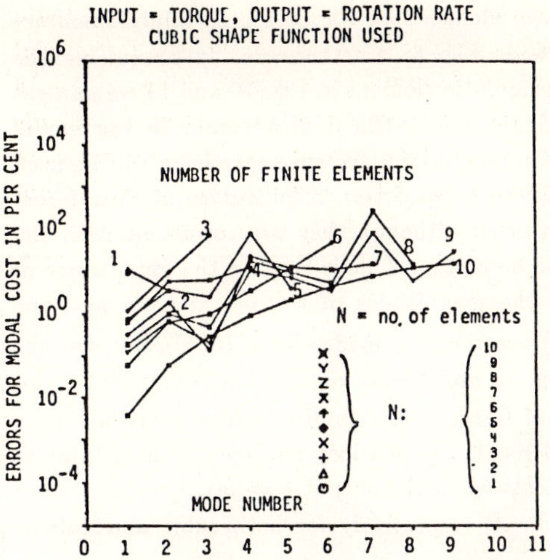

FIGURE 8 ERRORS IN MODAL COST VS. MODE NUMBER AS THE NUMBER OF ELEMENTS INCREASES

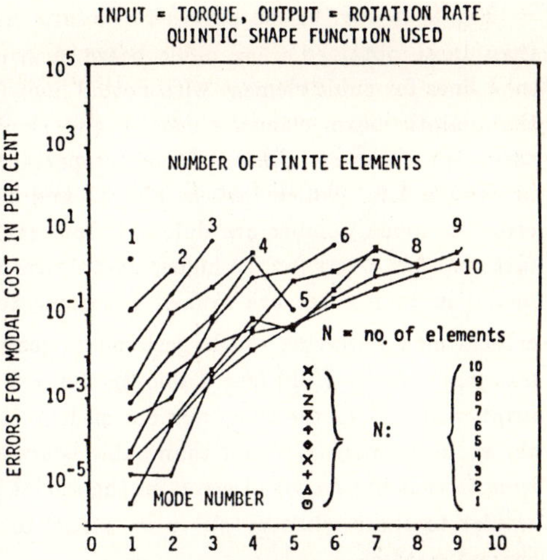

FIGURE 9 ERROR IN MODAL COST VS. MODE NUMBER AS THE NUMBER OF ELEMENTS INCREASES

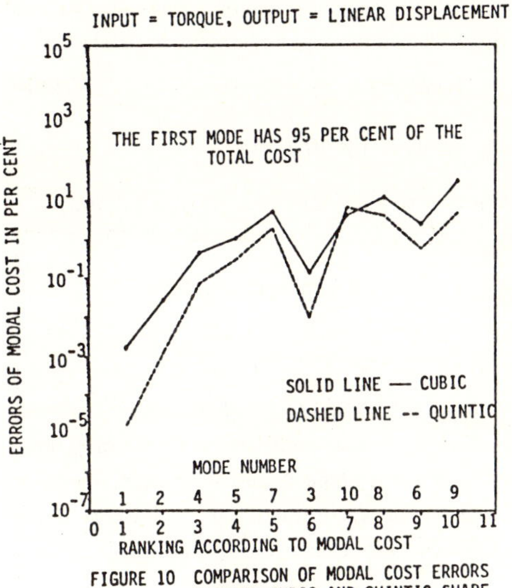

FIGURE 10 COMPARISON OF MODAL COST ERRORS BETWEEN CUBIC AND QUINTIC SHAPE FUNCTIONS WITH 20 D.O.F. FINITE ELEMENT MODELS

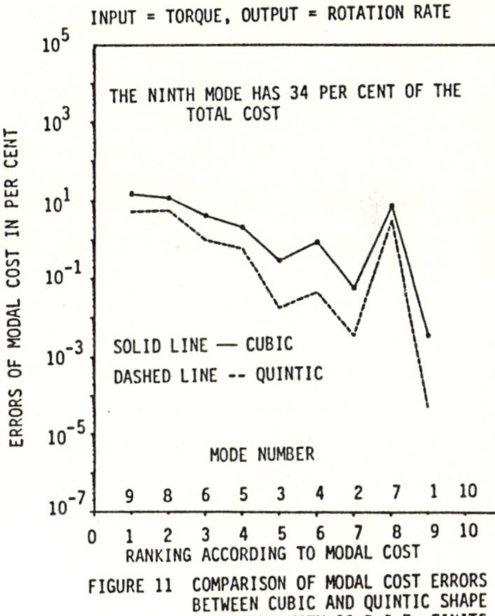

FIGURE 11 COMPARISON OF MODAL COST ERRORS BETWEEN CUBIC AND QUINTIC SHAPE FUNCTIONS WITH 20 D.O.F. FINITE ELEMENT MODELS

rately than a 10-element model ($e_{v_2} = 3x10^{-2}\%$). However, Fig. 7 shows that when using quintic beam elements the modal cost errors decrease *monotonically* for $y = \mu(r,t)$. Although the errors in the modal costs do not decrease monotonically for quintic beam elements when y

$= \theta(r_o, t)$ in Fig. 9, the curves for quintic beam elements in Fig. 9 are certainly *smoother* than those obtained using cubic beam elements in Fig. 8. Furthermore, by comparing the solid lines for cubic element with dotted lines for quintic element in Fig. 10 and 11 we can see that quintic beam elements usually give significant improvement of accuracy in the modal costs, the *accuracy of which* is very important in closed-loop system analysis as will be seen in Section 5.0. Notice that in Fig. 10 and 11, the general trends for curves of modal cost error vs. mode number are different from each other although they are consistent with the fact that the errors for the higher modes tend to be larger. This is because the importance of modes in both cases are ranked according to the magnitudes of the modal costs and the critical modes (having the largest modal costs) are the first modes for $V_i(\mu, T)$, but are the last modes for $V_i(\theta,T)$ (see Table 2). Since Fig. 10 and 11 are obtained based on finite element models with the same number of degrees of freedom, the quintic beam elments are thus shown to be *more efficient* than cubic beam elements in open loop analysis. It is of interest to note that the *trends* of errors in the modal costs when the output $y = \dot{\mu}(r_o,t)$ or $\theta(r_o,t)$ are *similar* to those when output $y = \mu(r_o,t)$ or $\theta(r_o,t)$ respectively for both cubic and quintic shape functions.

Moreover, the open loop modal cost analysis for beams with other boundary conditions are also performed [15], which shows the same trend as that for the simply-supported beam.

The reasons for the better convergence properties of the modal cost errors using the quintic beam elements can be stated as follows. Let $\{V_i, \omega_i, \zeta_i, \psi_i(r_o), \phi_i(r_c)\}$ be the exact modal cost, frequency, damping, mode shape and mode slope for mode i and let $\{\hat{V}_i, \hat{\omega}_i, \hat{\zeta}_i, \hat{\psi}_i(r_o), \hat{\phi}_i(r_c)\}$ represent the approximate quantities from finite element methods. Define

$$R_{\zeta_i} \triangleq \zeta_i/\hat{\zeta}_i \,, \quad R_{\omega_i} \triangleq (\omega_i/\hat{\omega}_i)^3 \,, \tag{31a}$$

$$R_{\phi_i} \triangleq \|\phi_i(r_c)\|^2/\|\hat{\phi}_i(r_c)\|^2 \,, \quad R_{\psi_i} \triangleq \|\psi_i(r_o)\|^2/\|\hat{\psi}_i(r_o)\|^2 \tag{31b}$$

Then the modal cost error defined by eqn (30) is related to eqn (31a) and (31b) as follows.

$$e_{v_i} = \frac{R_{\zeta_i} R_{\omega_i}}{R_{\phi_i} R_{\psi_i}} - 1 \tag{31c}$$

This result follows immediately by substitution of eqn (31a), (31b) into eqn (29). Eqn (31c) makes it clear why e_{v_i} may not be proportional to e_{ω_i} (as Fig. 6 illustrated), since the mode shape and mode slope errors R_{ϕ_i} and R_{ψ_i} can dominate the effects of frequency errors. Yedavalli [24] also showed for plates that the cost function V is most sensitive to errors in ζ_i, and V_i is next most sensitive to errors in mode shape ψ_i or mode slope ϕ_i and V_i is *least* sensitive to errors in frequency ω_i. Obviously we could improve the accuracy of the modal cost if we can better approximate ϕ_i and ψ_i. It is known that the exact solution for the displacement function $\mu(r,t)$ for the vibration of the simply supported beam is

$$\mu(r,t) = \sum_{i=1}^{\infty} \psi_i(r) q_i(t) \qquad \psi_i(r) = \sqrt{\frac{2}{\rho L}} \sin\left(\frac{i\pi}{L} r\right)$$

where

$$\sin r = r - \frac{r^3}{3!} + \frac{r^5}{5!} - \frac{r^7}{7!} + \cdots$$

therefore including two higher order terms r^4, r^5 in the approximate shape function helps better approximate ψ_i and ϕ_i and hence improve the accuracy of the modal costs. Obviously, this should be true for the entire domain of r for mode shape ψ_i and mode slope ϕ_i, which explains why the trends shown in Fig. 6-11 are quite representative of all input-output locations. This result is also consistent with the general observation in the finite element practice, that is, refining the element (the polynomial degrees of freedom is increased, while the mesh remains fixed) is superior than refining the mesh (the mesh size is approaching zero while the polynomial degrees of freedom of shape function remains fixed). Interested readers are referred to [23], which suggests that refining the element is more attractive than refining the mesh.

At this point, one would naturally expect to use the seventh (or even higher) order shape functions [22], since they are more efficient than quintic shape functions in the open loop modal cost analysis [15] and they enable the designer to consider all the possible boundary conditions for beam-like structures [22]. However, as we shall see in Fig. 12 since the quintic beam element models are already within 5% of the theoretically optimum performance for the closed loop system discussed in Section 5.0, the seventh order shape functions are not recommended here for *simply supported beam-like* structures.

We conclude that quintic elements are much more computationally efficient than cubic elements in the construction of models whose mean-squared value of positions and rates on beam-like structures are of paramount importance. Seventh order beam elements are not needed for the simply supported beam-like structures studied here, but may be important for other configurations. The number of modes needed in reduced order models is made available by the new theories of convergence in Section 3.0.

5.0 Closed Loop Performance Analysis

The aim of this section is to analyze the effects on closed loop regulator performance of model errors due to finite element approximations for the simply supported beam. However, as was the case of the open loop modal cost analysis, the results can be easily shown valid for more general cases such as beams with other boundary conditions and with multi-input and multi-output [15].

The input u of the closed loop system (Fig. 2) is a noisy torque applied at the left end and the output y is either linear displacement $\mu(r_o,t)$ or rotation rate $\theta(r_o,t)$ at $r_o = .35L$ where L is the beam length. It is desired to design control laws to minimize $V_y + \rho V_u$ as discussed in Section 2.0.

The *evaluation models*, considered to accurately represent the open loop plant, are 10-mode models derived from the exact shape functions. Thus, the *exact* frequencies, mode shapes and mode slopes were used to obtain the evaluation models which are used as the truth model in our simulations. When the output is the linear displacement, the first ten modes ranked by modal costs are 1, 2, 4, 5, 7, 3, 10, 8, 6 and 9 and they are all controllable and observable. Therefore the first ten modes are all retained in the evaluation model when the output is linear displacement. When the output is rotation rate, the first ten modes ranked by modal costs are 9, 8, 6, 5, 3, 4, 2, 7, 1, 10. But the tenth mode is unobservable and thus has zero cost. Therefore, this mode is discarded and only 9 modes are retained in the evaluation model when the output is rotation rate. Notice that when the output is $y = \mu(r_o,t)$, the first modal cost constitutes 95% of the total modal cost and that when the output is $y = \theta(r_o,t)$, the ninth modal cost constitutes 39% of the total modal cost.

Control design models are obtained by retaining the first 10 or 9 modes of finite element models derived using either cubic or quintic shape functions. Linear Quadratic Gaussian (LQG) theory [25] for either linear displacement or rotation rate output is used to design control laws based on the control design models. These control laws are then used to drive the evaluation models (as the control weighting ρ varies from 0 to ∞). However, although the control laws are optimal for the control design models, because of errors, they are suboptimal for the evaluation models.

In general the errors of a control design model are decreased by *increasing the degrees of freedom* of the initial finite element model. However, it is shown in this study that the *order* of polynomial shape function used in the finite elment modeling may also have a more *significant* effect on the errors of the control design model and thus the performance of the resultant regulator, depending upon the input and output combination. In particular, when the input is a torque and the output is rotation rate, quintic beam elements are shown to be more efficient than cubic beam elements for closed loop performance, judged in part by the minimum achievable output cost V_y. The results of this section are interpreted in terms of the convergence properties of the modal costs discussed in Section 3.0.

The analysis proceeds as follows. First, for either linear displacement or rotation rate output two control design models are obtained from 20 degrees of freedom finite element models derived using either cubic or quintic shape functions. Control laws are developed based on the design models and are used to drive the evaluation models. The results of this analysis can be found in Table 5-7 and Fig. 12. The performance graphs in Fig. 12 have the generic shape of the suboptimal plot shown in Fig. 1. The performance plots are compared with each other and with the optimal performance graphs (obtained by using the evaluation models themselves as the control design models). The smallest possible output cost V_y theoretically achievable is obtained as the control effort approaches infinity ($\rho \longrightarrow 0$) and is equal to either 1.6460 m^2 or 716.9 $(rad/sec)^2$ for either linear displacement or rotation rate output respectively. Note from Fig. 12 that the cubic element models lead to unstable controllers very quickly as the control effort increases beyond $V_u = 1$. (The 20 degree of freedom cubic element model leads to instability at $V_u = 1$.) Using less than 20 degrees of freedom with cubic elements the closed loop performance can be worse than the open loop performance, even when the designs are stable [15].

A) Linear displacement output $y = \mu(r_o,t)$

It is found from Table 5 that when the output is the linear displacement the performances of the controllers developed using both 20 degrees of freedom cubic beam element model and 20 degrees of freedom quintic beam element model are all very close to the theoretical optimal performance.

Now, referring to Table 5 let us compare the performances of the controllers derived using 20 degrees of freedom finite element models in which the output is the linear displacement $y = \mu(r_o,t)$. When using cubic shape functions the resultant controllers can achieve a minimum output of 1.6789 m^2, a performance degradation of only 1% from the theoretical minimum $(1.6460/m^2)$. When using the quintic shape function the resultant controllers are able to achieve a minimum output cost of 1.6462 m^2, a performance degradation of only 0.01% from the theoretical minimum. Both sets of controllers remain stable for very large control efforts and achieve minimum output costs very close to the theoretical minimum.

The equivalence of suboptimal and optimal performances of closed loop systems obtained when using both the exact and finite element models, may be related to the open loop modal cost errors found in Table 5. When the output is linear displacement at $r_o = .35L$ and the input is a white noise torque at the left end of the beam, the modal cost of the first mode constitutes 95% of the total cost. From Table 5 it is seen that the error in the first modal cost is 0.0017% using the 20 degrees of freedom cubic beam element model. Note that judging from closed loop performance this is accurate enough for acceptable closed loop performance even if the 20 degrees of freedom quintic beam model provides two-order of magnitude improvement in the modal cost accuracy (0.000017% vs. 0.0017%).

TABLE 5: Comparison of closed loop performance for 20 degrees of freedom cubic and quintic beam element models when $u = T$, $y = \mu(r_o t)$

	Minimum Achievable Output Cost v_y	Performance Degradation (From Theoretical Performance)
Cubic	1.6789	1%
Quintic	1.6462	0.01%

Note: v_y (theoretical performance) $= 1.6460$
$v_1/v = 95\%$, e_{v_1} (cubic) $= 0.0017\%$, e_{v_1} (quintic) $= 0.000017\%$

B) Rotation rate output $y = \dot{\theta}(r_o,t)$

It is also found from Table 6 and Fig. 12 that when the output is the rotation rate the performance of the controllers developed using the 20 degrees of freedom cubic beam element model is *significantly worse* than the controllers developed using the 20 degrees of freedom quintic beam element model. Thus we increase the degrees of freedom of the cubic beam element model (to 24 this time) and repeat the above procedure. This process is repeated several times and it is shown that 32 degrees of freedom are needed for the cubic beam element model in order to obtain performance equivalent to that obtained when using a 20 degrees of freedom quintic beam element model.

Now, referring to Table 6 and Fig. 12 let us compare the performances of the controllers derived using 20 degrees of freedom finite element models when the output is the rotation rate. When using cubic shape functions the resultant controllers can achieve a minimum output cost of 2694 $(rad/sec)^2$, a performance degradation of 275% from the theoretical minimum. However, when using the quintic shape functions the resultant controllers are able to achieve a minimum output cost of 757.0 $(rad/sec)^2$, a performance degradation of only 5.3% from the theoretical minimum.

The differences in the closed loop system performances obtained when using the cubic and quintic shape functions may again be *related* to the open loop modal cost errors found in Table 7. When the output is rotation rate at $r_o = .35L$ and the input is a white noise torque at the left end of the beam, the modal cost of the 9th mode constitutes 39% of the total cost, and hence is the dominant mode. From Table 7 it is seen that the error in the 9th modal cost obtained from a 20 degrees of freedom cubic beam element model is 15% while the corresponding error obtained from a 20 degrees of freedom quintic element beam model is 5.5%. This error in the modal costs accounts for the much improved closed-loop performance (Fig. 12 shows a factor of 4 improvement) of the quintic over the cubic element.

TABLE 6: Comparison of closed loop performance for 20 degrees of freedom cubic and quintic beam elements modes when u = T, y = $\theta(r_o,t)$

	Minimum achievable output cost v_y	Performance degradation (from theoretical performance)
cubic	2694	275%
quintic	757	5.3%

Note: V_y (Theoretical performance) = 717.7 (rad^2/sec^2)

TABLE 7: Comparison of critical modal cost errors using cubic and quintic beam elements (NDOF = degrees of freedom of finite element model) when u = T, y = $\theta(r_o,t)$

	Quintic	Cubic				
NDOF	20	20	24	28	30	32
$e_{v_9}(\%)$	5.5	15.1	9.0	4.9	3.6	5.0
$e_{v_7}(\%)$	3.3	7.7	46.2	26.9	26.2	14.4

Table 7 and Fig. 12 may also be used to compare the performances of the controllers derived using the 20 degrees of freedom quintic beam element model with those of the controllers derived using 24, 28, 30, 32 degrees of freedom cubic beam element models. The minimum achievable output cost obtained by using the 24 degrees of freedom cubic beam element model is still significantly higher than that obtained by using the 20 degrees of freedom quintic beam element model. By using the 28 degrees of freedom cubic beam element model, the performance is similar to that obtained by using the 20 degrees of freedom quintic beam element model. However, the controllers derived using the 28 degrees of freedom cubic beam element model drives the system unstable more quickly as the control effort is increased. The performance of the controllers using the 30 degrees of freedom cubic beam element model is worse than that using the 28 degrees of freedom cubic beam element model. Finally, the performance obtained by using the 32 degrees of freedom cubic beam element model is equivalent to that obtained by using the 20 degrees of freedom quintic beam element model. Both sets of controllers remain stable for large control efforts and achieve minimum output costs close to the theoretical minimum.

The stability performance of the controllers may also be related to errors in the open loop modal costs. Table 7 lists errors for the modal costs of the 7th and 9th modes when the output is rotation rate at .35L and the input is a white noise torque at the left end. The 9th mode is listed because, as previously mentioned, it has the highest modal cost. The 7th mode is included in the Table because there are large modal cost errors associated with the 24, 28, 30 and 32 degrees of freedom cubic beam element models. The 20 degrees of freedom cubic beam element model yields controllers with poor performance because of the large error in the 9th modal cost while the 24 degrees of freedom cubic beam element model gives poor controller performance because of large errors in the modal costs of both the 9th and 7th modes. The 28 and 30 degrees of freedom cubic beam element model yields controllers that are unstable for large control efforts because of the large errors in the cost of the 7th modes. Our *numerical* results indicate that in order to obtain good performance in the closed loop system, the modal cost error of the critical mode (highest modal cost) must be within about 5%, and the errors of the next critical mode must be within about 15% for the control design models. The relation between open-loop modal cost errors and closed loop stability performance is still an active research topic and more work needs to be done in this area.

The results in Fig. 12 have significance in terms of the *computer memory* required to develop control laws based on finite element models. The computer memory needed for a square matrix is proportional to the *square* of the order of the matrix. To compare the efficiency of the cubic and quintic beam element models, let us define β to be the square of the ratio of the degrees of freedom of the cubic beam element models to the degrees of freedom of the quintic beam element model which yields controllers with *equivalent* performance. For the problem considered in this section, $\beta = (32/20)^2 = 2.56$. Thus, *less than half* the computer memory is required when using the quintic beam element. Because the rotation rate is a quantity we are often interested in and because large flexible structures to be

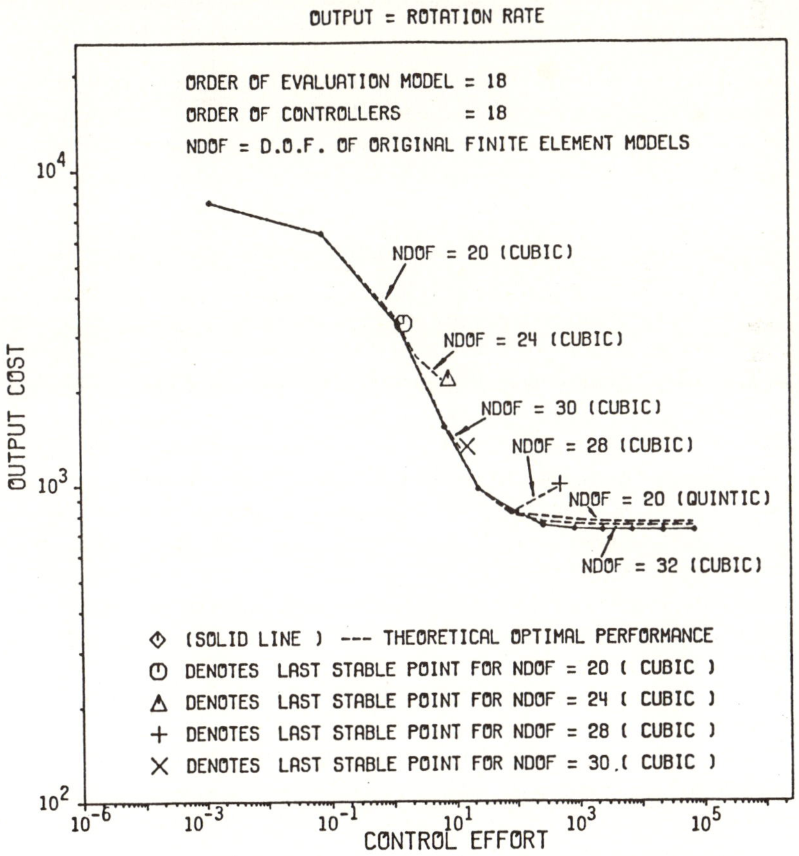

FIGURE 12 OUTPUT COST VS. CONTROL EFFORT FOR CONTROLLERS DERIVED USING EXACT AND FINITE ELEMENT MODELS

controlled may have thousands of degrees of freedom it is fair to say that the quintic beam element may provide great savings in computer memory requirements when developing models for successful control design.

6.0 Conclusion

Modal cost analysis is a decomposition of the control objective function in terms of each modal contribution, and this analysis is recommended in the selection of finite element models to be used for control design. The convergence theorems are developed for the flexible structures. For the Euler-Bernoulli beam, a quintic beam element is shown to be more efficient (per degree of freedom) than a cubic beam element for the convergence of the modal costs and for the convergence to a satisfactory closed loop control design. The open loop modal cost analysis was also related to the performance properties of feedback controllers designed by using finite element models.

It was shown that for beam-like structures:

(1) The convergence theorems for the modal cost are useful in determining the exact modal cost and finding the number of modes to attain a given "completeness" in the control of objective functions.

(2) The controllers designed from the quintic beam element model are able to achieve a *significantly lower* mean-squared rotational displacement rates than the controllers designed using a cubic beam element model with the same number of degrees of freedom and

(3) The quintic beam element model provides *substantial* savings in the computer memory requirements when compared to the requirements of a cubic beam element model which will yield *equivalent* closed loop performance.

The fundamental reason for these advantages in the quintic is that mode shapes and mode slopes are more accurate with quintic elements (compared to cubic) and that these shapes and slopes are generally more critical to successful control than accuracies in structural frequencies.

Generally, the control problems which seek to regulate rectilinear displacements using force or torque actuators do not require quintic elements; cubic is sufficient. But control problems which seek to regulate angular displacements or angular rates (using force or torque actuators) need quintic element models rather than cubic. These results make it clear that *the modeling and control problems are not independent,* and that the structural modeling methods and model reduction methods should be influenced by the specific control objectives. This paper has taken this step for beam-like structures.

ACKNOWLEDGMENTS

Part of this research was supported by NASA grant NAG1-642, under the direction of Dr. Howard Adelman.

References

[1] P. Likins 1977 *Proceedings of the fourth Symposium of International Federation of Automatic Control on Multivariable Technological Systems,* New Brunswick, N.B. Canada, July 1977. The Application of Multivariable Control Theory to Spacecraft Attitude Control.

[2] M. Balas 1977 *Symposium of American Institute of Aeronautics and Astronautics on Dynamics and Control of Large Flexible Spacecraft,* Blacksburg, Virginia, U.S.A., June 1977. Active Control of Flexible Systems.

[3] R. Skelton 1981 *Theory and Applications of Optimal Control in Aerospace Systems* (Chapter 8) ed. P. Kant, AGARD publication no. 251, KBN 92-835-1391-6, July 1981. Distributed in the United States, by NASA Langley Field, Virginia, 23365, attn: Report Distribution and Storage Unit. Control Design of Flexible Spacecraft.

[4] R. Skelton, P. Hughes, and H. Hablani 1982 *Journal of Guidance and Control* 5(4) 351-357. Order Reduction for Models of Space Structures Using Modal Cost Analysis.

[5] P. Hughes 1980 *Journal of Applied Mechanics* 47(1) 177-184. Modal Identities for Elastic Bodies with Application to Vehicle Dynamics and Control.

[6] J. Sesak, P. Likins, and T. Coradetti 1979 *Proceedings of the 2nd Symposium of American Institute of Aeronautics and Astronautics on Dynamics and Control of Large Flexible Spacecraft.* Blacksburg, Virginia, U.S.A., 1979. Flexible Spacecraft Control by Model Error Sensitivity Suppression.

[7] L. Meirovitch and H. Baruh 1980 *Journal of Guidance and Control* 3(2), 140-150. Optimal Control of Damped Flexible Gyroscopic Systems.

[8] L. Pinson, A. Amos, and V. Venkayya 1982 *Proceedings of Workshop of NASA Conference,* NASA Conference Publication 2258, Williamsburg, Virginia, U.S.A., May 1982. Modeling, Analysis and Optimization Issues for Large Space Structures.

[9] J. Oden and J. Reddy 1976 *An Introduction to the Mathematical Theory of Finite Elements.* New York: Wiley.

[10] O. Zienkiewicz 1971 *The Finite Element Method in Engineering Science.* New York: McGraw-Hill.

[11] J. Przemienicki 1968 *Theory of Matrix Structural Analysis.* New York: McGraw-Hill.

[12] J. Archer 1965 *Journal of American Institute of Aeronautics and Astronautics* 3(10), 1910-1918. Consistent Matrix Formulation for Structural Analysis Using Finite Element Techniques.

[13] R.E. Skelton and A. Hu 1984 *Proceedings of the First European Workshop on the Real Time Control of Large Scale Systems,* Patras, Greece, July 1984. Modeling and Control of Large Flexible Structures.

[14] R.E. Skelton and A. Hu 1985 *Computers and Structures* 20(1-3), 303-309. Modeling Structures for Control Design.

[15] A. Hu *Ph.D. Thesis, Purdue University,* U.S.A., (To appear in 1987). Modeling and Control of Large Flexible Space Structures.

[16] R.E. Skelton 1980 *Journal of Interdisciplinary Modeling and Simulation* 3(1), 47-62. Application of Disturbance-Accommodating Control in the Model Error Problem.

[17] R.E. Skelton 1985 *Proceedings of International Federation of Automatic Control Workshop on Model Error Concept and Compensation,* Boston, Massachusetts, U.S.A., June 1985. On the Structure of Modeling Errors and the Inseparability of the Modeling and Control Problems.

[18] R.E. Skelton and A. Yousuff 1983 *International Journal of Control* 37(2), 285-304. Component Cost Analysis of Large Scale Systems.

[19] R. Skelton and P.C. Hughes 1980 *Journal of Dynamic Systems, Measurement and Control,* 102, 151-158. Modal Cost Analysis for Linear Matrix-Second-Order System.

[20] R.J. Melosh 1983 *Journal of Astronautical Sciences* XXXI(3), 343-358. Finite Element Applications in Transient Analysis.

[21] C.T. Sun and S.N. Huang 1975 *Computers and Structures* 5, 297-303. Transverse Impact Problems by Higher Order Beam Finite Element.

[22] C.W.S. To 1979 *Journal of Sound and Vibration* 63(1), 33-50. Higher Order Tapered Beam Finite Elements for Vibration Analysis.

[23] J.K. Bennighof and L. Meirovitch *1985 Society of Engineering Science Conference,* State College, Pennsylvania, U.S.A., Oct. 1985. Eigenvalue Convergence in the Finite Element Method.

[24] R. Yedavalli and R. Skelton 1983 *Journal of Dynamic Systems, Measurement and Control,* 105, 238-244. Determination of Critical Parameters in Large Flexible Space Structures with Uncertain Modal Data.

[25] H. Kwakernaak and R. Sivan 1972 *Linear Optimal System.* New York: Wiley.

On the Transient Dynamics of Flexible Orbiting Structures

V.J. Modi and A.M. Ibrahim
Department of Mechanical Engineering
The University of British Columbia
Vancouver, B.C., Canada V6T 1W5

ABSTRACT

Complex interactions between deployment, attitude dynamics and flexural rigidity are reviewed using a rather general formulation applicable to a large class of space platforms with flexible, extensible members. The governing nonlinear, nonautonomous and coupled hybrid set of equations are extremely difficult to solve even with the help of a computer, not to mention the cost involved. Effectiveness of the versatile formulation is demonstrated through its application to several dynamical situations of contemporary interest involving beam type appendages. Both transient as well as post-deployment phases are considered. Results suggest significant influence of flexibility, inertia, deployment time history and orbital parameters on the system stability. The presence of free molecular and solar radiation induced environmental forces may further accentuate this tendency. The information has relevance to the design of control systems for the next generation of communications satellites with large solar panels; the Orbiter based experiments such as SAFE, COFS, NASA/CNR tethered subsatellite system, etc.; as well as constructional and operational phases of the proposed space station.

1. INTRODUCTION

In the early stages of space exploration, satellites tended to be relatively small, mechanically simple and essentially rigid. However, for a modern space vehicle carrying light deployable members, which are inherently flexible, this is no longer true. The space shuttle based experiments as well as construction of the proposed space station with gigantic trusses, power booms and solar panels, which will have to be erected in space, dramatically emphasize the role of deployment, slewing maneuvers and flexibility on dynamics, stability and control [1-7]. This being the case, flexibility effects on satellite attitude motion and its control have become topics of considerable importance. Over the years, a large body of literature pertaining to the various aspects of satellite system response, stability and control has appeared. A recent issue of the Journal of Guidance, Control, and Dynamics published by the AIAA (American Institute of Aeronautics and Astronautics) contains a series of articles reviewing the state of the art in the general area of large space structures [8]. Attention is also directed towards planning of on-orbit experiments such as COFS (Control Of Flexible Structures), the Orbiter Mounted Large Platform Assembler Experiment, NASA/Lockheed Solar Array Flight Experiment (SAFE) and a host of others to

Springer Series in Computational Mechanics
S. N. Atluri, A. K. Amos (Eds.)
Large Space Structures: Dynamics and Control

check, calibrate and improve algorithms. It is generally concluded that on-orbit information acquired during the construction phase of a space station is the only dependable procedure for its overall design. Obviously, this promises to open up an exciting area of in-flight measurements of structural dynamics, stability and control parameters necessary for design. With the U.S. commitment to a spce station in mid 1990's, the need for understanding structural response and control characteristics of such time varying, highly flexible systems is further emphasized.

With this as background, the paper briefly reviews a relatively general approach to the dynamical analysis of this class of structures and the results obtained. Details of the mathematical formulation and analysis, being extremely lengthy, are purposely omitted here, however, appropriate references cited. The emphasis throughout is on methodology, interpretation of results and corresponding conclusions.

2. DISTINCTIVE FEATURES OF THE FORMULATION

Essential features of this highly versatile Lagrangian matrix formulation of the nonlinear, nonautonomous, and coupled equations of motion describing dynamics of a large class of systems characterized by flexible interconnected structural members (Fig. 1) may be summarized as follows:

- arbitrary number, type (tether, membrane, beam, plate, shell) and orientation of flexible members, connected so as to form a topologicl configuration, open or closed, deploying independently at specified velocities and accelerations;
- the appendage is permitted to have variable mass density, flexural rigidity and cross-sectional area along its length;
- governing equations account for gravitational effects, shifting center of mass, changing inertia and appendage offset together with transverse, axial, and torsional oscillations;
- appendage as well as system rotations can be described using Euler or Rodrigues parameters, or any of the orientation angles;
- elastic deformations can be discretized using modal representation or admissible functions, finite elements, or lumped mass method;
- in general, joints between the flexible members are taken to be elastic and dissipative permitting relative rotation and translation between bodies;
- the system may contain momentum or reaction wheels, gimballed or fixed, as well as thrusters;
- the equations are applicable to earth-bound, underwater and space-based systems;
- in spacecraft dynamics studies, the generalized coordinates corresponding to librational degrees of freedom can be so chosen as to make the governing equations applicable to both spin stabilized and gravity gradient orientations;
- the equations are programmed in nonlinear as well as linearized forms to permit the study of:

  (i) large angle maneuvers;

  (ii) nonlinear effects.

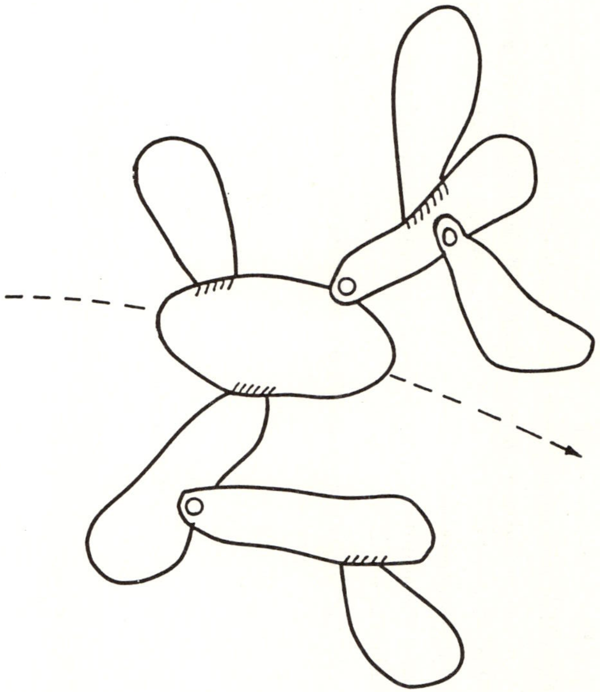

Fig. 1 A system of arbitrarily connected flexible deployable bodies.

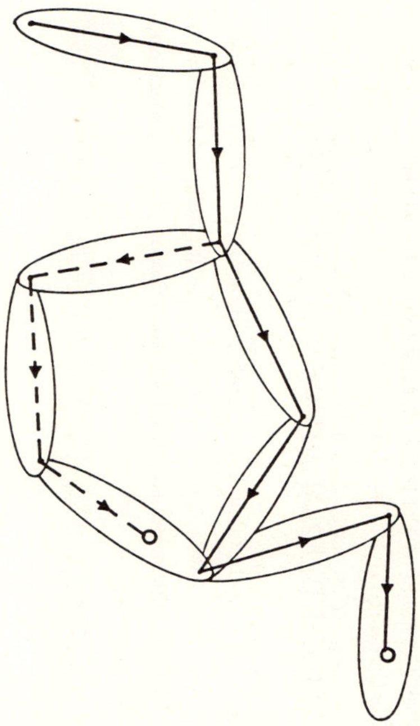

Fig. 2 A schematic diagram showing the direct path approach to the formulation.

The program is written in a modular fashion to help isolate the effects of flexibility, deployment, character and orientation of the appendages, inertia and orbital parameters, number and type of admissible functions, etc. Environmental effects due to solar radiation pressure, aerodynamic forces, Earth's magnetic field interaction, etc., can be incorporated easily through generalized forces. The same is true with internal and external dissipation mechanisms.

3. APPROACH TO FORMULATION OF THE PROBLEM

As can be expected, development of such a general formulation presented a challenging task and involved efforts spanning over several years. Details of the formulation form a part of the thesis in preparation [9]. Here the attention is focused on the philosophy of approach and physical appreciation of complex interactions.

3.1 Kinematics

Objective here is to obtain mathematical expressions for position and velocity of an arbitrarily located mass element. This is achieved by approaching the element in the direct way (Fig. 2) as against isolating each body and imposing reactions and constraints.

Consider a system of n elastic bodies connected arbitrarily to form a branched or closed loop topological geometry as shown in Fig. 3. Let X, Y, Z be an inertial coordinate system and x_0, y_0, z_0 with origin at O, a reference frame of coordinates. Now let body ① be referred to as the main body. Fixed to it, at a convenient location, is the coordinate system x_1, y_1, z_1. On the successive bodies, at their respective hinge points on the direct path connecting i and $i-1$ members, are located body coordinates x_i, y_i, z_i. Thus coordinates x_2, y_2, z_2 are fixed to body 2 at the hinge connecting bodies 1 and 2. Similarly, O acts as a hinge for body ①. The hinges are permitted to have three dimensional translational as well as rotational degrees of freedom.

Consider a mass element dm_i on link i defined by a position vector $\bar{r}^i$ with reference to the inertial coordinate system X, Y, Z, (Fig. 4). The geometry of the system is described by a matrix S_i^j hence numbering of bodies in sequence is not necessary,

$$\bar{r}^i = \sum_{j=0}^{n} S_{i+1}^j C_{j-1} \bar{a}_j \qquad \text{with} \qquad \bar{a}_{i+1} = \bar{\varsigma}_i . \tag{1}$$

Here: $S_i^j = \begin{cases} 0, & \text{if } j \text{ is not in the direct path to } i; \\ 1, & \text{if } j \text{ is in the direct path to } i; \end{cases}$

$C_j =$ relative rotation matrix relating coordinates j to the inertial set;

$n =$ number of connected bodies;

$\bar{a}_j =$ position of coordinates j with respect to coordinates $j-1$ on the direct path;

$\bar{\varsigma}_i =$ position vector to the element dm_i with respect to coordinates

x_i, y_i, z_i, (Fig. 4), $\rho_i + \phi_i \bar{\delta}_i$;

$\bar{\rho}_i =$ position vector to the element dm_i in nominal undeflected position;

$\phi_i =$ matrix of admissible functions;

$\bar{\delta}_i =$ generalized coordinates; $\phi_i \bar{\delta}_i$ represent deflection of dm_i from nominal equilibrium position.

Differentiating with respect to time, the expression for inertial velocity can be written as,

$$\dot{\bar{r}}^i = \check{\bar{r}}^i + \sum_{j=0}^{n} S_i^j C_{j-1} \tilde{\omega}_j \, {}^j\bar{r}^i , \tag{2}$$

where: $\check{\bar{r}}^i = \sum_{j=0}^{n} S_{i+1}^j C_{j-1} \check{\bar{a}}_j$;

$\tilde{\omega}_j =$ skew symmetric matrix representing rotation of coordinates j with respect to coordinates $j-1$;

${}^j\bar{r}^i =$ position of dm_i with respect to coordinates j. ${}^1\bar{r}^i$ is shown in Fig. 4.

Note, the first term on the righthand side represents translational velocity contributions due to deformations, deployment, hinge translations, etc., while the second term contains contributions due to rotations of the hinges.

For orbital applications, the reference point will be usually the centre of mass of the system. In that case:

$$\bar{a}_1 = -C_1 \sum_{i=1}^{n} \left(\frac{m_i}{m}\right) \left[\sum_{j=2}^{n} S_i^j C_{j-1} \bar{a}_j + \frac{C_i}{m_i} \int_{m_i} \bar{\varsigma}_i dm_i\right];$$

$$\check{\bar{a}}_1 = -\sum_{i=1}^{n} \left(\frac{\dot{m}_i}{m}\right) C_1^T \left[\sum_{j=1}^{n} S_i^j C_{j-1} \bar{a}_j - C_i \bar{e}_i\right]$$

$$- C_1^T \sum_{i=1}^{n} \left(\frac{m_i}{m}\right) \left[\sum_{j=2}^{n} S_i^j C_{j-1} \check{\bar{a}}_j + C_i \{E^i \check{\bar{e}}_i + D^i \check{\bar{\delta}}_i\}\right];$$

with E^i, D^i defined by the equation

$$\frac{d}{dt} \int_{m_i} \bar{\varsigma}_i dm_i = m_i \{E^i \check{\bar{e}}_i + D^i \check{\bar{\delta}}_i\} - \dot{m}_i \bar{e}_i .$$

Here m_i is the mass of the body i; m the total mass of the system; $\check{\bar{e}}_i$ represents velocity of deployment or ejected materials; and $\bar{e}_i$, the position vector to the center of mass of the deployed member or orifice for mass expulsion.

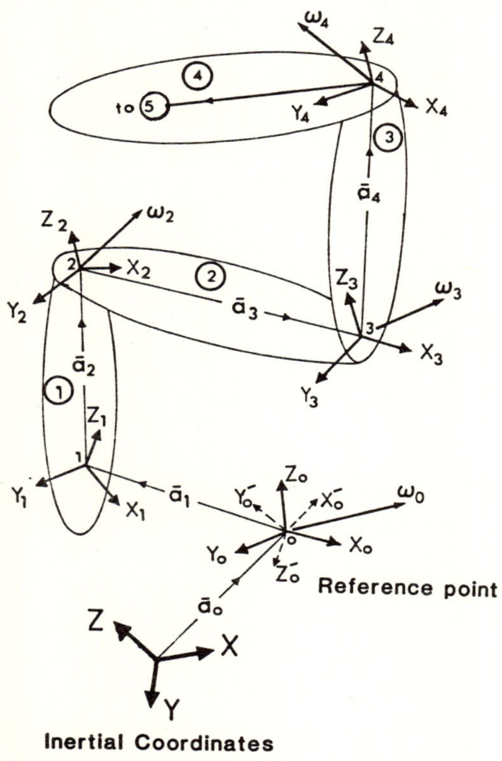

Fig. 3 System description and coordinate system. Note the reference point will be usually c.m. of the system in space dynamics application.

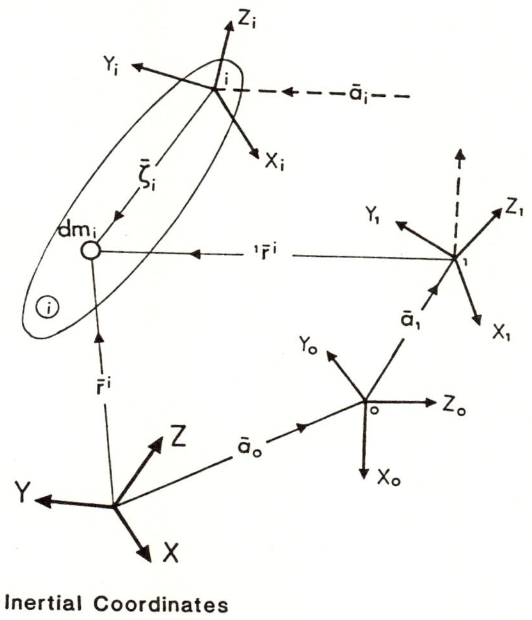

Fig. 4 Geometry of position vectors defining location of the mass element dm_i on body i.

3.2 Kinetics

For the Lagrangian approach, the next logical step would be to obtain expressions for kinetic, potential and elastic energies of the system. As can be expected, because of the complex and quite general character of the system, derivation of the expressions demands considerable time, space, and effort. For conciseness only more important steps relevant to the development are indicated here.

Kinetic energy.

$$T = \frac{1}{2}\sum_{i=1}^{n} \int_{m_i} \dot{\bar{r}}^{i} \cdot \dot{\bar{r}}^{i} dm_i .$$

Substituting from Eq. (2),

$$T = \frac{1}{2}\sum_{i=1}^{n} \int_{m_i} \Big(\check{\bar{r}}^{i} \cdot \check{\bar{r}}^{i} + 2\check{\bar{r}}^{i} \cdot \sum_{j=0}^{n} S_i^j C_{j-1} \tilde{\omega}_j \, {}^{j}\bar{r}^{i}$$

$$+ \big(\sum_{j=0}^{n} S_i^j C_{j-1} \tilde{\omega}_j \, {}^{j}\bar{r}^{i}\big) \cdot \big(\sum_{j=0}^{n} S_i^j C_{j-1} \tilde{\omega}_j \, {}^{j}\bar{r}^{i}\big)\Big) dm_i .$$

This is of the form,

$$T = \frac{1}{2}\sum_{i=1}^{n} e_i + \sum_{i=1}^{n}\sum_{j=0}^{n} \omega_j^T g_i^j + \frac{1}{2}\sum_{i=1}^{n}\sum_{j=0}^{n}\sum_{k=0}^{n} \omega_j^T H_i^{jk} \omega_k = \frac{1}{2}e + \omega^T g + \frac{1}{2}\omega^T H \omega, \quad (3)$$

where:
$$e_i = \int_{m_i} \check{\bar{r}}^{i} \cdot \check{\bar{r}}^{i} dm_i ;$$

$$\sum_{j=0}^{n} \omega_j^T g_i^j = \sum_{j=0}^{n} S_i^j \int_{m_i} \check{\bar{r}}^{i} \cdot (C_{j-1}\tilde{\omega}_j) \, {}^{j}\bar{r}^{i} dm_i = \sum_{j=0}^{n} S_i^j \omega_j \cdot C_{j-1}^T \cdot \int_{m_i} {}^{j}\bar{r}^{i} \times \check{\bar{r}}^{i} \, dm_i ,$$

therefore
$$g_i^j = S_i^j C_{j-1}^T \int_{m_i} {}^{j}\bar{r}^{i} \times \check{\bar{r}}^{i} \, dm_i ;$$

$$\sum_{j=0}^{n}\sum_{k=0}^{n} \omega_j^T H_i^{jk} \omega_k = \sum_{j=0}^{n}\sum_{k=0}^{n} S_i^j S_i^k \int_{m_i} (C_{j-1}\omega_j) \times {}^{j}\bar{r}^{i} \cdot \big((C_{k-1}\omega_k) \times {}^{k}\bar{r}^{i}\big) dm_i ,$$

$$= \sum_{j=0}^{n}\sum_{k=0}^{n} S_i^j S_i^k \omega_j^T C_{j-1}^T \Big[\int_{m_i} [{}^{j}\bar{r}^{i} \cdot {}^{k}\bar{r}^{i} U - {}^{k}\bar{r}^{i}\, {}^{j}\bar{r}^{i}] dm_i\Big] C_{k-1}\omega_k ,$$

therefore
$$H_i^{jk} = S_i^j S_i^k C_{j-1}^T \Big[\int_{m_i} [{}^{j}\bar{r}^{i} \cdot {}^{k}\bar{r}^{i} U - {}^{k}\bar{r}^{i} \cdot ({}^{j}\bar{r}^{i})^T] dm_i\Big] C_{k-1} ;$$

U = identity matrix;

ω = vector with angular velocity of joints as elements; $\omega^T = (\omega_0, \omega_1, \ldots, \omega_n)$;

$$g^T = (g^0, g^1, \ldots, g^n) ;$$
$$g^j = \sum_{i=1}^{n} g_i^j ;$$
$H = 3(n+1)$ square matrix with H^{jk} submatrices as elements;
$$H^{jk} = \sum_{i=1}^{n} H_i^{jk} .$$

Recognizing that the kinetic energy expression is composed of velocity-square terms, Eq. (3) can be rewritten as,

$$T = \frac{1}{2} v^T E v + \omega^T G v + \frac{1}{2} \omega^T H \omega = \frac{1}{2} \dot{x}^T M^* \dot{x} , \tag{4}$$

where: $\dot{x} = \begin{Bmatrix} v \\ \omega \end{Bmatrix}; \quad M^* = \begin{bmatrix} E & G^T \\ G & H \end{bmatrix} ;$

and v is defined by an array of all the linear velocities (i.e., time derivatives of the variables) appearing in the kinetic energy expression, $(\dot{m}_i, \dot{a}_i, \dot{e}_i, \dot{\delta}_i)$.

From the consideration of physical appreciation, it is important to recognize that E has the dimensions of mass; G the dimensions of the first mass-moment; and H the dimensions of the second mass-moment; Thus E represents mass associated with deployment, ejection, translation and vibration, while G corresponds to their first moment about the hinges. H matrix has more direct meaning. Let H^{jk} represent a submatrix. Then for $j = k$, H^{jj} represents a 3×3 matrix corresponding to inertias about the axis at j. For $j \neq k$, the submatrix has elements involving distances to joints j and k with common mass elements affected by rotations of both the hinges.

The next logical step is to introduce holonomic constraints and modify $\{\dot{x}\}, [M]$ making them consistent with the constraints. The kinetic energy can then be rewritten separating contributions from specified, s, and generalized coordinates, q, as follows:

$$T = \frac{1}{2} \begin{Bmatrix} \dot{s} \\ \dot{q} \end{Bmatrix}^T \begin{bmatrix} N & L^T \\ L & M \end{bmatrix} \begin{Bmatrix} \dot{s} \\ \dot{q} \end{Bmatrix} = \frac{1}{2} \dot{q}^T M \dot{q} + \dot{s}^T L^T \dot{q} + \frac{1}{2} \dot{s}^T N \dot{s}$$
$$= \frac{1}{2} \dot{q}^T M \dot{q} + \Gamma^T \dot{q} + T_0 . \tag{5}$$

Note here both $\dot{s}$ and $\dot{q}$, in general, will be arrays of the form $(\dot{m}_i, \dot{a}_i, \dot{e}_i, \dot{\delta}_i \omega_i)$. It is important to recognize that this general expression for kinetic energy consists of quadratic as well as linear terms in generalized velocities and a term independent of the generalized velocity. The quadratic term represents contribution from the generalized coordinates while T_0 is that from the specified coordinates. The linear term is due to a coupling between the system of two coordinates.

Potential energy. Contribution to the potential energy arises from two sources: gravitational field and strain energy due to flexibility.

Gravitational contribution (U).

$$U = -\frac{\mu_e m}{a_0} + \frac{\mu_e}{a_0^2}\{l\}^T\{L^o\} - \frac{\mu_e}{2a_0^3}\text{tr}[H^0] + \frac{3\mu_e}{2a_0^3}\{l\}^T[H^0]\{l\}, \tag{6}$$

where: $m =$ mass of the system, $\sum_{i=1}^{n} m_i$;

$$\{L^0\} = \sum_{i=1}^{n}\sum_{j=1}^{n} m_i S_i^j C_{j-1}\bar{a}_j + \sum_{i=1}^{n} C_i \int_{m_i} \varsigma_i dm_i;$$

$H^0 =$ inertia of the system about the reference point O (associated with ω_0);

$\mu_e =$ universal gravitational constant;

$\bar{a}_0 =$ position vector from the earth's center to the instantaneous center of mass;

$\{l\} =$ direction cosines of the unit vector along $\bar{a}_o$ w.r.t. x_0, y_o, z_0, coordinates.

Note, the first term represents gravitational energy due to the system treated as a point mass; the second term appears due to separation between the reference point and the center of mass and vanishes when they are coincident; while the remaining two terms are due to finite size of the system.

Strain energy (V). The expression for strain energy takes into account:

- large deformations in bending and torsion, i.e., rotary inertia, axial foreshortening, etc.;
- any arbitrary structural element, i.e., string, membrane, beam, plate, shell;
- variable mass and elastic properties;
- different discretization procedures such as admissible functions or finite elements.

For a given flexible body,

$$V = \frac{1}{2}\int_\tau \sigma^T \epsilon d\tau = \frac{1}{2}\int_\tau \epsilon^T C \epsilon d\tau. \tag{7}$$

where: $\tau =$ volume;

$C =$ matrix of elastic constants.

Note here stresses and strains are expressed in the form of vectors:

$$\sigma^T = (\sigma_{11}, \sigma_{22}, \sigma_{33}, \sigma_{12}, \sigma_{13}, \sigma_{23}).$$

$$\epsilon^T = (\epsilon_{11}, \epsilon_{22}, \epsilon_{33}, \epsilon_{12}, \epsilon_{13}, \epsilon_{23});$$

Considering the strain to have linear and nonlinear contributions from deformations δ, it can be written as:

$$\{\epsilon\} = \{B^0\delta\} + \{\delta^T B^j \delta\},$$

where B^0 and B^j depend on the chosen admissible functions ϕ. Substituting this expression for ϵ in Eq. (7), the strain energy expression can be written as

$$V = \delta^T[V_1 + 2\delta^T V_2^j + \delta^T V_3^{jk}\delta]\delta ,$$

where:

$$V_1 = \sum_{j=1}^{6}\sum_{k=1}^{6} C_{jk}\int_\tau B_{j-}^0 (B_{k-}^0)^T d\tau ;$$

$$V_2^j = \sum_{k=1}^{6}\sum_{l=1}^{6} C_{kl}\int_\tau B_{k\,j}^0 B^j d\tau ;$$

$$V_3^{jk} = \sum_{l=1}^{6}\sum_{m=1}^{6} C_{lm}\int_\tau B_{-j}^l B_{-k}^m d\tau .$$

Now the elastic force is given by,
$$\frac{\partial V}{\partial \delta} = V,_\delta = K\delta , \tag{8}$$

where:

$$K = \text{stiffness matrix} = K_1 + K_2 + K_3 ;$$

$$K_1 = V_1 + V_1^T ;$$

$$K_2 = 2[\delta^T V_2^j + (V_2^j)^T\delta + \sum_{j=1}^{6} \delta_{,j} V_2^j] ;$$

$$K_3 = [\delta^T[V_3^{jk} + (V_3^{jk})^T]\delta] + \sum_{j=1}^{6}\sum_{k=1}^{6} \delta_{,j}\delta_{,k}[V_3^{jk} + (V_3^{jk})^T] .$$

Note, the notation for partial differentiation used above for brevity. Global stiffness matrix can now be assembled with δ as generalized coordinates or nodal displacements.

3.3 Functions of Interest

Before proceeding further with the formulation, it would be useful to study more important key functions depending on the kinetic and potential energies. They are summarized below:

$$\text{Lagrangian} = L = T - (U + V)$$

$$= \frac{1}{2}\{\dot q\}^T[M]\{\dot q\} + \{\Gamma\}^T\{\dot q\} + T_0 - (U + V) ; \tag{9a}$$

$$\text{Hamiltonian} = H = \frac{1}{2}\{p - \Gamma\}^T[M]^{-1}\{p - \Gamma\} - T_0 + (U + V) ; \tag{9b}$$

$$\text{Linear Momentum} = \frac{\partial T}{\partial v} = [E]\{v\} + [G]^T\{\omega\}\,; \tag{9c}$$

$$\text{Angular Momentum} = \frac{\partial T}{\partial \omega} = [G]\{v\} + [H]\{\omega\}\,; \tag{9d}$$

$$\text{Generalized Momenta} = \left\{\frac{\partial L}{\partial \dot{q}}\right\} = \{p\} = [M]\{\dot{q}\} + \{\Gamma\}\,. \tag{9e}$$

4. EQUATIONS OF MOTION

4.1 Lagrangian Equations

Using the Lagrangian procedure in conjunction with kinetic and potential energy expressions (5), (6), and (8),

$$\frac{d}{dt}\left(\frac{\partial T}{\partial \dot{q}}\right) - \frac{\partial T}{\partial q} + \frac{\partial (U+V)}{\partial q} = Q + A^T\lambda\,,$$

gives $M\ddot{q} + [\dot{M} - \frac{\partial \Gamma^T}{\partial q_j}]\dot{q} - \frac{1}{2}\{\dot{q}^T\frac{\partial M}{\partial q_j}\dot{q}\} + \dot{\Gamma} + \{\frac{\partial}{\partial q_j}(U + V - T_0)\} = \{Q + A^T\lambda\}$

subject to the constraints $A\dot{q} + B = 0$. Substituting for:

$$\dot{M} = M_{,t} + \sum_j M_{,q_j}\dot{q}_j\,; \qquad \dot{\Gamma} = \Gamma_{,t} + \sum_j \Gamma_{,q_j}\dot{q}_j\,;$$

and rearranging the terms, the above equation can be written as,

$$M\ddot{q} + [M_{,t} + G]\dot{q} + \{\dot{q}^T\gamma_j\dot{q}\} + \Gamma_{,t} + Kq + (U - T_0)_{,q} = Q + A^T\lambda\,, \tag{10}$$

where: $M =$ square symmetric matrix of generalized mass, Eq. (5)

$G = [\{\Gamma_{j,q}\}^T - \Gamma^T_{,q_j}]$, skew symmetric matrix;

$\gamma_j = [M_{j-,q} - (1/2)M_{,q_j}]$, square matrix associated with the Coriolis force contribution;

$\Gamma =$ coefficient associated with the linear contribution in the kinetic energy, Eq. (5);

$K =$ stiffness matrix, Eq. (8);

$U =$ gravitational potential energy, Eq. (6);

$T_0 =$ kinetic energy contribution independent of generalized velocities, Eq. (5);

$Q =$ generalized forces.

Note, the Lagrangian character of the formulation is amenable to point transformation making it possible to obtain governing equations using alternate standard procedure of Hamilton. More importantly, the equations clearly isolate contribution of forces from different sources thus retaining physical insight into the problem. Furthermore, the form is ideally suited for implementing control strategies. It should be emphasized that most available formulations of multibody systems do not possess the above mentioned features. Terms representing contributions from various sources are explained next:

$M\ddot{q} =$ inertia forces;

$M_{,t}\dot{q} =$ reaction forces due to deployment and mass expulsion;

$G\dot{q} =$ gyroscopic forces;

$\{\dot{q}^T\gamma_j\dot{q}\} =$ Coriolis and centrifugal forces arising from generalized coordinates;

$\Gamma_{,t} - T_{0,q} =$ centrifugal forces arising from specified coordinates;

$Kq =$ elastic forces;

$U_{,q} =$ gravitational force;

$A^T\lambda =$ nonholonomic constraint forces as well as holonomic constraints not accounted for earlier.

4.2 Hamilton's Equations

Applying point transformation to the Lagrangian L, an expression for Hamiltonian can be obtained as given before (Eq. 9b). This leads to Hamilton's equations:

$$\dot{p}_k = -\frac{\partial H}{\partial q_k}; \qquad \dot{q}_k = \frac{\partial H}{\partial p_k};$$

$$\dot{p}_k = \frac{1}{2}\{p-\Gamma\}^T[M]^{-1}[\frac{\partial M}{\partial q_k}][M]^{-1}\{p-\Gamma\} + \{\frac{\partial \Gamma}{\partial q_k}\}^T[M]^{-1}\{p-\Gamma\} + \frac{\partial T_0}{\partial q_k} - \frac{\partial(U+V)}{\partial q_k};$$

$$\{\dot{q}\} = [M]^{-1}\{p-\Gamma\}.$$

4.3 Linearization About the Nominal Equilibrium

The highly nonlinear, nonautonomous and coupled equations of motions (Eq. 10) are ideally suited for studying librational as well as vibrational dynamics in the large. Furthermore, they account for coupling between orbital, librational, and vibrational degrees of freedom. Note, they can also be used for studying slewing dynamics.

However, for an actively controlled system, one would generally prefer to use the corresponding set of equations linearized about the nominal equilibrium configuration. A form of the linearized equations is presented below:

$$M\ddot{q} + [M_{,t} + G]\dot{q} + [S + K_1]q = Q + A^T\lambda$$

where $M, M_{,t}, G, S$ and K_1 are evaluated at the equilibrium. Here S is a geometric stiffness matrix arising from gravity and forces due to specified motions. $Q, A^T\lambda$ are now linearized forms of the original generalized and constraint forces.

5. APPLICATION OF THE GENERAL FORMULATION TO SPACE DYNAMICS PROBLEMS: AN ILLUSTRATIVE EXAMPLE

Consider a spacecraft with central rigid body, and a flexible deployable appendage attached to it, in a specified arbitrary orbit (Fig. 5a). Let the reference point O be the instantaneous center of mass of the system and the inertial coordinate X, Y, Z located at the center of the Earth. Thus O is the hinge point of body ①. Now in this special case, $\bar{a}_0$ represents position vector from the center of the Earth to the instantaneous center of mass and hence specifies the trajectory. Let 1 be the center of mass of the rigid body. Thus $\bar{a}_1$ represents instantaneous position of the moving center of mass with respect to point 1. $\bar{a}_2$ defines position of the appendage attachment point taken stationary in this case. The system variables now become, $\{\dot{\bar{x}}\}^T = (\dot{\bar{a}}_0, \dot{\bar{e}}_2, \dot{\bar{\delta}}_2, \bar{\omega}_0)$. Note, $\dot{m}_1$ and $\dot{m}_2$ do not appear explicitly because of the constraint relations between $\dot{m}_2$ and $\dot{\bar{e}}_2$. For numerical results, the rigid body is taken to be the Orbiter and the appendage a flexible beam (Fig. 5b). The configuration corresponds to the Orbiter Mounted Large Platform Assembler Experiment once proposed by Grumman Aerospace Corporation. It also resembles the SCOLE geometry proposed by the NASA Langley Research Center. The Grumman experiment aimed at establishing capability of manufacturing beams in space which would serve as one of the fundamental structural elements in construction of future space stations. Objective of the SCOLE experiment is to assess dynamics and control of large flexible structures in space when excited by external disturbances including slewing maneuvers. The formulation presented here can readily tackle both the problems. Three dynamical situations are considered for response studies:

Case 1: The Rigid Orbiter Without Any Flexible Appendage

$$\{\dot{\bar{x}}\}^T = (\dot{\bar{a}}_0, \bar{\omega}_0) .$$

Here $\bar{a}_0$ is the specified coordinate defining the Keplerian motion and the generalized co-ordinate $\bar{\omega}_0$ describes the attitude motion α (pitch), β (yaw), and γ (roll).

Case 2: The Orbiter Having a Deployable Flexible Beam with Libration Degrees of Freedom Held Zero

$$\{\dot{\bar{x}}\}^T = (\dot{\bar{a}}_0, \dot{\bar{e}}_2, \dot{\bar{\delta}}_2, \bar{\omega}_0) .$$

Here $\quad \dot{\bar{q}} = \dot{\bar{\delta}}_2$ and $\dot{\bar{s}} = \dot{\bar{a}}_0, \dot{\bar{e}}_2, \dot{\bar{\omega}}_0 = 0 .$

Case 3: The Orbiter Having a Deployable Flexible Beam and Undergoing Specified Librational Motion

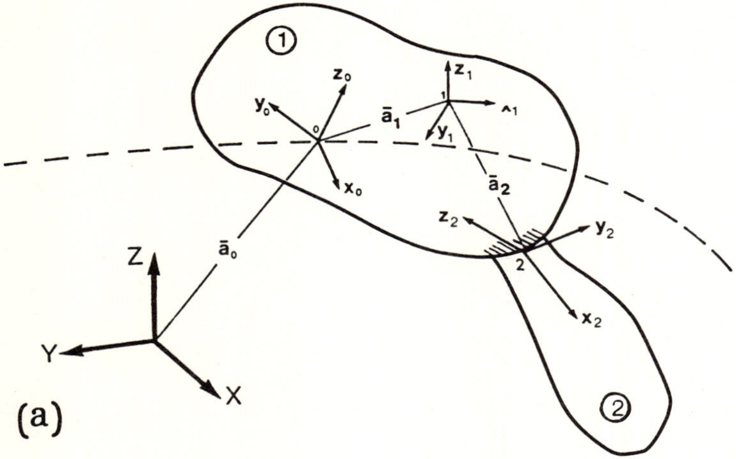

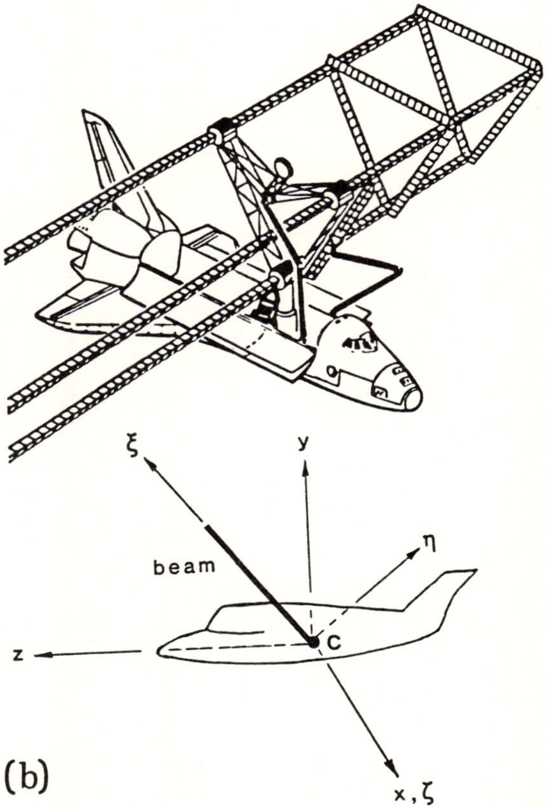

Fig.5 A particular case showing application of the general formulation:

(a) an orbiting rigid body with a flexible deployable appendage attached to it;

(b) artist's view of the Orbiter based manufacture of the beam type structural components for construction of space platforms. Principal body coordinates x, y, z with their origin at the instantaneous center of mass and beam coordinates ς, η, ξ with the origin at the attachment point are also indicated. In general the two origins are not coincident.

$$\{\dot{\bar{x}}\}^T = (\dot{\bar{a}}_0, \dot{\bar{e}}_2, \dot{\bar{\delta}}_2, \dot{\bar{\omega}}_0) .$$

Here $\dot{\bar{q}} = \dot{\bar{\delta}}_2$ and $\dot{\bar{s}} = \dot{\bar{a}}_0, \dot{\bar{e}}_2, \dot{\bar{\omega}}_0$.

5.1 Results and Discussions

For analysis, the flexibility and deployment rate parameters were taken to be of the same order of magnitude as used or likely to be employed in practice. In the diagrams e represents orbital eccentricity; Ψ, Λ, Φ (roll, yaw, pitch, respectively) are the librational angles; EI is the beam flexural rigidity, assumed constant over the length in this particular example; and $\dot{L}$ corresponds to the deployment rate. λ_{in} and λ_{out} denote beam inclinations to the local vertical in and normal to the orbital plane, respectively. The perigee was taken to be 331 km. The truss or beam vibrations were represented by a maximum of the first four modes, Ψ_i, of a cantilever. P_l, Q_l represent generalized coordinates associated with the admissible functions used to represent beam-type appendage oscillations in the l th mode in ς and η directions, respectively. $\overline{P}_l$ and $\overline{Q}_l$ represent transverse generalized coordinates normalized with respect to the total length.

Numerical values for some of the more important parameters used in the computation are given below:

Orbiter: Mass $= 79,710\,kg$; $I_{xx} = 8,286,760\,kgm^2$; $I_{xy} = 27,116\,kgm^2$;
$I_{yy} = 8,646,050\,kgm^2$; $I_{yz} = 328,108\,kgm^2$;
$I_{zz} = 1,091,430\,kgm^2$; $I_{zx} = -8,135\,kgm^2$.

Here x, y, z are the principal body coordinates of the Orbiter with the origin coinciding with the center of mass. In the nominal configuration x is along the orbit normal, y coincides with the local vertical and z is aligned with the local horizontal in the direction of motion (Fig. 5b). γ (roll), β (yaw), and α (pitch) refer to rotations about the local horizontal, local vertical, and orbit normal, respectively.

Beam: Mass $(M_b) = 129\,kg$;
Length $(L) = 33\,m$;
Flexural Rigidity $(EI) = 436\,kgm^2$.

To get some appreciation as to the system dynamics during transition to instability, the Lagrange configuration was subjected to pitch, yaw, and roll disturbances separately (Fig. 6). With a pitch disturbance as large as 30° (Fig. 6a), the roll and yaw remain unexcited and the system is stable. The same is essentially true with a yaw disturbance (Fig. 6b). However, even with a relatively small roll disturbance (Fig. 6c), the diverging yaw oscillations set-in tending towards instability. Thus roll control seems to be a key to ensure stability of the Orbiter in the Lagrangian configuration.

Effects of beam deployment on the tip dynamics is studied in Fig. 7. Initial tip deflection is 4% of the beam length. Two time histories with the same duration of deployment are considered. As can be expected, the frequency of oscillation in and out of the orbital plane gradually decreases

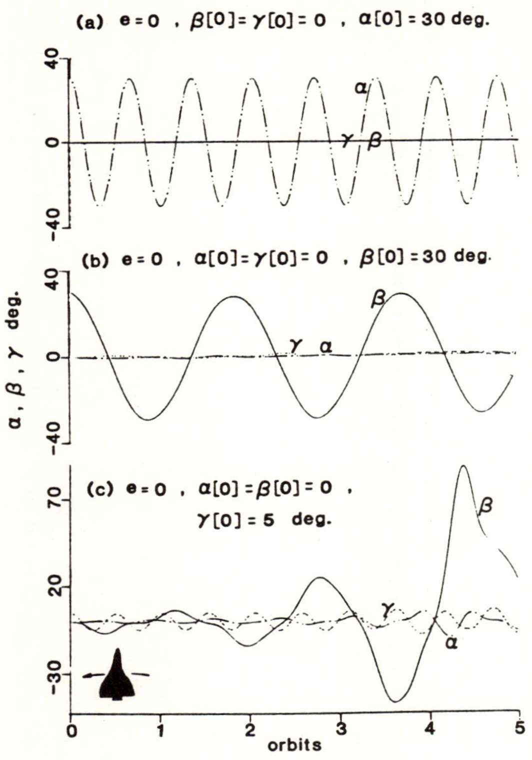

Fig. 6 Librational response of the Orbiter to an independent excitation in pitch, yaw, and roll. Note the pitch and yaw disturbances lead to essentially uncoupled motions. The system appears to become unstable in yaw through its coupling with roll.

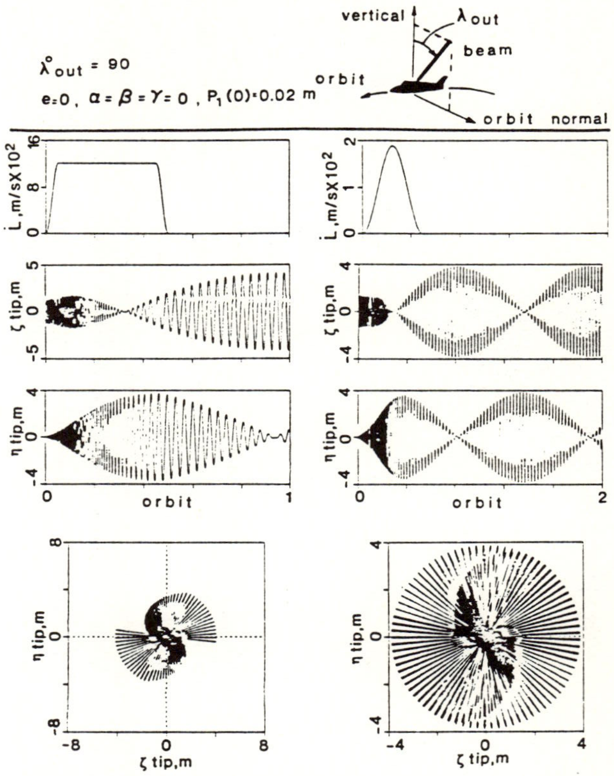

Fig. 7 Effect of deployment strategies on tip response of a beam deploying normal to the orbital plane. Note a reduction in beam frequency during deployment. The steady state amplitude is essentially independent of the strategy for a given time of deployment.

with deployment finally attaining a steady state value upon its termination. It is of interest to recognize that they reach the same steady state amplitude although it is much larger during deployment compared to the deployed case.

In practice, the Orbiter's librations will be controlled to a specified tolerance limit. A typical time history [10] of the controlled Space Shuttle librations during an orbit is shown in Fig. 8. In the following results attention is focused on response of the deployed beam during such forced excitation of the Orbiter in the Lagrangian configuration.

Figure 9 shows the forced tip response as well as the first two modes contributing to it for a beam deployed along the orbit normal with the Orbiter in the Langrangian configuration. At the outset it should be recognized that, for this out-of-plane configuration of the beam, the out-of-plane motion ς and inplane response η are coupled. Hence one would expect the Orbiter's yaw and roll to be reflected in both η (inplane) and ς (out-of-plane) motions. The response shown here precisely reveals these trends. However, the roll disturbance being at a higher frequency, and hence with a higher acceleration, appears to be dominant as apparent from the amplitude modulation of the response at the roll frequency (around 13 cycles per orbit).

It was thought useful to study the effect of truncation in terms of the number of modes used to represent the beam dynamics. To this end the Orbiter was taken in the Lagrange configuration with the beam normal to the orbital plane ($\lambda_{out} = 90^{\circ}$). The 4% tip deflection in ς direction and in the first mode was applied with the Orbiter free of librational motion. In absence of any external forces one would expect the beam to vibrate in the first mode alone. However, due to orbital motion giving rise to centrifugal, Coriolis and other forces, the beam deflection deviates from the first mode requiring higher modes to represent its dynamics. To assess contribution of the higher modes, the beam response was analyzed using the first, the first two and the first four modes. The results are presented in Fig. 10. Two important conclusions can be drawn: (i) time history of the beam deflection remains essentially unaffected by the number of modes used; (ii) amplitudes of the generalized coordinates associated with the first mode (P_1, Q_1) are around three orders of magnitude higher than P_2, Q_2 of the second mode and, of course, still larger compared to the higher modes.

In case of the beam excited by the controlled librations of the Orbiter (Fig. 8) the dominant mode will be governed by the resonance condition. It is apparent that over the range of frequency of the forcing functions the first mode will have the major contribution.

6. CONCLUDING REMARKS

With a relatively general formulation in hand and its validity assessed through application to specific configurations, efforts are in progress to develop a comprehensive data bank for spacecraft with flexible appendages. Not only will it prove useful to design engineers involved in planning of future communications satellites but also help in assessing dynamical, stability, and control considerations associated with the Orbiter based construction of space-platforms.

The entire field is wide open to innovative contributions. Dynamics and control of such nonlinear, time-dependent systems accounting for damping and environmental forces remains vir-

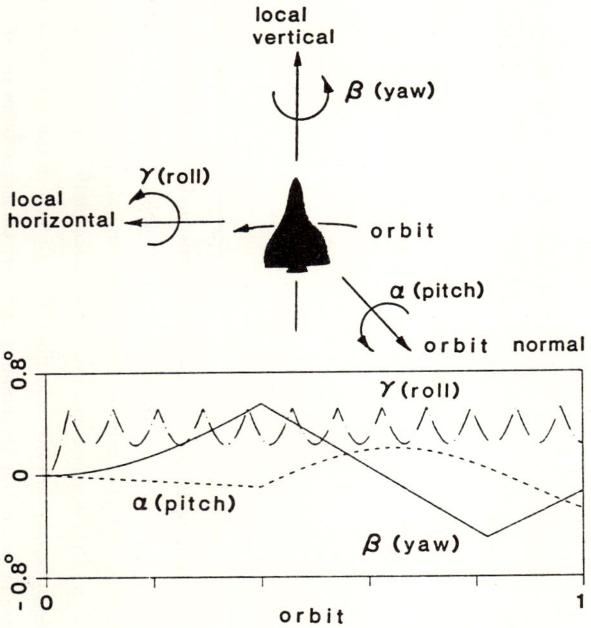

Fig. 8 A representative controlled motion of the Orbiter during a typical orbit. Note the roll, yaw, and pitch motions are with reference to the local horizontal, local vertical, and orbit normal, respectively.

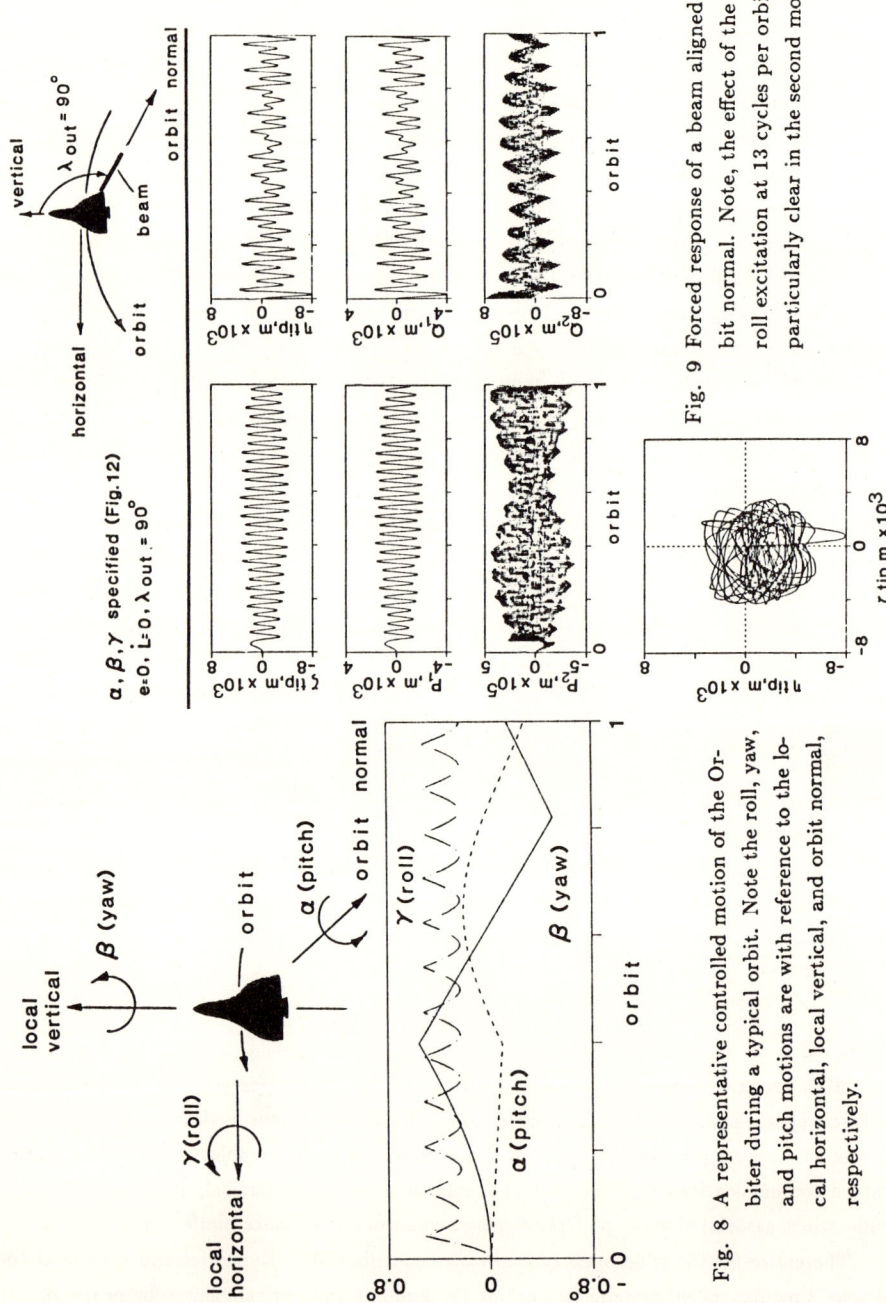

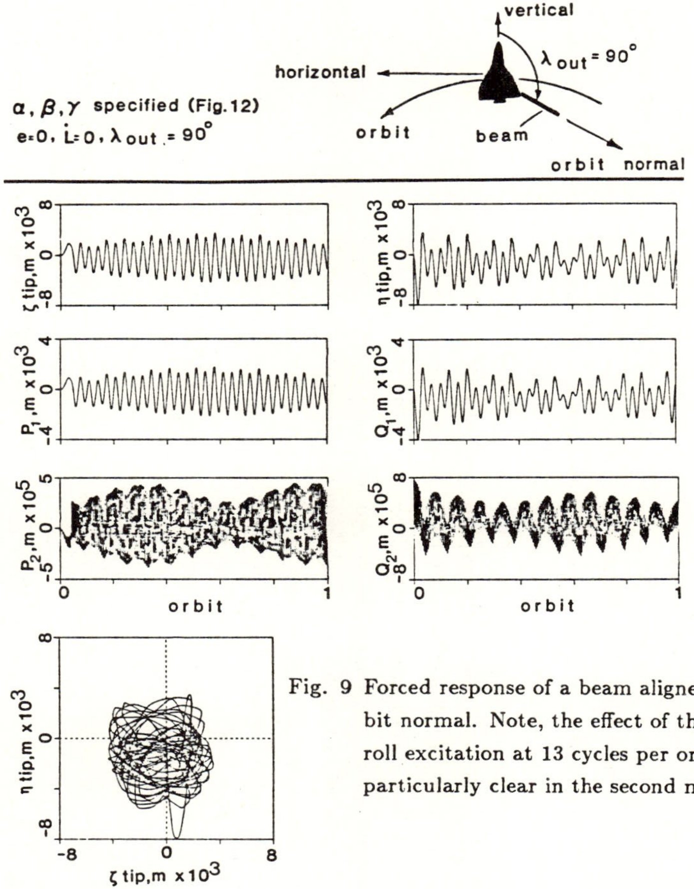

Fig. 9 Forced response of a beam aligned with the orbit normal. Note, the effect of the dominant roll excitation at 13 cycles per orbit which is particularly clear in the second mode.

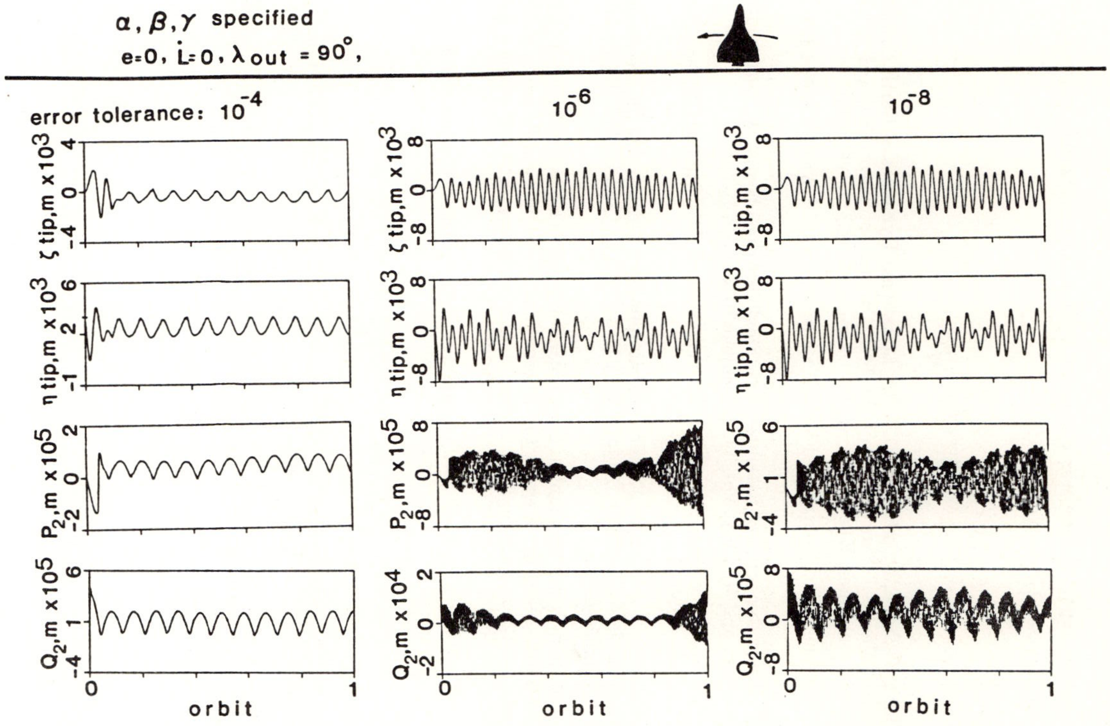

Fig. 10 A typical case of beam response showing the effect of mode truncation. It appears that the higher modes have very little effect on the time history of the beam deflection.

tually untouched. Their applications to the construction of space station has received attention only recently. Development of an algorithm to predict the effect of mass, inertia, and stiffness of the station as it evolves on the dynamics and control paramaters represents an exciting challenge. Application of the tether and associated dynamics presents an area of enormous potential.

ACKNOWLEDGEMENT

The investigation reported here was supported by the Natural Sciences and Engineering Research Council of Canada, Grant No. G-1547.

REFERENCES

[1] Shrivastava, S.K., Tschann, C., and Modi, V.J., "Librational Dynamics of Earth Orbiting Satellites - A Brief Review," Proceedings of the 14th Congress of Theoretical and Applied Mechanics, ISTAM Publisher, Kharagpur, India, December 1969, pp. 284-306.

[2] Likins, P.W., "Dynamics and Control of Flexible Space Vehicles," NASA, TR-32-1329, January 1970.

[3] Modi, V.J., "Attitude Dynamics of Satellites with Flexible Appendages - A Brief Review," Journal of Spacecraft and Rockets, Vol. 11, November 1974, pp. 743-751.

[4] Roberson, R.E., "Two Decades of Spacecraft Attitude Control," Journal of Guidance and Control Vol. 2, January-February, 1979, pp. 3-8.

[5] Lips, K.W., and Modi, V.J., "Three Dimensional Response Characteristics for Spacécraft with Deploying Flexible Appendages," Journal of Guidance and Control, Vol. 18, November-December, 1981, pp. 650-656.

[6] Misra, A.K., and Modi, V.J., "Dynamics and Control of Tether Connected Two-Body Systems," Invited Address, 33rd Congress of the International Astronautical Federation, Paris, France, September 1982; also Space 2000, Editor: L.G. Napolitane, AIAA Publisher, pp. 473-514.

[7] Modi, V.J., and Shrivastava, S.K., "Satellite Attitude Dynamics and Control in Presence of Environmental Torques - A Survey," Journal of Guidance, Control, and Dynamics, AIAA, Vol. 6, November-December 1983, pp. 461-471.

[8] Special Section, "Large Space Structure Control: Early Experiments," Journal of Guidance, Control, and Dynamics, Vol. 7, September-October 1984, pp. 513-562.

[9] Ibrahim, A.M., "Mathematical Modelling of Flexible Multibody Dynamics with Application to Orbiting Systems," Ph. D. Thesis (in preparation), University of British Columbia, 1987.

[10] Budica, R.J., and Tong, K.L., "Shuttle On-orbit Attitude Dynamics Simulation," AIAA/AAS Astrodynamics Conference, San Diego, California, August 1982, Paper No. AIAA-82-1452.

Computational Issues in Control-Structure Interaction Analysis

K. C. Park
Center for Space Structures and Controls
University of Colorado, Campus Box 429
Boulder, Colorado 80309

ABSTRACT

This paper surveys computational issues for large-scale simulation of dynamics and controls of space structures, which involve structural elements that are capable of large combined rigid and flexible motions, accurate and efficient treatment of constraints, robust integration of both translational and large rotational motions in the equations of motion, modular interface with active control synthesis packages, and a capability from wave motions to slowly varying transient responses. A particular feature of the present survey is a partitioned solution (or divide-and-conquer) procedure that can handle the numerical solution of multidisciplinary simulation problems by relying on individual modular solution packages that treat each aspect of simulation requirements.

1. Introduction

Computer simulation of large space structural systems requires a concerted integration of several physical and computational aspects. The physical considerations include the modeling of structural systems, control systems, thermal systems, liquid systems and environmental disturbances as well as manuevering thrusts and torques[1]. The attendant computational aspects include: a data management system that facilitates the necessary interprocessor communications; a simple yet versatile data structure for describing the system topology; and, computational modules for generating the governing structural equations of motion, for control synthesis that can easily interfaced with the structural analysis module, for performing transient analyses ranging from wave to slowly varying responses, for accurate robust treatment of constraints, for fluid sloshing interactions with structures, for thermal analysis and for disturbance characterizations.

Of the computational aspects listed above, this paper will concentrate on some of the computational modules. Specifically, we will address a partitioned analysis procedure by which one can utilize exisiting analysis modules to conduct the interaction analysis primarily involving structures, controls and constraints, nevertheless a procedure that can be extendable to accommodate any additional system interactions such as thermal and fluid analyzers.

In Section 2, we overview the partitioned procedure to simulate the control-structure interaction problems along with a set of the governing equations of motion. In Section 3, we present a new constraint treatment technique that can solve the constraint equations in a separate module from the solution module for the equations ofmotion. A major feature of this technique is to preserve the system topology for a variety of multibody dynamics systems. Solution techniques for large

Springer Series in Computational Mechanics
S. N. Atluri, A. K. Amos (Eds.)
Large Space Structures: Dynamics and Control

totational motions are discussed in Section 4 while the solution of the translational motions is treated by a conventional explicit direct integration method. Applications of these techniques are presented in Section 5. We conclude the present survey with future directions in methods development for the structure-interaction problems.

2. Equations of Motion for Control-Structure Interaction (CSI) Problems

We propose to express the discrete equations of motion for control-structure interaction systems in the form[2-7]:

$$\mathbf{M}\ddot{\mathbf{q}} + \mathbf{C}(\dot{\mathbf{q}}) + \mathbf{S}(u) + \mathbf{B}^T\boldsymbol{\lambda} = \mathbf{f}(t) - \mathbf{D}\mathbf{u} \qquad (2\cdot1)$$

$$\mathbf{u} = \mathbf{A}\dot{\mathbf{q}} + \mathbf{E}\mathbf{q} \qquad (2\cdot2)$$

$$\boldsymbol{\Phi}(\dot{\mathbf{q}},\ \mathbf{q},\ t) = 0 \qquad (2\cdot3)$$

In (2.1), $\mathbf{M}$ is the mass matrix, $\mathbf{C}(\cdot)$ is the generalized velocity-dependent force operator, $\mathbf{S}(\cdot)$ is the internal force operator due to member flexibility, $\mathbf{B}$ is the gradients of the nonholonomic and holonomic constraints (2.3), $\boldsymbol{\lambda}$ is the constraint forces, $\mathbf{f}(t)$ is the applied force, $\mathbf{D}$ is the projection matrix for the control force $\mathbf{u}$ that is acting on the structure as an applied force, q is the generalized displacement vector, $(\dot{\ })$ denotes time differentiation and $(\ )^T$ designates the matrix transposition. In (2.2), $\mathbf{A}$ and $\mathbf{E}$ are obtained as a combination of sensor and actuator characteristics, open-loop control and closed-loop control performance requirements. Finally, $\boldsymbol{\Phi}$ in (2.3) represents system constraints.

3. Partitioned Solution Procedure for CSI Problems

There are three ways of solving (2.1)-(2.3). In most conventional approaches, one eliminates both the constraint force, $\boldsymbol{\lambda}$, and the control force, $\mathbf{u}$, and solve for $\mathbf{q}$. The recovery of $\boldsymbol{\lambda}$ and $\mathbf{u}$ is realized by substitutions via (2.2) and (2.3). While such approaches appear straightforward for small-scale problems, there arise several computational and software difficulties when applied to large-scale problems.

First, the embedding of the control laws(2.2) into the structural equations of motion becomes a difficult task if the structural system exhibit strong nonlinearities as in slewing motions and/or deployment stages.

Second, the above embedding process often destory the sparsity of the associated matrices such that the solution task can burden not only the computational aspects but also the attendent software requirements.

Third, most important of all, the embedding appraoch inhibits a balanced interaction of two experts, viz, the dynamist and the control scientist. This is because, if the control laws are embedded into the structural dynamics equations, there is an unavoidable attempt by the dynamist to simplify the control laws as much as possible. On the other hand, if the structural dynamics equations are to be subsummed into the control laws, the control scientist would like to simply the physics of structural dynamics as much as possible. One evidence of such tendency is that most existing control laws are based on linearized modal characteristics.

As an alternative to the prevailing embedding approach, we propose a partitioned solution procedure[8-10]. The proposed partitioned procedure is designed to meet the following requirements.

Structural design and analysis programs have been developed over the last two decades with substantial financial investment. It is best thus for us to adopt the structural analysis programs to conduct CSI problems by preserving the integrity of the programs. This can be possible if one views the control laws acting on the structure as an external force, even though it may be state-dependent as is the case.

Likewise, there have been considerable software development efforts to analyze and conduct various control systems problems. It is thus highly desirable to adopt such software packages to conduct the CSI problems as well. As a means to accommodate the preceding practical considerations, of several possibilities one may employ the so-called equation augmentation which will be illustrated below.

If we differentiate (2.2) with respect to time, we obtain

$$\dot{\mathbf{u}} = \mathbf{A}\ddot{\mathbf{q}} + \mathbf{E}\dot{\mathbf{q}} \tag{3·1}$$

Substituting $\ddot{\mathbf{q}}$ from (2.1) into the above equation one obtains

$$\dot{\mathbf{u}} + \mathbf{A}\mathbf{M}^{-1}\mathbf{D}\mathbf{u} = \mathbf{A}\mathbf{M}^{-1}\dot{\mathbf{p}} + \mathbf{E}\dot{\mathbf{q}} \tag{3·2}$$

where $\mathbf{A}$ and $\mathbf{E}$ are assumed to be time-invariant for present purposes and $\dot{\mathbf{p}}$ is the generalized rate of momentum given by

$$\dot{\mathbf{p}} = \mathbf{f}(t) - \mathbf{C}(\dot{\mathbf{q}}) - \mathbf{S}(u) - \mathbf{B}^T\boldsymbol{\lambda} \tag{3·3}$$

Thus, (3.2) constitutes a set of differential equations for the control laws. We note that, if (2.2) involve an integral of $\mathbf{q}$, the resulting differential equations for the control laws will be of second-order in time.

We now observe that the solution of (2.1) (2.3) and (3.2) can be carried out by a judicious employment of three software packages: the structural analyzer to obtain $\mathbf{q}$, the solver for constraint force, λ, which we will described later in the paper, and the solver for the control force, $\mathbf{u}$. We will now describe some detailed computational procedrues associated with the three solution packages.

4. Staggered Stabilized Procedure for Constraints

In this Section, we review the staggered stabilized technique[11-12] by which one can solve the constraint force in a separate solution package from that for the structural displacement. We recall (2.1) and (2.3) as

$$\mathbf{M}\ddot{\mathbf{q}} + \mathbf{C}(\dot{\mathbf{q}}) + \mathbf{S}(\mathbf{q}) + B^T\boldsymbol{\lambda} = \mathbf{f} - \mathbf{D}\mathbf{u} \tag{4·1}$$

$$\Phi(\mathrm{q}, \quad \dot{\mathrm{q}}, \quad t) = 0 \tag{4·2}$$

We invoke the well-known penalty technique to approximate the constraint force in the form

$$\lambda = \frac{1}{\epsilon}\Phi(\mathrm{q}, \quad \dot{\mathrm{q}}, \quad t), \qquad 0 < \epsilon << 1 \tag{4·3}$$

Differentiation of λ with respect to time then yields

$$\epsilon\dot{\lambda} = B\ddot{\mathrm{q}} + \dot{B}\dot{\mathrm{q}} + \frac{\partial\Phi}{\partial t} \tag{4·4}$$

Substituting $\ddot{\mathrm{q}}$ from (4.1) into (4.4), we obtain

$$\epsilon\dot{\lambda} + BM^{-1}B^T\lambda = BM^{-1}(\dot{\mathrm{p}} - \mathrm{Du}) + \dot{B}\dot{\mathrm{q}} + \frac{\partial\Phi}{\partial t} = \mathrm{r}_\lambda \tag{4·5}$$

Hence, the solution of (2.1) and (2.3) has been replaced by (2.1) and (4.5). In addition to its modular solution feature, the staggered stabilized technique(4.5) possesses one favorable algorithmic characteristic. We note that when $\mathbf{BM}^{-1}\mathbf{B}$ becomes computationally signular as occurs when two or more constraint conditions become computationally dependent, i.e., due to the same geometric or motion state, most conventional technique including the Baumgarte technique[13-14] as given by

$$\begin{bmatrix} \mathrm{G} & \mathrm{B}^T \\ \mathrm{B} & 0 \end{bmatrix} \begin{Bmatrix} \mathrm{q} \\ \lambda \end{Bmatrix} = \begin{Bmatrix} \mathrm{b}_q \\ \mathrm{b}_\lambda \end{Bmatrix} \tag{4·6}$$

will fail to converge unless a sophisticated reduction technique[15-17] is invoked which can complicate the solution process.

On the other hand, the staggered stabilized technique overcomes the above mentioned numerical difficulty since its solution iteration matrix, $(\epsilon + \delta\mathrm{BM}^{-1}\mathrm{B}^T)$, remains nonsigular where δ is dependent on the time stepsize.

5. Methods for Large Structural Motions [18]

A basic difficulty in direct integration of (2.1) is that the angular velocity ω that is a part of $\dot{\mathrm{q}}$ is not directly integrable, except for some special kinematic configurations. This motivates us to partition $\ddot{\mathrm{q}}$ into the translational acceleration vector, $\ddot{\mathrm{x}}$, which is directly integrable and the angular acceleration vector, $\dot{\omega}$, which is not, and treat them by two different algorithms, viz

$$\ddot{\mathrm{q}} = \begin{Bmatrix} \ddot{\mathrm{x}} \\ \dot{\omega} \end{Bmatrix}, \qquad \dot{\mathrm{q}} = \begin{Bmatrix} \dot{\mathrm{x}} \\ \omega \end{Bmatrix} \tag{5·1}$$

The equations of motion (2.1) can be partitioned according to the above acceleration partitioning:

$$\begin{bmatrix} M_d & 0 \\ 0 & M_\omega \end{bmatrix} \begin{Bmatrix} \ddot{\mathrm{x}} \\ \dot{\omega} \end{Bmatrix} + \begin{Bmatrix} Q_d \\ Q_\omega \end{Bmatrix} = \begin{Bmatrix} f_x - \mathrm{D}_x\mathrm{u} \\ f_\omega - \mathrm{D}_\omega\mathrm{u} \end{Bmatrix} \tag{5·2}$$

where

$$\left\{\begin{matrix} Q_x \\ Q_w \end{matrix}\right\} = \left\{\begin{matrix} C_x(\dot{q}) + S_x(q) - B_x^T\lambda \\ C_w(\dot{q}) + S_w(q) - B_w^T\lambda \end{matrix}\right\} \tag{5·3}$$

To effect the node-by-node integration for the rotational degrees of freedom, we partition $\dot{\omega}$ further into

$$\dot{\omega} = \lfloor \dot{\omega}^1, \dot{\omega}^2, \ldots, \dot{\omega}^P \rfloor^T \tag{5·4}$$

where $\dot{\omega}^{(j)}$ is a (3×1) angular acceleration vector for the j-th node,

$$\dot{\omega}^{(j)} = \lfloor \dot{\omega}_1^{(j)}, \dot{\omega}_2^{(j)}, \dot{\omega}_3^{(j)} \rfloor \tag{5·5}$$

5.1 Euler Parameters and Angular Accelerations

In order to update the angular orientations that correspond to the angular velocity states in the equations of motion, we introduce the well-known four-parameter Euler representation of the angular velocity for each node as (see, e.g., Wittenburg[19]):

$$\dot{e} = \frac{1}{2}\begin{bmatrix} 0 & -\omega^T \\ \omega & -\tilde{\omega} \end{bmatrix} e = G(\omega)q, \qquad e = \lfloor e_0 \;\; e_1 \;\; e_2 \;\; e_3 \rfloor^T, \quad e^T e = 1 \tag{5·6}$$

where

$$\tilde{\omega} = \begin{bmatrix} 0 & -\omega_3 & \omega_2 \\ \omega_3 & 0 & -\omega_1 \\ -\omega_2 & \omega_1 & 0 \end{bmatrix}, \qquad \omega = \lfloor \omega_1 \;\; \omega_2 \;\; \omega_3 \rfloor \tag{5·7}$$

and the nodal-designating superscript is omitted for notational simplicity.

Time differentiation of (5.6a) once more yields

$$\ddot{e} = G(\dot{\omega}) \cdot e + G(\omega) \cdot \dot{e}, \quad e^T \dot{e} = 0 \tag{5·8}$$

where $G(\dot{\omega})$ is obtained by substituting $\dot{\omega}$ for ω in $G(\omega)$.

5.2 Mid-Point Integration of Euler Parameters

Suppose that we know the state variables, $\dot{\omega}^k$ and ω^k, at the k-th time step and we want to solve for e^{k+1}. Because of the special properties of $G(\omega)$ and $G(\dot{\omega})$, one can take advantage of the following set of mid-point rules:

$$\left\{\begin{aligned} e^{k+\frac{1}{2}} &= e^k + \delta \dot{e}^{k+\frac{1}{2}} \\ \dot{e}^{k+\frac{1}{2}} &= \dot{e}^k + \delta \ddot{e}^{k+\frac{1}{2}} \\ e^{k+1} &= 2e^{k+\frac{1}{2}} - e^k, \qquad \delta = \tfrac{h}{2} \end{aligned}\right. \tag{5·9}$$

where h is the stepsize.

Substituting (5.6) and (5.8) into (5.9), one obtains

$$\left\{\begin{aligned} &[I - \delta G(\omega^{k+\frac{1}{2}}) - \delta^2 G(\dot{\omega}^{k+\frac{1}{2}})]e^{k+\frac{1}{2}} = [I - \delta G(\omega^{k+\frac{1}{2}}) + \delta G(\omega^k)]e^k \\ &[I - \delta G(\omega^{k+\frac{1}{2}})]\dot{e}^{k+\frac{1}{2}} = \dot{e}^k + \delta G(\omega^{k+\frac{1}{2}})e^{k+\frac{1}{2}}, \quad (e^{k+\frac{1}{2}})^T \dot{e}^{k+\frac{1}{2}} = 0 \end{aligned}\right. \tag{5·10}$$

Since $G(\omega^{k+\frac{1}{2}})$ and $G(\dot{\omega}^{k+\frac{1}{2}})$ are not available, we approximate them by

$$G(\omega^{k+\frac{1}{2}}) \simeq G(\omega^k) \quad , \quad G(\dot{\omega}^{k+\frac{1}{2}}) \simeq G(\dot{\omega}^k) \tag{5·11}$$

which are equivalent to a Newton-like tangent approximation. Hence, we have from (5.10a) and (5.11)

$$[I - \delta G(\omega^k) - \delta^2 G(\dot{\omega}^k)] \cdot \mathbf{e}^{k+\frac{1}{2}} = \mathbf{e}^k \tag{5·12}$$

Once $\mathbf{e}^{k+\frac{1}{2}}$ is obtained from (5.12), we can obtain $\dot{\mathbf{e}}^{k+\frac{1}{2}}$ from (5.10b) by

$$[I - \delta G(\omega^{k+\frac{1}{2}})]\dot{\mathbf{e}}^{k+\frac{1}{2}} = \dot{\mathbf{e}}^k + \delta G(\omega^{k+\frac{1}{2}})\mathbf{e}^{k+\frac{1}{2}} \tag{5·13}$$

Finally, one can update ω from the following formula:

$$\omega = 2(e_0\dot{\mathbf{g}} - \tilde{\mathbf{g}}\dot{\mathbf{g}} - \dot{e}_0\mathbf{g}), \quad \mathbf{g} = \lfloor e_1 \quad e_2 \quad e_3 \rfloor, \quad e_0^2 + \mathbf{g}^T\mathbf{g} = 1 \tag{5·14}$$

where $\tilde{\mathbf{g}}$ has the same form as $\tilde{\omega}$ as given by (5.7). Hence, if necessary, one can obtain $\dot{\omega}$ from the equations of motion and iterate on $\mathbf{e}$ and $\dot{\mathbf{e}}$ by (5.10).

The skew-symmetric solution matrices in the above difference equations can be explicitly inverted via the formula:

$$(\mathrm{I} - \delta G(\omega))^{-1} = \begin{bmatrix} 1 & -a & -b & -c \\ a & 1 & c & -b \\ b & -c & 1 & a \\ c & b & -a & 1 \end{bmatrix}^{-1} = \frac{1}{1 + a^2 + b^2 + c^2} \begin{bmatrix} 1 & a & b & c \\ -a & 1 & -c & b \\ -b & c & 1 & -a \\ -c & -b & a & 1 \end{bmatrix} \tag{5·15}$$

Hence, the solution of (5.12) and (5.13) becomes straightforward.

5.3 Update of New Angular Orientation

Once $\mathbf{e}^{k+1}$ is computed from (5.12) and (5.9c), it is often required to compute the body-fixed basis vector, $\mathbf{b} = \lfloor \mathbf{b}_1 \quad \mathbf{b}_2 \quad \mathbf{b}_3 \rfloor^T$ in terms of the inertial basis vectors, $\mathbf{a} = \lfloor \mathbf{a}_1 \quad \mathbf{a}_2 \quad \mathbf{a}_3 \rfloor^T$. These two vectors are related by

$$\mathbf{b} = \mathbf{R}\mathbf{a} \tag{5·16}$$

where

$$\mathbf{R} = \begin{bmatrix} 2(e_0^2 + e_1^2) - 1 & 2(e_1e_2 + e_0e_3) & 2(e_1e_3 - e_0e_2) \\ 2(e_1e_2 - e_0e_3) & 2(e_0^2 + e_2^2) - 1 & 2(e_2e_3 + e_0e_1) \\ 2(e_1e_3 + e_0e_2) & 2(e_2e_3 - e_0e_1) & 2(e_0^2 + e_3^2) - 1 \end{bmatrix} \tag{5·17}$$

In order to satisfy the orthonormality of $\mathbf{R}$, it is crucial to satisfy the two constraints, viz, $\mathbf{e}^T\mathbf{e} = 1$ in computing $\mathbf{e}$ and $\mathbf{e}^T\dot{\mathbf{e}} = 0$ in computing $\dot{\mathbf{e}}$, respectively. This can be accomplished for $\mathbf{e}^{k+\frac{1}{2}}$ by augmenting the constraint and solving the following equation by a Newton-like procedure:

$$\begin{bmatrix} \mathbf{E}_e & \mathbf{e} \\ \mathbf{e}^T & 0 \end{bmatrix} \begin{Bmatrix} \mathbf{e} \\ \lambda \end{Bmatrix} = \begin{Bmatrix} \mathbf{b}_e \\ 1 \end{Bmatrix} \tag{5·18}$$

where $\mathbf{E}_e$ is the (4×4) solution matrix in the lefthand side of (5.10a) and $\mathbf{b}_e$ is the righthand side vector of (5.10a), respectively.

Similarly, the constraint $\mathbf{e}^T\dot{\mathbf{e}} = 0$ can be satisfied by

$$\begin{bmatrix} \mathbf{E}_{\dot{e}} & \mathbf{e} \\ \mathbf{e}^T & 0 \end{bmatrix} \begin{Bmatrix} \dot{\mathbf{e}} \\ \lambda \end{Bmatrix} = \begin{Bmatrix} \mathbf{b}_{\dot{e}} \\ 0 \end{Bmatrix} \tag{5·19}$$

where $\mathbf{E}_{\dot{e}}$ is the (4×4) solution matrix in the lefthand side of (5.13) and $\mathbf{b}_{\dot{e}}$ is the righthand side vector of (5.13), respectively.

In addition, after the solution has converged at the $(k+\frac{1}{2})$-timestep, we must enforce the same two constraint conditions in updating $\mathbf{e}^{k+1}$ and $\dot{\mathbf{e}}^{k+1}$ by a similar procedure as for (5.18) and (5.19).

6. Solution Procedures for Control Laws

We recall the agmented equation for the control laws from (3.2):

$$\dot{\mathbf{u}} + \mathbf{A}\mathbf{M}^{-1}\mathbf{D}\mathbf{u} = \mathbf{A}\mathbf{M}^{-1}\dot{\mathbf{p}} + \mathbf{E}\dot{\mathbf{q}} \tag{6·1}$$

where

$$\dot{\mathbf{p}} = \mathbf{f}(t) - \mathbf{C}(\dot{\mathbf{q}}) - \mathbf{S}(\mathbf{q}) - \mathbf{B}^T\boldsymbol{\lambda} \tag{6·2}$$

It is noted that, if $\mathbf{A} = c\mathbf{D}$ with c being a scalar constant, the solution matrix for $\mathbf{u}$ becomes symmetric, hence achieving substantial computational simplicity.

To gain insight into the above partitioned equation for the control laws, we consider the linearized undamped $(\mathbf{C} = 0)$ structural dynamics case and express the coupled equations without any system constraint force to yield

$$\begin{bmatrix} \mathbf{M} & 0 \\ 0 & 0 \end{bmatrix} \begin{Bmatrix} \ddot{\mathbf{q}} \\ \ddot{\mathbf{u}} \end{Bmatrix} + \begin{bmatrix} 0 & 0 \\ -\mathbf{E} & \mathbf{I} \end{bmatrix} \begin{Bmatrix} \dot{\mathbf{q}} \\ \dot{\mathbf{u}} \end{Bmatrix} + \begin{bmatrix} \mathbf{S} & \mathbf{D} \\ \mathbf{A}\mathbf{M}^{-1}\mathbf{S} & \mathbf{A}\mathbf{M}^{-1}\mathbf{D} \end{bmatrix} \begin{Bmatrix} \mathbf{q} \\ \mathbf{u} \end{Bmatrix} = \begin{Bmatrix} \mathbf{f} \\ \mathbf{A}\mathbf{M}^{-1}\mathbf{f} \end{Bmatrix} \tag{6·3}$$

whose original eigensystem reads:

$$\begin{bmatrix} \mu^2\mathbf{M} + \mathbf{S} & \mathbf{D} \\ -\mu\mathbf{A} + \mathbf{E} & \mathbf{I} \end{bmatrix} \begin{Bmatrix} \mathbf{q} \\ \mathbf{u} \end{Bmatrix} = 0 \tag{6·4}$$

or in its augmented form:

$$\begin{bmatrix} \mu^2\mathbf{M} + \mathbf{S} & \mathbf{D} \\ -\mu\mathbf{E} + \mathbf{A}\mathbf{M}^{-1}\mathbf{S} & \mu\mathbf{I} + \mathbf{A}\mathbf{M}^{-1}\mathbf{D} \end{bmatrix} \begin{Bmatrix} \mathbf{q} \\ \mathbf{u} \end{Bmatrix} = 0 \tag{6·5}$$

It is emphasized that the eigenvalues μ are not just for the strutural systems only but for the combined structural and control system. It is also interesting to observe that in the original eigensystem the rate feeback gain matrix $\mathbf{A}$ is associated the eigenvalue whereas in the augmented form the position feedback gain matrix $\mathbf{E}$ is associated with the eigenvalue even though they are equivalent. Clearly, the latter form suggests that one must first attain stability for the block-diagonal system $(\mu\mathbf{I}+\mathbf{A}\mathbf{M}^{-1}\mathbf{D})$, which implies that stability for the rate feedback system requires higher priority than the position feedback scheme unless it is absent ($\mathbf{A} = 0$).

The augmented form of eigenproblem can also provide a natural perturbation vehicle for the sensitivity analysis of stability margin of closed-loop systems. For example, one may start with the two sets of uncoupled eigenvalues:

$$(\mu^2\mathbf{M}+\mathbf{S})\mathbf{q}=0, \quad (\mu\mathbf{I}+\mathbf{A}\mathbf{M}^{-1}\mathbf{D})\mathbf{u}=0 \tag{6·6}$$

and iterate on μ using the augmented eigenproblem.

We note that the augmented form of equation for the control laws(6.1) provides a natural means of obtaining the control laws by direct integration of the augmented differential equation. On the other hand, the original form of control laws (2.2) can be obtained only after both $\mathbf{q}$ and $\dot{\mathbf{q}}$ are obtained. Finally, it should be mentioned that the computation of $\mathbf{u}$ as the primary variable from the augmented eqaution yields more accurate results than by post-processing with the original equation.

7. Summary of Partitioned Solution Procedures

We summarize the present overall computational procedures.

- For the flexible structural system we have

$$\begin{cases} \mathbf{M}_x\ddot{\mathbf{x}}=\mathbf{f}_x-\mathbf{C}_d(\dot{\mathbf{q}})-\mathbf{S}_d(\mathbf{q})-\mathbf{B}_d^T\lambda-\mathbf{D}_q\mathbf{u} \\ \mathbf{M}_\omega\dot{\omega}=\mathbf{f}_\omega-\mathbf{C}_\omega(\dot{\mathbf{q}})-\mathbf{S}_\omega(\mathbf{q})-\mathbf{B}_\omega^T\lambda-\mathbf{D}_\omega\mathbf{u} \end{cases} \tag{7·1}$$

with the computational sequences as follows:

$$\begin{cases} \dot{\mathbf{x}}^{n+\frac{1}{2}}=\mathbf{x}^{n-\frac{1}{2}}+h\ddot{\mathbf{x}}^n \\ \mathbf{x}^{n+1}=\mathbf{x}^n+h\dot{\mathbf{x}}^{n+\frac{1}{2}} \end{cases} \tag{7·2}$$

$$\begin{cases} \mathbf{G}(\omega^{k+\frac{1}{2}})\simeq\mathbf{G}(\omega^k) \quad \mathbf{G}(\dot{\omega}^{k+\frac{1}{2}})\simeq\mathbf{G}(\dot{\omega}^k) \\ [I-\delta\mathbf{G}(\omega^{k+\frac{1}{2}})-\delta^2\mathbf{G}(\dot{\omega}^{k+\frac{1}{2}})]\mathbf{e}^{k+\frac{1}{2}}=[I-\delta\mathbf{G}(\omega^{k+\frac{1}{2}})+\delta\mathbf{G}(\omega^k)]\mathbf{e}^k \\ [I-\delta\mathbf{G}(\omega^{k+\frac{1}{2}})]\dot{\mathbf{e}}^{k+\frac{1}{2}}=\dot{\mathbf{e}}^k+\delta\mathbf{G}(\omega^{k+\frac{1}{2}})\mathbf{e}^{k+\frac{1}{2}}, \quad (\mathbf{e}^{k+\frac{1}{2}})^T\dot{\mathbf{e}}^{k+\frac{1}{2}}=0 \\ \mathbf{e}^{k+1}=2\mathbf{e}^{k+\frac{1}{2}}-\mathbf{e}^k \\ \dot{\mathbf{e}}^{k+1}=2\dot{\mathbf{e}}^{k+\frac{1}{2}}-\dot{\mathbf{e}}^k, \quad (\mathbf{e}^{k+1})^T\dot{\mathbf{e}}^{k+1}=0 \\ \omega=2(e_0\dot{\mathbf{g}}-\tilde{\mathbf{g}}\dot{\mathbf{g}}-\dot{e}_0\mathbf{g}) \end{cases} \tag{7·3}$$

It should be mentioned that the above procedures require only (4×4)-matrices which can be inverted explicitly via (5.15).

- For the control laws, we have the following differential equation:

$$\begin{cases} \dot{\mathbf{u}}+\mathbf{A}\mathbf{M}^{-1}\mathbf{D}\mathbf{u}=\mathbf{A}\mathbf{M}^{-1}\dot{\mathbf{p}}+\mathbf{E}\dot{\mathbf{q}} \\ \dot{\mathbf{p}}=\mathbf{f}(t)-\mathbf{C}(\dot{\mathbf{q}})-\mathbf{S}(\mathbf{q})-\mathbf{B}^T\lambda \end{cases} \tag{7·4}$$

with the solution procedure

$$\begin{cases} (\mathbf{I}+\delta\mathbf{A}\mathbf{M}^{-1}\mathbf{D})\mathbf{u}^{k+\frac{1}{2}}=\delta(\mathbf{A}\mathbf{M}^{-1}\dot{\mathbf{p}}^{k+\frac{1}{2}}+\mathbf{E}\dot{\mathbf{q}}^{k+\frac{1}{2}}+\mathbf{u}^k \\ \mathbf{u}^{k+1}=2\mathbf{u}^{k+\frac{1}{2}}-\mathbf{u}^k \end{cases} \tag{7·5}$$

- For the system constraints we have

$$\epsilon\dot{\lambda} + BM^{-1}B^T\lambda = \mathrm{r}_\lambda = BM^{-1}(\dot{\mathrm{p}} - \mathrm{Du})\dot{B}\dot{\mathrm{q}} + \frac{\partial\Phi}{\partial t} \tag{7·6}$$

with the resulting solution equation:

$$(\epsilon\mathbf{I} + \delta\mathbf{B}\mathbf{M}^{-1}\mathbf{B}^T)\lambda^{k+\frac{1}{2}} = \delta\mathrm{r}_\lambda^{k+\frac{1}{2}} + \epsilon\lambda^k \tag{7·7}$$

We will now apply (7.1) - (7.7) to some sample problems.

8. Example Problems

The structural equation solver and the constraint force solver have been implemented as stand-alone modules and the control law module is presently under implementation. We will now illustrate some simple examples for partitioned solution of the constrained dynamical systems. We hope to report on partitioned solution examples involving all the three solution modules in the near future.

The first example is a classical crank mechanism whose governing equations of motion are characterized by the following matrices and constraints[20].

$$M = \begin{bmatrix} J_1 & & & \\ & J_2 & & \\ & & m & \\ & & & m \end{bmatrix} \tag{8·1}$$

$$\Phi = \begin{Bmatrix} r\cos\theta - (x - l_1\cos\phi) \\ r\sin\theta - (y - l_1\sin\phi) \\ (l - l_1)\sin\phi + y \end{Bmatrix} = 0 \tag{8·2}$$

$$B^T = \begin{bmatrix} -r\sin\theta & r\cos\theta & 0 \\ -l_1\sin\theta & l_1\cos\phi & (l - l_1)\cos\phi \\ -1 & 0 & 0 \\ 0 & -1 & 1 \end{bmatrix} \tag{8·3}$$

and

$$\mathrm{q} = \lfloor \theta \quad \phi \quad x \quad y \rfloor^T, \quad \lambda = \lfloor \lambda_1 \quad \lambda_2 \quad \lambda_3 \rfloor^T$$

$$\mathrm{S(q)} = 0, \quad \mathrm{f} = \{0 \quad 0 \quad 0 \quad -mg\}^T \tag{8·4}$$

Figure 1 shows the problem definition and the numerical performance of the staggered stabilized technique along with Baumgarte's technique. In carrying out the computations, the mid-point rule has been used to time-discretize the equations of motion(2.1), the constraints(2.3) and their stabilized form (6.7). For the initial condition applied, we chose the time increment, $h = 0.01$, for the period, $T = 0.82$. During the computation no iteration was performed at each integration step so that we could measure the performance of the two techniques directly in terms of violation of the constraint conditions vs time during one complete cycle. In each technique, the three constraint conditions exhibited a same order of accuracy level. Hence, we illustrate only one constraint violation history, i.e., the pin joint constraint between the crank and the connecting

iterations. It is noted that the staggered stabilized technique as given by (4.5) overcomes this singularity difficulty since $\dot{\lambda}$ still exist.

The third example is a sphere rolling on a flat surface with the constraint

$$y = sinx \qquad (8 \cdot 5)$$

This problem was investigated by Huston et al [22], whose equations do not involve the constraint force, λ. In the present analysis, we employ a formulation that incorporates the constraint force as part of the system variables. Fig. 3 illustrates the ball with its radius a and an offset center r_0 that is to follow the sine curve ($y = \sin x$).

The various matrices and vector quantities for (6.1)-(6.3) can be derived as

$$\mathbf{M} = \begin{bmatrix} m & 0 & -mr_0\mathbf{a}_1 \cdot \mathbf{b}_2 & mr_0\mathbf{e}_1 \cdot \mathbf{b}_1 & 0 \\ 0 & m & -mr_0\mathbf{a}_2 \cdot \mathbf{b}_2 & mr_0\mathbf{a}_2 \cdot \mathbf{b}_1 & 0 \\ & & J_1 & 0 & 0 \\ & \text{sym.} & & J_2 & 0 \\ & & & & J_3 \end{bmatrix}$$

$$B = \begin{bmatrix} 1 & 0 & -a\mathbf{b}_1 \cdot \mathbf{a}_2 & -a\mathbf{b}_2 \cdot \mathbf{a}_2 & -a\mathbf{b}_3 \cdot \mathbf{e}_2 \\ 0 & 1 & a\mathbf{b}_1 \cdot \mathbf{a}_1 & a\mathbf{b}_2 \cdot \mathbf{a}_1 & a\mathbf{b}_3 \cdot \mathbf{a}_1 \\ \cos x & -1 & 0 & 0 & 0 \end{bmatrix} \qquad (8 \cdot 6)$$

rod. Note that the constraint error level by Baumgarte's technique remains about two digits above that by the staggered stabilized technique. For the staggered stabilized technique, the penalty parameter chosen was $\epsilon = 10^{-6}$, which yielded an accuracy level about 10^{-5} for the technique.

The second problem tested is a simplified version of the seven-link manipulator deployment problem[21]. The three links are initially folded as shown in Figure 2. For modeling simplicity, attached between the two joints is a coil spring which resists a constant deploying force at the tip of the third link. Also, the left-hand end of the first link is fixed through the same coil spring to the wall. These three coil springs are to be *locked up* once the links are deployed straight. The time-discretized difference equations both for Baumgarte's technique and the staggered stabilized technique have been solved at each time increment by a Newton-type iterative procedure to meet a specified accuracy level. Hence, the perfomance of the two techniques can be assessed by the average number of iterations taken per time increment. This is presented in Fig. 2 for the accuracy of 10^{-4}. Notice that the staggered stabilized technique requires on the average about 4.5 iterations per step whereas Baumgarte's technique requires about 22 iterations per step.

When the deployed links become straight, a state of singularity is experienced because at least two or more lows in $\mathbf{B}$ become numerically linearly dependent. Consequently, the matrix, $\mathbf{BM}^{-1}\mathbf{B}^T$, that is needed in solving the constraint forces by Baumgarte's technique (see Eq.(4.6)) becomes singular. When this condition is approached, Baumgarte's technique requires an excessively large number of iterations, whereas the staggered stablized technique still converges within 50

$$\mathbf{D}_d(\dot{\mathbf{q}}) = -mr_0 \begin{Bmatrix} \omega_1\omega_3\mathbf{a}_1\cdot\mathbf{b}_1 + \omega_2\omega_3\mathbf{a}_1\cdot\mathbf{b}_2 - (\omega_1^2+\omega_2^2)\mathbf{a}_1\cdot\mathbf{b}_3 \\ \omega_1\omega_3\mathbf{a}_2\cdot\mathbf{b}_1 + \omega_2\omega_3\mathbf{a}_2\cdot\mathbf{b}_2 - (\omega_1^2+\omega_2^2)\mathbf{a}_2\cdot\mathbf{b}_3 \end{Bmatrix} \tag{8·7}$$

$$\mathbf{D}_\omega(\dot{\mathbf{q}}) = - \begin{Bmatrix} \omega_2\omega_3(J_2-J_3) \\ \omega_3\omega_1(J_3-J_1) \\ \omega_1\omega_2(J_1-J_2) \end{Bmatrix} \tag{8·8}$$

$$\mathbf{f}_d = 0, \qquad \mathbf{f}_\omega = mgr_0 \begin{Bmatrix} \mathbf{a}_3\cdot\mathbf{b}_2 \\ -\mathbf{a}_3\cdot\mathbf{b}_1 \\ 0 \end{Bmatrix} \tag{8·9}$$

$$\mathbf{x} = \lfloor x, y \rfloor^T, \dot{\omega} = \lfloor \omega_1 \quad \omega_2 \quad \omega_3 \rfloor^T, \lambda = \lfloor \lambda_1 \quad \lambda_2 \quad \lambda_3 \rfloor \tag{8·10}$$

$$S(\mathbf{q}) = 0, \quad \mathbf{f} = 0 \tag{8·11}$$

$$\mathbf{b} = \mathbf{R}\mathbf{a} \tag{8·12}$$

$$\omega = 2(e_0\mathbf{g} - \tilde{\mathbf{g}}\dot{\mathbf{g}} - \dot{e}_0\mathbf{g}), \quad \mathbf{g} = \lfloor e_1 \quad e_2 \quad e_3 \rfloor^T \tag{8·13}$$

There is a total of eight variables in the foregoing equations of motion as given by (7.10). However, in adopting the present solution procedure—viz, (6.1)-(6.7)—we solve for nine variables.

Numerical solutions of the rolling of sphere on a flat sinusoidal curve have been obtained with the data summarized in Table 1.

Table 1 Physical Dimensions and Initial Conditions for a Rolling Sphere

$$m = 71.32N, \quad a = 10.9cm, \quad r_0 = 0 \text{ or } 0.15cm$$

$$J_1 = J_2 = J_3 = 2/5ma^2, \epsilon = 10^{-6}$$

$$x^0 = y^0 = 0, \quad \{\omega_2^0 = -\omega_1^0 = 1, \ \omega_3^0 = 0\}$$

$$\dot{x}^0 = \dot{y}^0 = a\omega_1^0, \qquad \{e_0^0 = 1, e_1^0 = e_2^0 = e_3^0 = 0\}$$

The time histories of the three constraint forces are shown in Fig. 4 for the case of the off-set center, where $(\lambda_1, \ \lambda_2)$ correspond to the x and y- component of the constraint force to maintain the rolling contact condition, and λ_3 is to maintain the sinusoidal trajectory as imposed by (7.5). Hence, the first two constraints are indicative of skidding phenomenon and the third corresponds to the steering force required in the ball manuevering. Notice that they exhibit highly nonlinear behavior while still periodic. The angular velocities for the two cases are shown in Fig. 5 and 6. Note that the angular velocities are periodic for the no-offset case whereas they no longer exhibit periodic response with the off-set center.

8. Concluding Remarks

A partitioned procedure for an accurate and efficient solution of control-structure interaction involving system constraints have been presented. The proposed partitioned procedure incorporates two novel solution algorithms for large angular motions and constraint stailization. The staggered stabilized constraint algorithm overcomes ill-conditioning of the constraint matrix, $\mathbf{B}$ and consistently has performed more robust than Baumgarte's technique. The angular orientation updating procedure treats each nodal angular orientation separately in terms of the four-parameter Euler formula. Hence, one deals with only (4×4)-matrices whose inversions are done explicitly. It has been found from our numerical experiments that it is essential to enforce the constraint conditions on the Euler parameters as given by (5.18) and its time derivatives (5.19) at each integration step.

The full potential of the proposed partitioned procedure, however, must wait until we have performed an extensive numerical experiments involving all the three field problems. We are currently pursuing such experiments and hope to report on our results shortly.

Acknowledgements

The work reported herein was supported in part by Air Force Office of Scientific Research under Grant F49620-87-C-0074 and in part by NASA/Langley Research Center under Grant NAG - 1 - 756. The authors wish to thank Dr. Anthony K. Amos of AFOSR and Drs. Jerry Housner and Jeff Stroud of NASA/Langley Research Center for their keen interest and encouragement during the course of the present work.

References

1. Wright, R. L.(ed.), *NASA/DOD Control/Structure Interaction Technology*, NASA Publication 2447, 1986.

2. Bodley, C. S., Devers, A. D., Park, A. C. and Frish, H. P., "A Digital Computer Program for the Dynamic Interaction Simulation of Controls and Structures (DISCOS)," NASA Technical Paper 1219, May 1978.

3. Balas, M. J., "Active Control of Flexible Systems", *Journal of Optimization Theory and Applications*, **25**, July 1978, 415-436.

4. Hale, A. L., Lisowski, R. J. and Dahl, W. E., "Optimal Simultaneous Structural and Control Design of Maneuvering Flexible Spacecraft," *Journal of Guidance, Control and Dynamics*, **8**, Jan.-Feb. 1985, 86-93.

5. Junkins, J. L., and Rew, D. W.,"Unified Optimization of Structures and Controllers", *Large Space Structures: Dynamics and Controls*, Springer-Verlag, to appear in 1987.

6. Skelton R. E. and Hughes, P. C., "Modal Cost Analysis for Linear Matrix Second-Order Systems," *J. Dyn. Syst. and Control*, **102**, 1980, 151-180.

7. Lõtstedt, P., "On a penalty function method for the simulation of mechanical systems subject to constraints," TRITA-NA-7919, 1979, Royal Institute of Technology, Stockholm, Sweden.

8. Park, K. C., Felippa, C. A. and DeRuntz, J. A., "Stabilization of staggered solution procedures for fluid-structure interaction analysis, " in: *Computational Methods for Fluid-Structure Interaction Problems*, Belytschko, T. and Geers, T. L. (editors), ASME, AMD Vol. 26, New York, N. Y., 1977, 95-124.

9. Park, K. C., "Partitioned Analysis Procedures for Coupled-Field Problems: Stability Analysis," *Journal of Applied Mechanics*, **47**, 1980, 370-378.

10. Park, K. C. and Felippa, C. A., "Partitioned Analysis of Coupled Systems," in: *Computational Methods for Transient Analysis*, T. Belytschko and T. J. R. Hughes (eds.), Elsevier Pub. Co., 1983, 157-219.

11. Park, K. C. and Chiou, J. C., "Evaluation of Constraint Stabilization Procedures for Multibody Dynamical Systems," *Proc. the 28th Structures, Structural Dynamics and Materials Conf.*, Part 2A, 1987, Monterey, CA, AIAA Paper No. 87-0927, 769-773.

12. Park, K. C. and Chiou, J. C., "Stabilization of Computational Procedures for Constrained Dynamical Systems," Report No. CU-CSSC-87-05, Center for Space Structure and Controls, College of Engineering, University of Colorado, Boulder, June 1987.

13. Baumgarte, J. W., "Stabilization of constraints and integrals of motion in dynamical systems," *Comp. Meth. Appl. Mech. Engr.*, **1**, 1-16 (1972).

14. Baumgarte, J. W., "A New Method of Stabilization for Holonomic Constraints," *Journal of Applied Mechanics*, **50**, 869-870 (1983).

15. Wehage, R. A. and Haug, E. J., "Generalized coordinate partitioning for dimension reduction in analysis of constrained dynamic systems," *ASME J. of Mech. Design*, **104**, 1982, 247-255.

16. Huston, R. L. and Kamman, J. W., "A discussion on constraint equations in multibody dynamics," *Mech. Res. Comm.*, **9**, (1982) 251-256.

17. Fuehrer, C. and Wallrapp, O., "A computer-oriented method for reducing linearized multibody system equations by incorporating constraints," *Comp. Meth. Appl. Mech. Engr.*, **46**, (1984) 169-175.

18. Park, K. C. and Chiou, J. C., "A Computational Procedure for Large Rotational Motions in Multibody Dynamics," Report No. CU-CSSC-87-08, Center for Space Structure and Controls, College of Engineering, University of Colorado, Boulder, August 1987.

19. Wittenburg, J., *Dynamics of Systems of Rigid Bodies*, B. G. Teubner, Stuttgart, 1977.

20. Haug, E. J., "Elements and Methods of Computational Dynamics," in: *Computer Aided Analysis and Optimization of Mechanical System Dynamics (Haug, E. J., Editor)*, Springer-Verlag, 1984, 3-38.

21. Housner, J. M., "Convected Transient Analysis for Large Space Structure Maneuver and Deployment," AIAA-84-1023-CP, *Proc. 25th Structures, Structural Dynamics and Material Conference*, Part 2, 14-16 May 1984, Palm Springs, 616-619.

22. Huston, R. L., Passerello, C., Winget, J. M. and Sears, J., "On the Dynamics of a Weighted Bowling Ball," *Journal of Applied Mechanics*, **46**, 1979, 937-943.

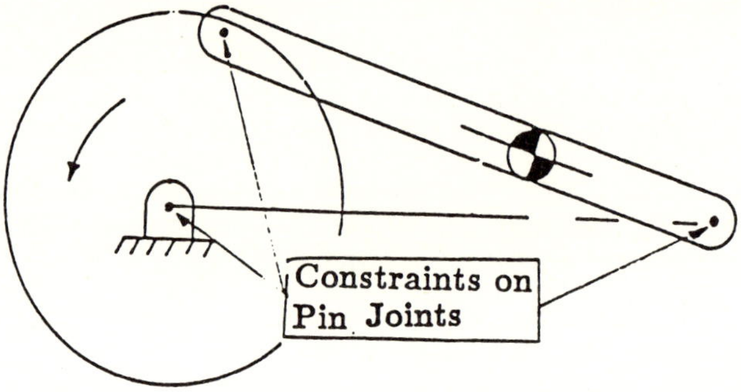

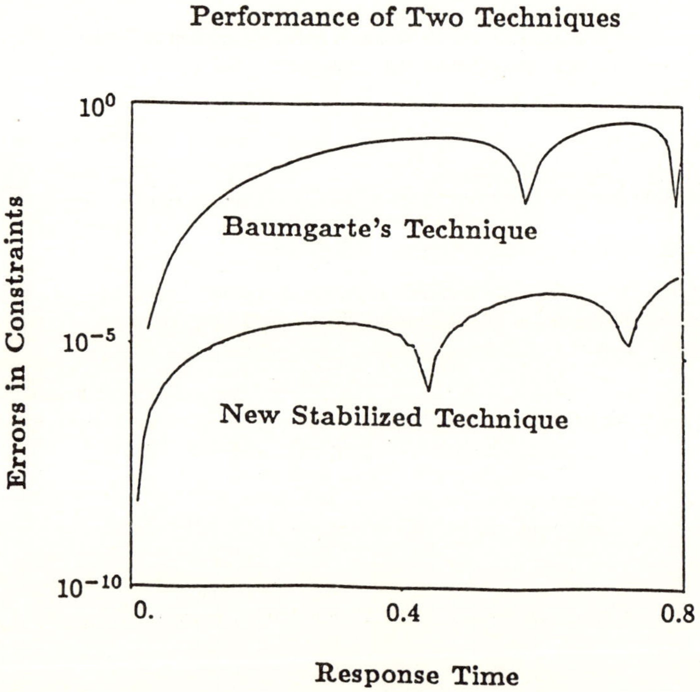

Fig. 1 Errors in Pin-Joint Constraint with No Iteration

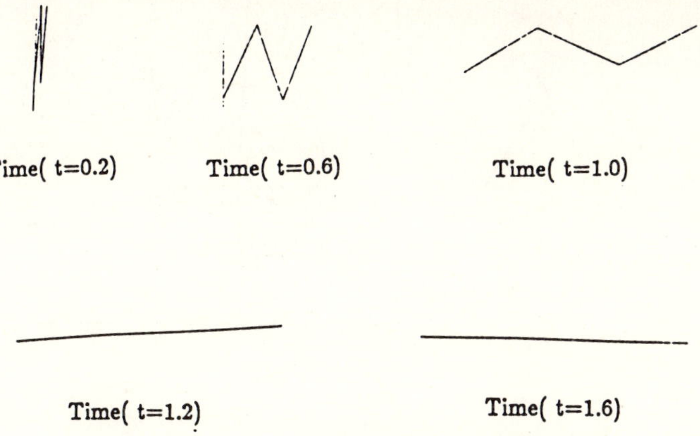

Performance of Two Stabilization Techniques
(Solution Accuracy=10^{-6})

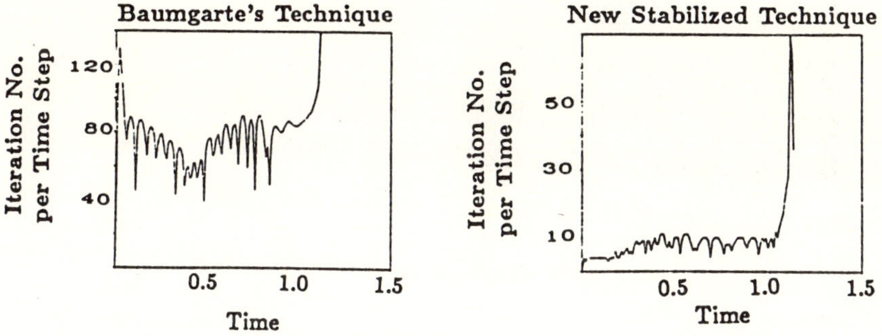

Fig. 2 Deployment of Three-Link Remote Manipulator

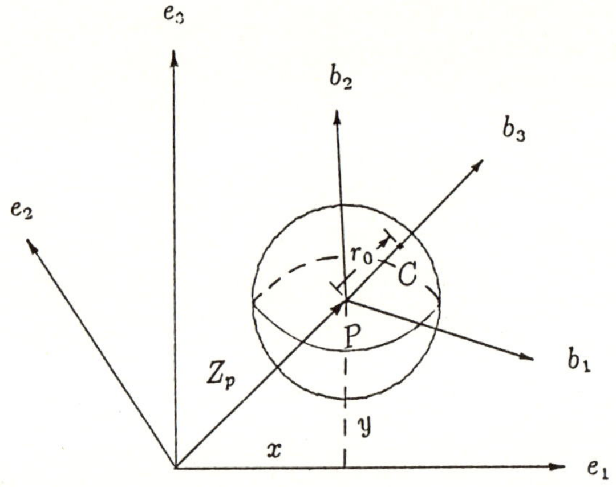

Fig. 3 Solid spherical ball rolling on a flat surface

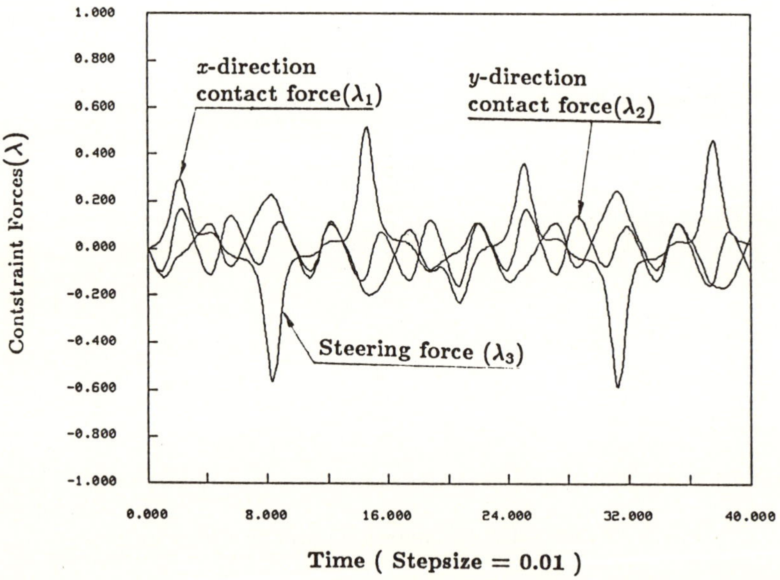

Fig. 4 Time histories of constraint forces with offset center

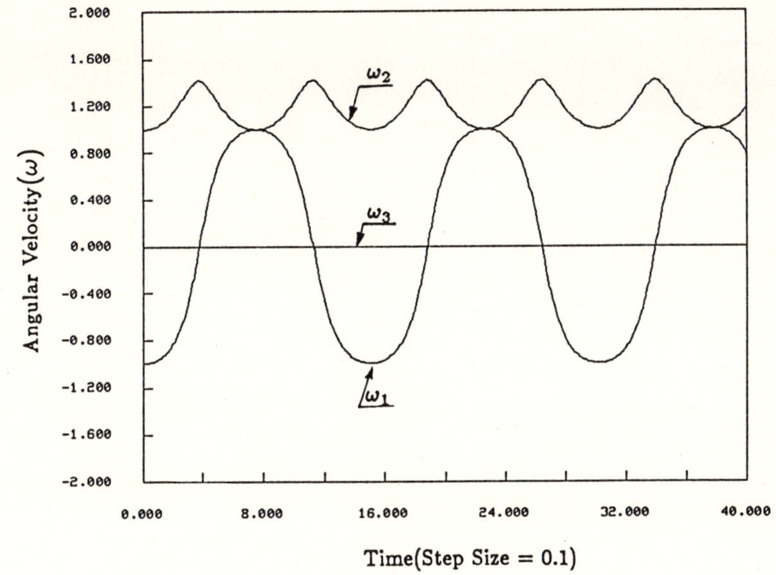

Fig. 5 Angular velocities of the sphere with no offset

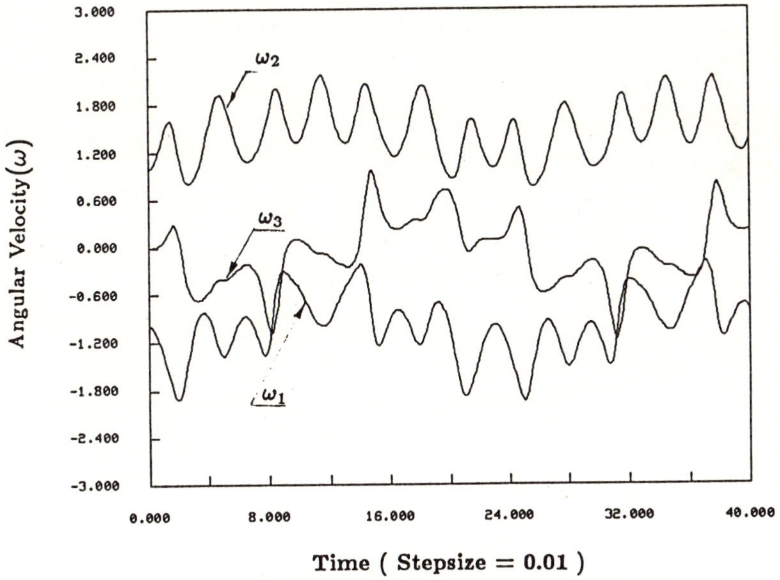

Fig. 6 Angular velocities of the sphere with offset center.

Dynamical Response to Pulse Excitations in Large Space Structures

Michail Zak
Applied Technology Section
Jet Propulsion Laboratory
Pasadena, CA 91109

Abstract

Finite dimensional approximations of large space structures as distributed parameter systems may lead to a loss of contribution of high frequencies to the dynamic response in case of impulsive or concentrated loads. It is shown that the unmodelled part of this response can be represented by a system of thin pulses which propagate as characteristic waves. It is demonstrated that the dynamical response to such a system of pulses can be modelled by a system of equations with delay argument. Fundamental dynamical properties of this system such as Liapunov stability, structural stability, loss of periodicity and transition to ergodicity are analyzed in this study. The results are illustrated by examples.

Introduction

The spectral (Fourier) approach to structural dynamics has attained widespread application as a result of the comparatively simple linear analysis of the behavior of each of the harmonic modes, and the possibility of the expansion of any relatively smooth dynamic process into harmonic modes[1] or waves[2]. However, along with the physical clearness and simplicity of engineering applications spectral methods have some disadvantages if they applied to distributed parameter systems such as large space structures. In this case due to finite dimensional approximation which is usually performed by finite element technique the contribution of high frequencies to structural response is lost. However, in many cases this contribution is significant and cannot be ignored. In general, the accuracy of the finite dimensional approximation depends on the degree of smoothness of the disturbances. Suppose that the load as a function of the length x of a structural member is described by a function f(x) and let f(x) as well as its derivatives f', f''... $f^{(k-1)}$ be continuous in $(-\pi, \pi)$ interval excluding the points $\xi_\mu (\mu=0,1,\ldots m)$ where all the functions have jumps: $\delta_\mu{}^{(0)}$, $\delta_\mu{}^{(1)}, \ldots . \delta\mu^{(k-1)}$, $(\mu=0,1,\ldots . m)$, respectively. Then the Fourier coefficients of this function will be:

Springer Series in Computational Mechanics
S. N. Atluri, A. K. Amos (Eds.)
Large Space Structures: Dynamics and Control

$$a_n = \frac{A_n}{n} - \frac{B_n'}{n^2} - \frac{A_n''}{n^3} + \frac{B_n'''}{n^4} + \ldots\ldots \text{ etc.}$$

$$b_n = \frac{B_n}{n} + \frac{A_n'}{n^2} - \frac{B_n''}{n^3} + \frac{A_n'''}{n^4} + \ldots\ldots \text{ etc.}$$

in which

$$A_n^{(i)} = \frac{1}{\pi} \sum_{\mu=1}^{m} \delta_\mu^{(i)} \operatorname{Sin} n\xi\mu$$

$$B_n^{(i)} = \frac{1}{\pi} 2 \sum_{\mu=0}^{m} \delta_\mu^{i} \quad \operatorname{Cos} n\xi\mu$$

$(i=0,1\ldots\ k-1;\ n=1,2\ldots)$

Hence, an approximation of the same dimension n leads to different accuracies depending upon the discontinuities of the function f(x) and its derivatives characterized by the constants $A_n^{(i)}$ and $B_n^{(i)}$. For instance, the Fourier coefficients of a concentrated disturbance considered as δ-function are all equal which means that any truncation may lead to a significant error.

It is important to notice that the discontinuities of the derivatives f', f''...etc. only slow down the convergence of the Fourier series while in addition the discontinuities of the function f itself lead to divergence of the series at the points of the discontinuities: the series tend to the average values of the function f. The difference between the original function f and its Fourier decomposition at $n \to \infty$ in this case has a form of isolated thin pulses located at the points of discontinuities, Fig. 1.

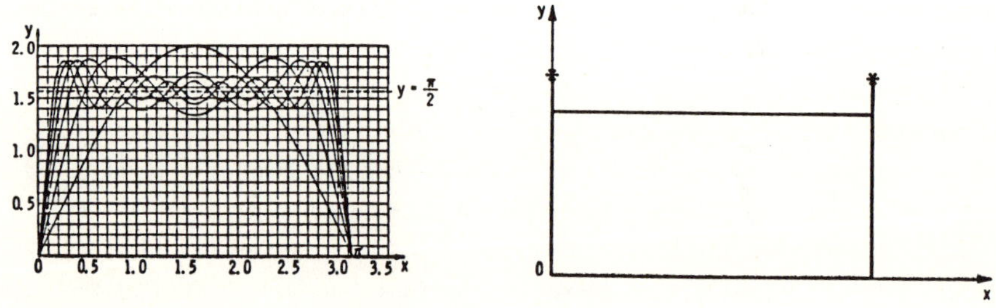

FIGURE 1

The widths of these pulses are vanishing:

$$\Delta\lambda = \frac{\pi}{2n} \to 0 \text{ at } n\to\infty$$

However, their heights remain finite. Surprisingly, they are 18% greater than the original discontinuities. This defect of the convergence is known as the Gibbs effect. Obviously, in the course of finite dimensional approximations these pulses are lost. This loss may lead to a significant discrepancy between predicted and real dynamical response especially in the case of controlled large space structures. Indeed, if a control law is based on finite dimensional approximation the energy imparted through the actuators may be added to unmodelled modes of a structure, i.e., to the pulses and cause destabilization of the system. This effect is known as control spillover, [3].

There is at least two possible sources of pulse excitations in large space structures.

Firstly, since large space structures will likely be composed of networks of long slender members, all the disturbances with a relatively small contact zone and short time interval (for instance, coming from actuators) can be considered as pulse excitations.

Secondly, all the disturbances coming from vehicle docking, crew motion, reboost, etc. are usually described by functions with discontinuities, and therefore, pulse excitations occur as a complement to finite dimensional modal approximation.

Since any discontinuity propagates with the characteristic speed, it is reasonable to turn to the characteristic wave approach in treating the pulse excitations. The advantage of this approach is in the fact that characteristic speeds depend only on the coefficients at the highest (second order) derivatives in the governing equation of structural members which significantly simplify the analysis of characteristic waves.

Thus, it appears that the application of the characteristic wave approach is the most beneficial in the domains where spectral methods fail. This approach also can be considered as a supplement to modal methods for linear analysis of controlled structures when loads can be decomposed into "smooth" and impulsive components.

In this paper some aspects of characteristic wave propagation, reflection and transmission in structures with one-dimensional structural members as well as possible engineering tools for this analysis are discussed.

1. Propagation of pulse excitations in one-dimensional structural members. We will start with a one-dimensional structural member subjected to a concentrated or impulsive load assuming that

$$\Delta\lambda << L, \text{ and } \Delta t << \frac{L}{c} \tag{1}$$

in which L is the length of the structural member, $\Delta\lambda$ is the width of the contact zone of the pulse, Δt is the duration of the concentrated load, and c is the characteristics speed of wave propagation. The characteristic speed c depends upon the type of deformation transmitted by the wave as follows:

$$c_B^2 = \frac{E}{\rho},\ c_T^2 = \frac{G}{\rho},\ c_V^2 = \frac{GA_s}{\rho A},\ c_M^2 = \frac{EI}{\rho I},\ c_S^2 = \frac{T}{\rho} \tag{2}$$

Here c_B, c_T, c_V, c_M, and c_S are the characteristic speeds for longitudinal, torsional, shear, bending and transverse string waves, respectively, E is Young's modulus, G is the shear modulus, A is the cross-sectional area, A_s is an effective shear area of a Timoshenko beam, I is the cross-sectional area moment of inertia, ρI is the rotatory inertia per unit length, ρ is the mass per unit length[4], T is the string tension, Fig. 2.

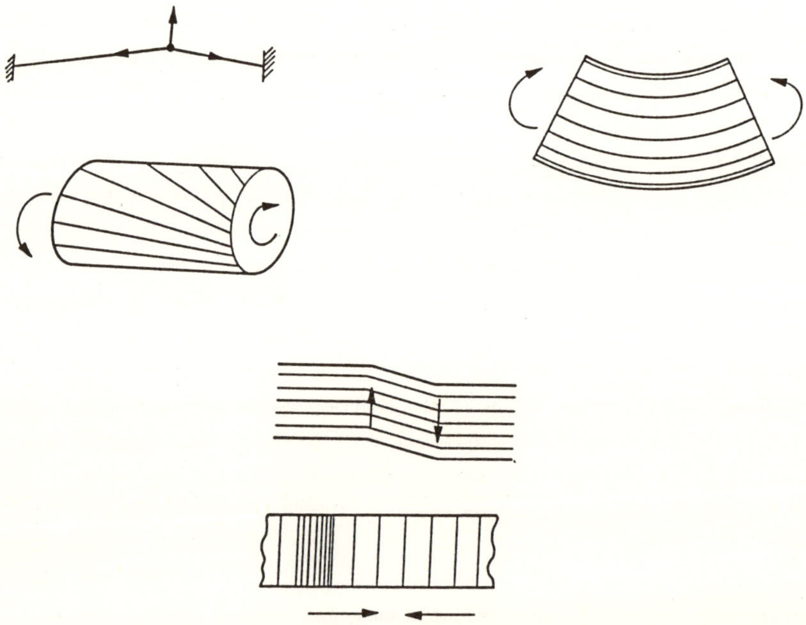

FIGURE 2

For homogeneous structural members all the characteristic speeds in Eq. 2 are constant, and consequently, the width Δl as well as the duration Δt of the pulse will be constant too. However, the initial configuration of the pulse will be preserved only for the simple wave equation without damping. In all other cases due to the dispersion phenomenon the configuration of the pulse will not be preserved. Nevertheless the dispersion can be ignored if the conditions (1) are satisfied. For further convenience we will introduce an equivalent rectangular pulse of the same length and energy. For such a rectangular pulse all the waves listed in (2) are decoupled even if they are simultaneously propagated in a structural member, and this is the important advantage of the characteristic wave approach to propagation of impulsive loads.

2. Modeling of periodic structures.

Repetitive lattice-type structures have recently been identified as candidates for space applications. We will describe the global picture of pulse propagation, transmission and reflection in truss-or beam-type periodic structures. For simplicity each structural member will be considered as homogeneous, and therefore, the propagation speed of the pulse (no matter, whether it is longitudinal, bending, shear, torsional pulse) will be constant (see Eq. 2).

Joints between structural members will be characterized by reflection and transmission coefficients

$$\xi = \frac{\sigma_R}{\sigma_I}, \quad \zeta = \frac{\sigma_T}{\sigma_I} \tag{3}$$

in which σ_I, σ_R and σ_T are the incident, reflected and transmitted stress pulses, respectively.

For the sake of simplicity in this study we will consider only symmetric joints whose reflection and transmission coefficients are the same for waves coming from the right and from the left.

For an ith joint (Fig. 3) the stress pulses $\sigma_1{}^i$ and $\sigma_2{}^i$ are partly reflected and partly transmitted so that

$$\sigma_3{}^i = \xi\,\sigma_1{}^i + \zeta\,\sigma_2{}^i, \quad \sigma_4{}^i = \zeta\,\sigma_1{}^i + \xi\,\sigma_2{}^i \tag{4}$$

The same stresses for the $(i-1)^{th}$ and $(i+1)^{th}$ joints will differ from σ^i only by a time delay τ:

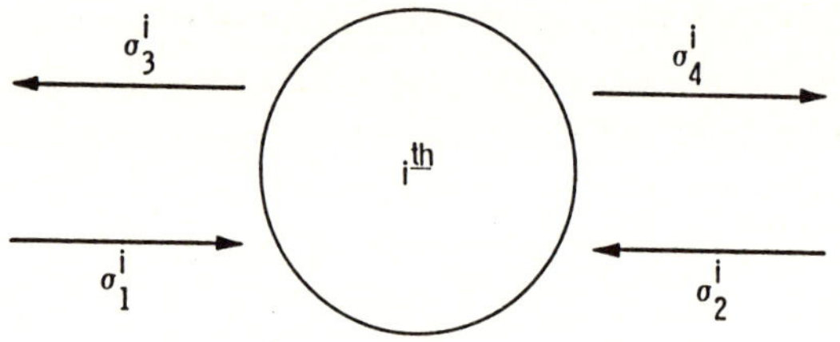

FIGURE 3

$$\tau = \frac{L}{c}$$

Introducing the dimensionless time $t \to t\frac{c}{L}$ one obtains:

$$\sigma_1^{\,i}(t+1) = \sigma_4^{\,i-1}(t),\ \sigma_2^{\,i}(t+1) = \sigma_3^{\,i+1}(t) \tag{5}$$

or after eliminating σ_3 and σ_4:

$$\begin{aligned} \sigma_1^{i}(t+1) &= \zeta\sigma_1^{i-1}(t)+\xi\, \sigma_2^{i-1}(t) \\ \sigma_2^{i}(t+1) &= \xi\sigma_1^{i+1}(t)+\zeta\sigma_2^{i+1}(t) \end{aligned} \qquad i=1,2,\ldots n \tag{6}$$

in which n is the number of joints.

Thus, the evolution of an initial stress pulse in a periodic structure is described by the system of difference Eqs. (5) and (6) which by using the conversional abbreviation for difference equation can be presented in the following matrix form:

$$\sigma_{k+1} = A\sigma_k \tag{7}$$

where

$$\sigma = \{\sigma_1^{1},\ \sigma_2^{1},\ \sigma_1^{2},\ \sigma_2^{2},\ \ldots.\ \sigma_1^{n},\ \sigma_2^{n},\}^T \tag{8}$$

and A is the mxm matrix whose elements are defined by the reflection and transmission coefficients ξ and ζ, while $m \leq 2(n-1)$. For $n=2$ (one isolated structural member with $\xi=1$, $\zeta=0$) the system (6) reduces to:

$$\begin{aligned} \sigma_1^{2}(t+1) &= \sigma_2^{1}(t) \\ \sigma_2^{1}(t+1) &= \sigma_1^{2}(t) \end{aligned} \tag{9}$$

$$\text{and } A = \begin{pmatrix} 0 & 1 \\ 1 & 0 \end{pmatrix} \tag{10}$$

For n=3 (two jointed structural members with isolated ends)

the system (6) and the matrix A, are:

$$
\begin{aligned}
\sigma_1{}^2(t+1) &= \sigma_2{}^1(t)\\
\sigma_2{}^1(t+1) &= \xi\sigma_1{}^2(t)+\zeta\sigma_2{}^2(t)\\
\sigma_1{}^3(t+1) &= \zeta\sigma_1{}^2(t)+\xi\sigma_2{}^2(t)\\
\sigma_2{}^2(t+1) &= \sigma_1{}^3(t)
\end{aligned}
\qquad (11)
$$

$$
A = \begin{pmatrix} o & 1 & o & o\\ \xi & o & o & \zeta\\ \zeta & o & o & \xi\\ o & o & 1 & o \end{pmatrix}
$$

Hence, formally the reflection - transmission matrix A plays here the same role as the stiffness matrix does in finite element technique.

If the joints between the structural members are subjected to external stress pulses then instead of Eq. (4) one obtains:

$$
\sigma_3{}^i = \xi\sigma_1{}^i+\zeta\sigma_2{}^i+\sigma_\ell{}^i,\ \sigma_4{}^i = \zeta\sigma_1{}^i+\xi\sigma_2{}^i+\sigma_r{}^i \qquad (12)
$$

in which $\sigma_\ell{}^i$ and $\sigma_r{}^i$ are the pulses applied to the ith joints from the left and the right sides, respectively. Eq. (6) is written now in the form:

$$
\begin{aligned}
\sigma_1{}^i(t+1) &= \zeta\sigma_1{}^{i+1}(t)+\xi\sigma_2{}^{i-1}(t)+\sigma_\ell{}^{i-1}(t)\\
\sigma_2{}^i(t+1) &= \xi\sigma_1{}^{i+1}(t)+\zeta\sigma_2{}^{i+1}(t)+\sigma_r{}^{i-1}(t)
\end{aligned}
\qquad (13)
$$

and instead of Eq. (7) one arrives at

$$
\vec{\sigma}_{k+1} = A\vec{\sigma}_k + \vec{\overset{o}{\sigma}}_k \qquad (14)
$$

in which

$$
\vec{\overset{o}{\sigma}}_k = \{\sigma\ell^1,\ \sigma_r{}^1,\ \sigma_\ell{}^2,\ \sigma_r{}^2,\ldots.\sigma_\ell{}^n,\sigma_r{}^n\}^T \qquad (15)
$$

For n=3 (Fig. 4) one obtains:

$$
\vec{\overset{o}{\sigma}}_k = \{0,\ \sigma_\ell{}^2\ 0,0\}^T \qquad (16)
$$

if the external stress pulses (repeated through each t=1 time interval) are applied to the second joint from the left side.

If external pulses are applied to a point which does not coincide with a joint one has to consider this point as a fictitious joint with no reflection ($\xi=0$) and fully transmission ($\zeta=1$).

Expressions for ξ and ζ as functions of the prameters of idealized models of joints are derived in Appendix [see Eqs. (121) and (123)].

3. Analysis of Solutions

The solution to Eq. (7) can be written in the following form[6]:

$$\vec{\sigma}_k = A^k\vec{\sigma}_o + \sum_{j=0}^{k-1} A^j \overset{\to}{\overset{o}{\sigma}}_{k-j-1} \tag{17}$$

Hence, it can be obtained from the initial pulse vector $\vec{\sigma}_0$ and the external pulse rector $\overset{\to}{\overset{o}{\sigma}}_k$ by successive matrix multiplications which makes a numerical strategy very simple.

All the qualitative properties of the solution (12) are defined by the eigenvalues λ_j of the fundamental matrix A, i.e., by the roots of the characteristic polynomial:

$$|A-\lambda I| = 0 \tag{18}$$

Indeed, as follows from Sylvester's formula:

$$A^m = \sum_j^n \lambda_j^m Z_j(A), \quad Z_r = \frac{\prod_{p\neq r}^n (A-\lambda_p I)}{\prod_{p\neq r}^n (\lambda-\lambda_p)} \tag{19}$$

if all λ_i are distinct, and

$$Z_r = [\lambda_1^m I + m\lambda_1^{m-1}(A-\lambda I)] \frac{\prod_{p\neq 3}^n (A-\lambda_p I)}{\prod_{p\neq 3}^n (\lambda-\lambda_p)} \tag{20}$$

if λ_1 is a characteristic root of multiplicity two. This formula clarifies the structure of the matrix A^m, and particularly, the effect of multiple characteristic roots.

The solution (17) is stable if

$$|\lambda_i| \leq 1 \tag{21}$$

By applying a linear fractional transformation:

$$\lambda = \frac{\lambda+1}{\lambda-1} \tag{22}$$

to Eq. (18) one reduces the stability analysis of Eq. (17) to conventional methods for ordinary differential equations.

As an example illustrating the theory we will consider a two-member structure with isolated ends at x=1 and x=3 ($\xi_1=1$, $\zeta_1=0$, $\xi_3=1$, $\zeta_3=0$)and a joint at x=2 with the reflection and transmission coefficients ξ and ζ, respectively, Fig. 4.

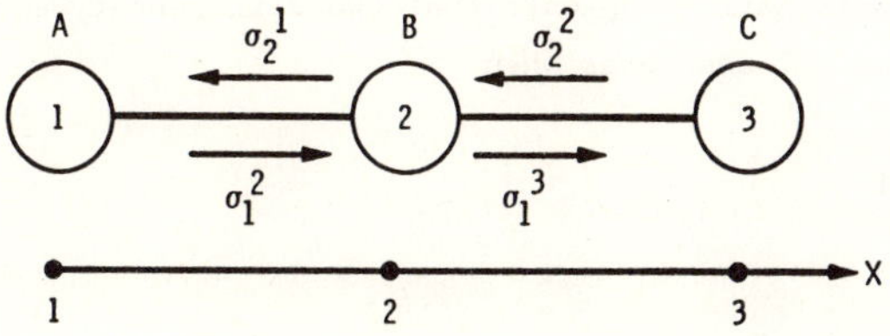

FIGURE 4

The fundamental matrix A for this case is presented by Eq. (11), and therefore, the characteristic polynomial (18) reduces to:

$$\lambda^4 - 2\xi\lambda^2 + \xi^2-\zeta^2 = 0 \tag{23}$$

while the characteristic roots are:

$\lambda = \pm\sqrt{\xi \pm \zeta}$, i.e.

$$\lambda_{1,2} = \pm 1, \quad \lambda_{3,4} = \pm \frac{1}{1+\eta} \sqrt{(1-\eta)^2-4\eta} \tag{24}$$

with ξ and ζ expressed by Eqs. (121 App) and (123 App), respectively.

Thus, all the root modulii do not exceed unity since

$$\frac{1}{1+\eta} \sqrt{(1-\eta)^2-4\eta} \quad <1 \text{ if } \eta> 0 \tag{25}$$

Hence,, the motion is neutrally stable with respect to small changes in initial conditions.

The roots $\lambda_{3,4}$ are imaginary in the domain

$$3 - \sqrt{8} < \eta < 3 + \sqrt{8} \tag{26}$$

The solution to the governing Eq. (14) can be written in the form (17) while the formulas (19) defining the matrix A^m reduces to:

$$A^m = \left\{\frac{1-(\sqrt{\xi-\zeta})^m}{2\zeta}[1-(-1)^m]\right\} A^3+$$

$$+\left\{\frac{1-(\sqrt{\xi-\zeta})^{m+2}}{2\zeta}[1+(-1)^m]\right\} A^2+$$

$$+\left\{\frac{\zeta-\xi+(\sqrt{\xi-\zeta})^{m+1}}{2\zeta}[1-(-1)^m]\right\} A+$$

$$+\left\{\frac{\zeta-\xi+(\sqrt{\xi-\zeta})^{m+1}}{2\zeta}[1+(-1)^m]\right\} I \qquad (27)$$

The structure of this matrix suggests that two different type of solutions can occur. Indeed, for all the cases when

$$|\xi-\zeta| < 1 \qquad (28)$$

the terms

$$(\sqrt{\xi-\zeta})^m,\ (\sqrt{\xi-\zeta})^{m+1},\ (\sqrt{\xi-\zeta})^{m+2} \qquad (29)$$

decay, and the solution tends to the form in which

$$\xi=\zeta,\ \text{i.e.,}\ \eta = 3\pm\sqrt{8} \qquad (30)$$

Hence, the solution obtained from Eqs. (17) and (27) at the condition (30) can be considered as an attractor: it is stable with respect to small changes in the reflection and transmission coefficients ξ and ζ.

Another type of solutions can be obtained if the condition (28) is replaced by

$$|\xi-\zeta| = 1 \qquad (31)$$

In particular, this case occurs if $\xi = 0$, $\zeta=1$, i.e., when the joint at $\chi=2$ (Fig. 4) can be ignored, and therefore, one deals with one-member structure. Unlike the previous case, here the terms (29) will not decay, and the solution will not tend to the attractor (30). However, it is easy to conclude that this isolated solution is unstable with respect to small changes of ξ or ζ since such changes replace the equality (31) by the inequality (28), and the perturbed solution can be considered as a repeller.

Let us pose now the following initial conditions:

$$\vec{\sigma}_0=(1,0,0,0)^T,\ \text{i.e.,}\ \sigma_1{}^2=1,\ \sigma_2{}^1=0,\ \sigma_1{}^3=0,\ \sigma_2{}^2=0\ \text{at}\ t=0 \qquad (32)$$

Hence, we will analyze now the free motion caused by a stress pulse $\sigma_1^2=1$ applied to the point x=1 (Fig. 4). Turning to Eq. (17) at $\sigma=0$ and utilizing Eq. (19) one obtains:

$$[\sigma_1^2]_k = (1,0,0,0)\, A^k (1,0,0,0)^T = [1+(-1)^k]\, \frac{\xi[1-(\sqrt{\xi-\zeta})^{K+2}] + \zeta - \xi + (\sqrt{\xi-\zeta})^{k+1}}{2\zeta} \quad (33)$$

It is obvious that

$$[\sigma_1^2]_k \to 1/2[1+(-1)^k] \text{ if } |\xi-\zeta| < 1 \quad (34)$$

while the solution (43) is the attractor, and

$$[\sigma_1^2]_k \to 1/2[1+(-1)^k](1+i^{k+1}) \text{ if } \xi=0,\ \zeta=1, \text{ i.e,}$$

$$[\sigma_1^2]_{2k} = 1,\ [\sigma_1^2]_{2k+1} = 0 \quad (35)$$

while the solution (45) is the repeller.

One should notice that for multi-member structures which are characterized by higher order of difference equations the method of successive matrix multiplications may be more convenient than the analytical method performed above. For simplicity we will apply this method to the same example. Since

$$\vec{\sigma}_{k+1} = A\vec{\sigma}_k$$

one obtains:

$$\vec{\sigma}_o = (1,0,0,0)^T,\ \vec{\sigma}_1 = A\vec{\sigma}_o = (0,\xi,\zeta,0)^T,$$
$$\vec{\sigma}_2 = A\vec{\sigma}_1 = (\xi,0,0,\zeta)^T,\ \vec{\sigma}_3 = A\vec{\sigma}_2 = (0,\xi^2+\zeta^2,2\xi\zeta,0)^T$$
$$\vec{\sigma}_4 = A\vec{\sigma}_3 = (\xi^2+\zeta^2,0,0,2\xi\zeta)^T,$$
$$\vec{\sigma}_5 = A\vec{\sigma}_4 = (0,\xi(\xi^2+\zeta^2)+2\xi\zeta^2,\ \zeta(\xi^2+\zeta^2)+2\xi^2\zeta,0)^T \quad (36)$$
$$\vec{\sigma}_6 = A\vec{\sigma}_5 = (\xi(\xi^2+\zeta^2)+2\xi\zeta^2,0,0,\zeta(\xi^2+\zeta^2)+2\xi^2\zeta)^T \quad \text{etc.}$$

Here the matrix A is given by Eq. (11).

As follows from Eq. (36) the function $\sigma(t)$ is changed by jumps at each unit dimensionless time-step. The functions $\sigma_1^2(t)$ at $\xi=0.1$, $\zeta=0.9$ are plotted in Fig. 5.

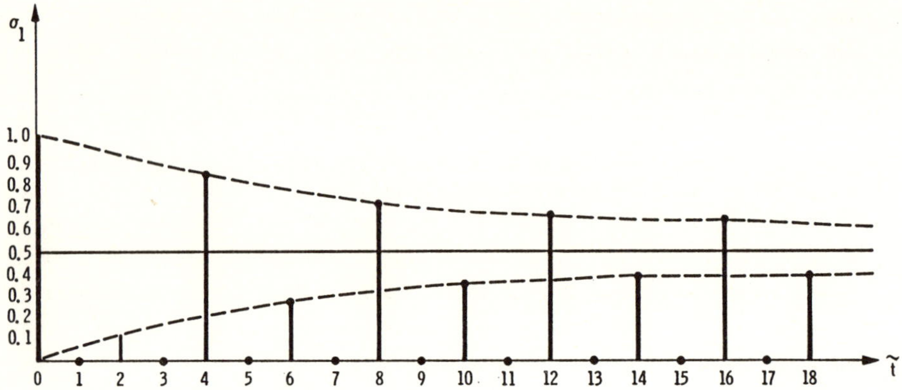

FIGURE 5

4. Modal Representation of Solutions

Let us transfer to a new state vector in Eq. (7):

$$\vec{\sigma}_k = \theta\vec{\sigma}_k \tag{37}$$

and therefore:

$$\vec{\sigma}_{k+1} = \theta^{-1}A\theta\vec{\sigma}_k \tag{38}$$

while selecting the matrix θ such

$$\theta^{-1}A\theta=\tilde{A} \tag{39}$$

is a diagonal matrix

The representation (39) is possible since all the characteristic roots of the matrix A are distinct. However since A is not symmetric [see Eq. (11)] it is more convenient to rewrite Eq. (37) and (38) in the form:

$$\sigma_{k+2} = A^2\sigma_k, \quad \sigma_{k+2} = A^2\sigma_k \tag{40}$$

in which A^2 is symmetric (for instance, if A is given by Eq. (11). Then

$$A^2 = \begin{pmatrix} \xi & o & o & \zeta \\ o & \xi & \zeta & o \\ \\ o & \zeta & \xi & o \\ \zeta & o & o & \xi \end{pmatrix} \tag{41}$$

Since all the characteristic roots λ^2 of A^2 or $\widetilde{A}^2$ are real, the system (38) can be rewritten in a decouple form:

$$\sigma^{\ell}_{k+2} = \lambda^2_{\ell}\sigma^{\ell}_{k} \tag{42}$$

in which σ_k^{ℓ} are components of the eigenvector σ_k, and λ_{ℓ}^2 are the corresponding characteristic roots. Like natural modes of vibrtion, the eigenvectors σ_k are orthogonal and lineraly independent. They form a complete system in the sense that any vector of the same dimension representing a possible state of the system can be constructed as a linear combination of σ_k. Each eigenvector can be excited independently of the other while its motion will be described by the corresponding Eq. (42). The solution to this equation has one of two possible forms:

$$\sigma_k = \begin{cases} |\lambda_{\ell}|^{k}[C_1+(-1)^{K}C_2] & \text{if } \lambda_{\ell}^2>0 \\ |\lambda_{\ell}|^{k}(C_1 \operatorname{Cos} \pi k/2 + C_2 \operatorname{Sin} \pi k/2) & \text{if } \lambda^2_{\ell}<0 \end{cases} \tag{43}$$

while the constants C_1 and C_2 are defined by initial conditions.

The solution to the orginal difference Eq. (7) now can be obtained by linear superposition of the solution (43).

5. Transition to Ergodicity

So far only structures with identical members (characterized by the dimensionless time delay 1) were considered. It was demonstrated that there exists a formal analogy between the matrix techniques for treating these structures under impulsive loads and for conventional modal analysis (although the matrices have different physical meanings). However, in reality the identity of structural members even in periodic structures is an exception rather than a rule. Indeed, different time delays for different structural members can be caused not only by different lengths, but also by different characteristic speeds. In turn, different characteristic speeds may occur if joints convert one type of deformation into another [see Eq. (2)]. Another source of different time delays is associated with external forces if they were applied between joints. In this case the points of their application must be considered as additional joints, but without reflection and this will lead to additional structural members with different time delays. As will be shown below, different time delays lead to new qualitative effects which cannot be detected in modal methods.

For simplicity we will start with the two-member structure given in Fig. 6 and assume the following (dimensionless) lengths:

$AB=p$, $BC=q$, $q>p>0$

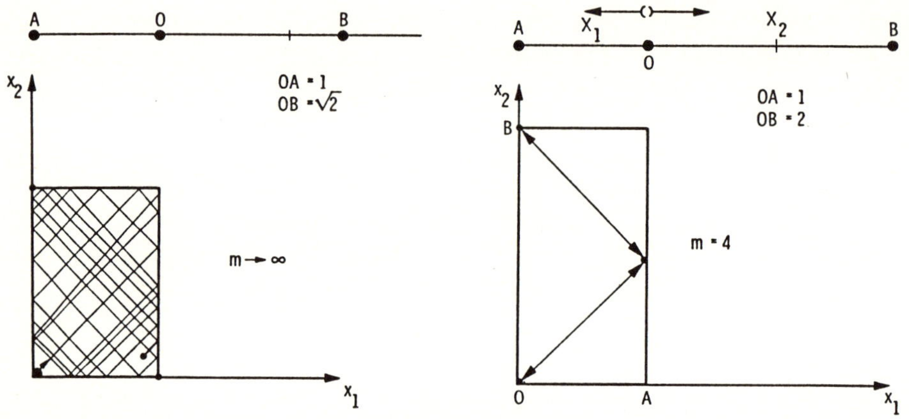

FIGURE 6

Obviously, the dimensionless time step now should be chosen as

$$t = \frac{1}{q} \tag{44}$$

Then the system of difference Eq. (11) should be replaced by the following:

$$\begin{aligned}
\sigma_1{}^2(t) &= \sigma_2{}^2(t-p) \\
\sigma_2{}^1(t) &= \xi\sigma_1{}^2(t-p) + \zeta\sigma_2{}^2(t-q) \\
\sigma_1{}^3(t) &= \zeta\sigma_1{}^2(t-p) + \xi\sigma_2{}^2(t-q) \\
\sigma_2{}^2(t) &= \sigma_1{}^3(t-q)
\end{aligned} \tag{45}$$

with the characteristic equation:

$$\lambda^{2(p+q)} - \xi\lambda^{2p} - \xi\lambda^{2q} + \xi^2 - \zeta^2 = 0 \tag{46}$$

Let us assume first that

$$\frac{q}{P} = \alpha \tag{47}$$

is an integer.

Then the degree of Eq. (46) can be reduced by changing the time scale:

$$\lambda^{2(1+\alpha)} - \xi\lambda^2 - \xi\lambda^{2\alpha} + \xi^2 - \zeta^2 = 0 \tag{48}$$

If α is a fraction

$$\alpha = \frac{\beta}{\gamma} > 1 \tag{49}$$

in which β and γ are relatively prime integers, one obtains:

$$\lambda^{2(\beta+\gamma)} - \xi\lambda^{2\gamma} - \xi\lambda^{2\beta} + \xi^2 - \zeta^2 = 0 \tag{50}$$

Obviously, this effect does not have an analogy in modal approach where the order of the governing differential equation depends only on the number of modes (or degrees-of-freedom) considered.

In order to clarify the physical meaning of such a phenomenon let us start with the following question: during what time interval T an initial pulse will return to its original location in "one piece" Simple geometrical consideration show that $T = 2$ if AB=1, BC=1, $T = 10$ if AB=4, BC=6, $T = \infty$ if AB=1, BC=$\sqrt{2}$. In other words, this (dimensionless) interval is equal to the order m of the governing difference equation divided by two. Thus, if at least two time delays in an n-member structure are not commensurate, the system will never return to its initial position, i.e., the motion will loose its periodicity (Fig. 6).

Hence, the degree of the characteristic Eq. (46) depends not only on the number of structural members, but also on the ratio of their lengths. For instance, if q=10, and p=5, the degree will be 6, but if q=10, and p=7, the degree will be 34.

Now it is easy to conclude that if α in Eq. (49) is irrational, then it can be represented by a non-terminating non-repeating decimal fraction, and therefore, the degree of Eq. (46) will tend to infinity since

$$\beta + \gamma \to \infty \tag{51}$$

For instance, if p=1, q=$\sqrt{2}$, the successive rational approximations for Eq. (46) will have degrees which tend to infinity as: 4, 24, 482, 4828, etc.

In classical mechanics such systems are known as ergodic systems. For infinite number of times they pass through every state of motion (which is consistent with constraints) spending equal time intervals near each state. As however the rational numbers are a set of measure zero, practically every motion of this type sooner or later becomes ergodic.

Nevertheless, in engineering applications there always can be found such a characteristic time interval within which the motion is approximately periodic, while the transition to ergodicity can be ignored due to damping.

Let us return to the characteristic Eq. (50) and analyze it roots when $(\beta+\gamma) \to \infty$. First of all, using Eq.s (121 App) and (123 App) one can rewrite it as

$$\lambda^{2(\beta+\gamma)} - \xi\lambda^{2\gamma} - \xi\lambda^{2\beta} + \xi^2 - \zeta^2 = 0, \quad \xi+\zeta = 1 \tag{52}$$

Then it is easily verifiable that

$$\lambda = \pm 1 \tag{53}$$

are among the roots of Eq. (52).

Let us assume now that Eq. (52) has roots which modulii exceed unity:

$$|\lambda| > 1 \tag{54}$$

Then the second and the third terms in Eq. (52) can be ignored since

$$\lambda^{2(\beta+\gamma)} >> \lambda^{2\gamma}, \lambda^{2\beta}, \text{ if } \gamma, \beta \to \infty \tag{55}$$

Hence, if the roots (55) exist they must be:

$$\lambda = (\zeta - \xi)^{\frac{1}{2(\beta+\gamma)}} \tag{56}$$

But as follows from Eq. (55)

$$|\zeta - \xi| < 1 \tag{57}$$

which means that the roots (54) do not exist.

Consequently, all the root modulii do not exceed unity, i.e., the motion is neutrally stable with respect to small changes in initial conditions.

In order to find the roots which modulii are less then unity

$$|\lambda| < 1 \tag{58}$$

on can replace Eq. (52) by the following:

$$1 - \xi\delta^{2\beta} - \xi\delta^{2\gamma} + (\xi - \zeta)\delta^{2(\beta+\gamma)} = 0, \quad \delta = \frac{1}{\lambda}, \quad |\delta| > 1, ..$$

Then

$$\lambda_n = (\zeta - \xi)^{\frac{1}{2(\beta+\gamma)}} \tag{60}$$

This formula gives $n=2(\beta+\gamma)$ roots (58) of Eq. (52), while

$$|\lambda| \to 1 \text{ if } \beta,\gamma \to \infty \tag{61}$$

Now the general solution to the problem in terms of the stress pulse $\sigma_1{}^2$ can be presented in the form:

$$[\sigma_1{}^2]_k = \sum^{n} C_n \lambda_n{}^k, \quad K=1,2,\ldots\infty,\ n \to \infty \tag{62}$$

in which C_n are constants to be found from initial conditions, and λ_n are the characteristic roots determined by Eq. (60).

It is interesting to notice that for each fixed time interval (k=Const)

$$|\lambda_n{}^k| \to 1,\ n \to \infty \tag{63}$$

but for each fixed n

$$|\lambda_n{}^k| \to 0,\ n=\text{Cosnt},\ k \to \infty \tag{64}$$

Thus, if $n=2(\beta+\gamma)$ is large but finite, i.e., α in Eq. (49) is rational, the solution (62) has the same qualitative properties as the solution (33): it possesses an attractor (30), as well as a repeller (31). All the motions described by this solution are periodic, although the period will be large:

$$\tilde{T} = n = 2(\beta+\gamma) \tag{65}$$

in which $\tilde{T}$ is the dimensionless period.

As follows from Eq. (44), the dimensionless time-step at which the solution changes by jumps is now very small:

$$t = \frac{1}{\gamma},\ \text{while } k = \frac{t}{\Delta t} = \gamma t \tag{66}$$

If $n=2(\beta+\gamma)$ is infinitely large, i.e, α in Eq. (49) is irrational, the solution (62) loses it periodicity and becomes ergodic, while the time-step (66) tends to zero. Hence, the solution is changed by jumps within each infinitesimal period of time, and therefore, it may be discontinuous at any instant. That is why the limit of the sum in Eq. (62) at $n \to \infty$ can be expressed through the Lebesque rather than the Riemann integral.

Let us analyze now the behavior of each term in the sum (62). First of all, at $\xi=\zeta$ there are only two non-zero characteristic roots (54) [see Eq. (60)], i.e., the solution (62) contains only two non-zero terms. Obviously, this isolated solution is not ergodic, but is unstable with respect to small changes of ξ or ζ.

At

$$|\xi-\zeta| = 1 \tag{67}$$

the solution is ergodic while

$$\lambda_n = \sqrt[n]{1}, \quad n\to\infty \tag{68}$$

and therefore

$$|\lambda_n{}^k| = 1 \tag{69}$$

The structure of this solution is very clear: all the characteristic roots λ_n in Eq. (62) are found as nth roots of unity, while the part of the solution corresponding to imaginary roots $\pm ib$ are expressed in the real form as:

$$b^k(\mathrm{Cos}\ \frac{\pi k}{2} + \mathrm{Sin}\ \frac{\pi k}{2}) \tag{70}$$

If

$$|\xi-\zeta| < 1 \tag{71}$$

then, as follows from Eq. (60)

$$\lambda_n{}^k \to (\zeta-\xi)^{\frac{k}{2(\beta+\gamma)}} \tag{72}$$

According to Eq. (66) for any finite time interval t<1

$$k=\gamma t \tag{73}$$

Consequently

$$\left|\lambda_n{}^k\right| = \left|(\zeta-\xi)^{\frac{\gamma t}{2(\beta+\gamma)}}\right| \to 1 \text{ at } \beta,\gamma \to \infty,\ \gamma>\beta \tag{74}$$

Thus, any ergodic solution at the condition (72) tends to the ergodic solution at the condition (70), which means that the latter can be considered as an "ergodic attractor".

It is worth emphasizing that the transition to ergodicity is not "inevitable" if one takes into account non-linear properties of real structures. Non-linearities may provide some mechanisms (such as dynamical synchronization effects) which suppress the disorder and lead to periodical motion. In this connection it is relevant to mention the experiment with coupled chain of inharmonic oscillators performed by Fermi, Pasta and Ulam. Instead of ergodicity which was expected they found periodic oscillations. However, if dynamical synchronization effects do not suppress ergodicity, and if the characteristic time during which the motion can be approximated as periodic is too short, one has to apply methods of statistical mechanics.

6. <u>Periodic structures with wave conversions</u>.

Previous discussions were restricted to the case when the type of the characteristic wave transmitting a pulse is not changed during its propagation. But in practice a conversion from one type of deformation to another is very likely. For instance, an originally longitudinal wave, after passing through a joint between two structural members forming some angle, can be partially converted into a transverse wave which has different characteristic speed of propagation. The rate of such a conversion depends not only on the properties of the joint, but also on the angle between the structural members. The relationships between the reflected, transmitted and converted parts of a propagating pulse are expressed by the Reflection-Conversion (RC) and the Transmission-Conversion (TC) matrices:

$$\begin{pmatrix} \sigma_R \\ \tilde{\sigma}_R \end{pmatrix} = \begin{pmatrix} \xi_{\ell\ell} & \xi_{\ell T} \\ \xi_{T\ell} & \xi_{TT} \end{pmatrix} \begin{pmatrix} \sigma_I \\ \tilde{\sigma}_I \end{pmatrix}$$

$$\begin{pmatrix} \sigma_T \\ \tilde{\sigma}_T \end{pmatrix} = \begin{pmatrix} \zeta_{\ell\ell} & \zeta_{\ell T} \\ \zeta_{T\ell} & \zeta_{TT} \end{pmatrix} \begin{pmatrix} \sigma_I \\ \tilde{\sigma}_I \end{pmatrix} \qquad (75)$$

in which σ_I, $\tilde{\sigma}_I$, $\tilde{\sigma}_R$, σ_R, $\tilde{\sigma}_T$, σ_T are the longitudinal and transverse components of incident, reflected and transmitted stresses, respectively, while $\xi_{\ell\ell}$, $\xi_{\ell T}$, $\xi_{T\lambda}$ and ξ_{TT} are the reflection - conversion coefficients, and $\zeta_{\ell\ell}$, $\zeta_{\ell T}$, $\zeta_{T\ell}$ and ζ_{TT} are the transmission-conversion coefficients.

For an ith joint (Fig. 7)

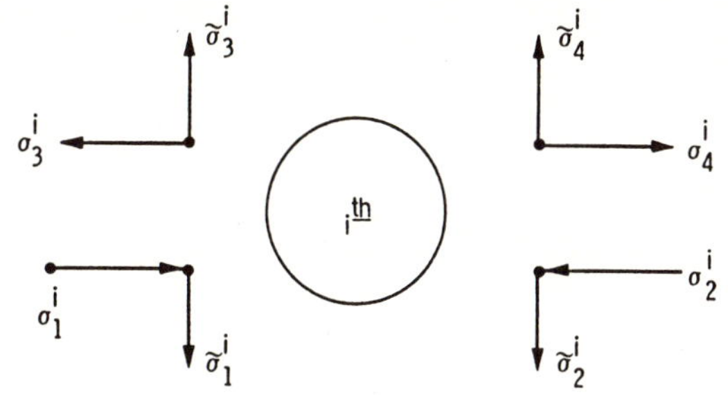

FIGURE 7

the stress pulses $\tilde{\sigma}_1{}^i$, $\sigma_1{}^i$, $\sigma_2{}^{i\sim}$ and $\sigma_2{}^i$ are partly reflected and partly transmitted, while each reflected and transmitted longitudinal stress is partly converted into the transverse stress, and vice versa. Hence,

$$\sigma_3{}^i = \xi_{\ell\ell}\sigma_1^i + \zeta_{\ell\ell}\sigma_2^i + \xi_{T\ell}\tilde{\sigma}_1^i + \zeta_{T\ell}\tilde{\sigma}_2^i \tag{76}$$

$$\sigma_4{}^i = \zeta_{\ell\ell}\sigma_1{}^i + \xi_{\ell\ell}\sigma_2{}^i + \zeta_{T\ell}\tilde{\sigma}_1{}^i + \xi_{T\ell}\tilde{\sigma}_2{}^i \tag{77}$$

$$\tilde{\sigma}_3{}^i = \xi_{TT}\tilde{\sigma}_1{}^i + \zeta_{TT}\tilde{\sigma}_2{}^i + \xi_{\ell T}\sigma_1{}^i + \zeta_{\ell T}\sigma_2{}^i \tag{78}$$

$$\tilde{\sigma}_4{}^i = \zeta_{TT}\tilde{\sigma}_2{}^i + \xi_{TT}\tilde{\sigma}_1{}^i + \zeta_{\ell T}\sigma_1{}^i + \xi_{\ell T}\sigma_2{}^i \tag{79}$$

The same stresses for the $(i-1)^{th}$ and $(i+1)^{th}$ joints will differ from σ^i only by a time delay

$$\tau = \frac{L}{c} \tag{80}$$

and from $\tilde{\sigma}^i$ by a time delay

$$\tilde{\tau} = \frac{L}{\tilde{c}} \tag{81}$$

in which L is the length of the structural member, while c and $\tilde{c}$ are the longitudinal and the transverse characteristics speeds, respectively.

Introducing the dimensionless time $t \to t\frac{c}{L}$ and the ratio

$$\frac{c}{\tilde{c}} = \chi \tag{82}$$

one obtains:

$$\sigma_1{}^i(t+1) = \sigma_4{}^{i-1}(t) \tag{83}$$

$$\sigma_2{}^i(t+1) = \sigma_3{}^{i+1}(t) \tag{84}$$

$$\sigma_1{}^i(t+\chi) = \sigma_4{}^{i-1}(t) \tag{85}$$

$$\sigma_2{}^i(t+\chi) = \sigma_3{}^{i+1}(t) \tag{86}$$

or after eliminating $\sigma_3{}^i$, $\tilde{\sigma}_3{}^i$, $\sigma_4{}^i$, and $\tilde{\sigma}_4{}^i$:

$$\sigma_1{}^i(t+1) = \zeta_{\ell\ell}\sigma_1{}^{i-1}(t) + \xi_{\ell\ell}\sigma_2{}^{i-1}(t) + \zeta_{T\ell}\tilde{\sigma}_1{}^{i-1}(t) + \xi_{T\ell}\tilde{\sigma}_2{}^{i-1}(t) \tag{87}$$

$$\sigma_2{}^i(t+1) = \xi_{\ell\ell}\sigma_1{}^{i+1}(t) + \zeta_{\ell\ell}\sigma_2{}^{i+1}(t) + \xi_{T\ell}\tilde{\sigma}_1{}^{i+1}(t) + \zeta_{T\ell}\tilde{\sigma}_2{}^{i+1}(t) \tag{88}$$

$$\tilde{\sigma}_1^{\,i}(t+\chi) = \xi_{TT}\tilde{\sigma}_1^{\,i-1} + \zeta_{TT}\tilde{\sigma}_2^{\,i-1} + \zeta_{\ell T}\sigma_1^{\,i-1}(t) + \xi_{\ell T}\sigma_2^{\,i-1}(t) \tag{89}$$

$$\tilde{\sigma}_2^{\,i}(t+\chi) = \xi_{TT}\tilde{\sigma}_1^{\,i+1}(t) + \zeta_{TT}\tilde{\sigma}_2^{\,i+1}(t) + \xi_{\ell T}\sigma_1^{\,i+1}(t) + \zeta_{\ell T}\sigma_2^{\,i+1}(t) \tag{90}$$

$i=1,2,\ldots n$

in which n is the number of joints.

Thus, the evolution of an initial stress pulse in a periodic structure with joints causing reflection, transmission and conversion from longitudinal to transverse stress and vice versa is described by the system of difference Eqs. (87-90) which can be presented in the matrix form (7):

$$\vec{\sigma}_{k+1} = A\vec{\sigma}_k \tag{91}$$

where

$$\vec{\sigma} = (\sigma_1^{\,1}, \tilde{\sigma}_1^{\,1}, \sigma_2^{\,1}, \tilde{\sigma}_2^{\,1}, \sigma_1^{\,2}, \tilde{\sigma}_1^{\,2}, \sigma_2^{\,2}, \tilde{\sigma}_2^{\,2}, \ldots \sigma_1^{\,n}, \tilde{\sigma}_1^{\,n}, \sigma_2^{\,n}, \tilde{\sigma}_2^{\,n})^T \ldots \tag{92}$$

and A is the mxm matrix which elements are defined by the reflection and transmission coefficients. The order of this matrix depends not only on the joints n, but also on the ratio χ (see Eq. 82). If χ is an integer, this order is

$$m \leq 2\chi(n-1) \tag{93}$$

If χ is a fraction:

$$\chi = \frac{\beta}{\gamma} \tag{94}$$

in which β and γ are relatively prime integers, instead of Eq. (93) one obtains [see Eqs. (44) and (50)]:

$$m \leq 2(\beta+\gamma)(n-1) \tag{95}$$

Finally, if χ is irrational, then $m \to \infty$, i.e., pulse motions loose their periodicity.

Hence, the transition to ergodicity in the case of stress conversion can occur not only due to different lengths of structural members, but also due to noncommensurating characteristic speeds for longitudinal and transverse waves. Since formally Eq. (91) is of the same type as Eq. (7), all the analysis of this case can be performed in the same way as it has been presented in the previous cases.

Example. Let us consider a two membertruss structure subjected to an initial vertical pulse (Fig. 8).

Suppose that $0_1 0_2 = 0_2 0_3 = 1$, $\alpha = 90^0$, $\chi = 2$, and the stress pulse $\sigma_0 = \sqrt{2}$ is applied to the truss $0_2 0_3$. Selecting:

$$\xi_{\ell\ell} = \xi_{TT} = 1, \ \xi_{\ell T} = \xi_{T\ell} = 0, \ \zeta_{\ell\ell} = \zeta_{TT} = \zeta_{\ell T} = \zeta_{T\lambda} = 0 \tag{96}$$

we will not specify these coefficients for the joint 0_2, assuming that in general they are non-zeros. Now the system (87)-(90) can be written as:

$$\sigma_1^{\ 2}(t+1) = \sigma_2^{\ 1}(t) \tag{97}$$

$$\sigma_2^{\ 1}(t+1) = \sigma_{2*}^{\ 1}(t) \tag{98}$$

$$\sigma_1^{\ 3}(t+1) = \sigma_{1*}^{\ 3}(t) \tag{99}$$

$$\tilde{\sigma}_1^{\ 2}(t+1) = \tilde{\sigma}_{1*}^{\ 2}(t) \tag{100}$$

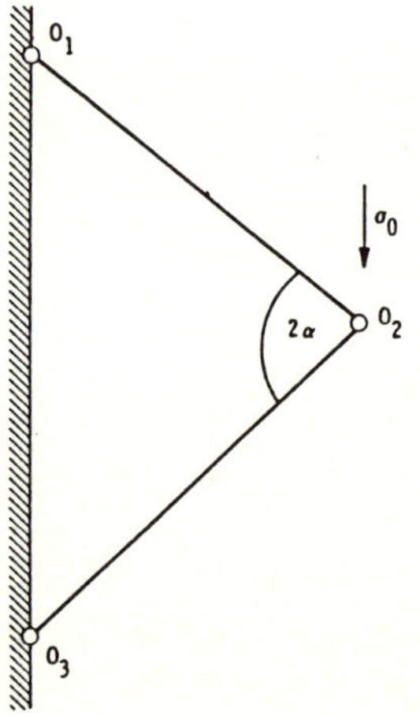

FIGURE 8

$$\tilde{\sigma}_1^{\,2}(t+1) = \tilde{\sigma}_{2*}^{1}(t) \tag{101}$$

$$\tilde{\sigma}_1^{\,3}(t+1) = \tilde{\sigma}_{1*}^{3}(t) \tag{102}$$

$$\tilde{\sigma}_2^{\,2}(t+1) = \tilde{\sigma}_{2*}^{2}(t) \tag{103}$$

$$\tilde{\sigma}_{1*}^{2}(t+1) = \tilde{\sigma}_2^{\,1}(t) \tag{104}$$

$$\sigma_{2*}^{1}(t+1) = \xi_{\ell\ell}\sigma_{1*}^{2}(t) + \xi_{T\ell}\tilde{\sigma}_1^{\,2}(t) + \zeta_{\ell\ell}\sigma_{2*}^{2}(t) + \zeta_{T\ell}\tilde{\sigma}_2^{\,2}(t) \tag{105}$$

$$\tilde{\sigma}_{2*}^{1}(t+1) = \xi_{\ell T}\sigma_{1*}^{2}(t) + \xi_{TT}\tilde{\sigma}_1^{\,2}(t) + \zeta_{\ell T}\sigma_{2*}^{2}(t) + \zeta_{TT}\tilde{\sigma}_2^{\,2}(t) \tag{106}$$

$$\sigma_{1*}^{3}(t+1) = \zeta_{\ell\ell}\sigma_{1*}^{2}(t) + \zeta_{T\ell}\tilde{\sigma}_1^{\,2}(t) + \xi_{\ell\ell}\sigma_{2*}^{2}(t) + \xi_{T\ell}\tilde{\sigma}_2^{\,2}(t) \tag{107}$$

$$\tilde{\sigma}_{1*}^{3}(t+1) = \zeta_{\ell T}\sigma_{1*}^{2}(t) + \zeta_{TT}\tilde{\sigma}_1^{\,2}(t) + \xi_{\ell T}\sigma_{2*}^{2}(t) + \xi_{TT}\tilde{\sigma}_2^{\,2}(t) \tag{108}$$

$$\tilde{\sigma}_{2*}^{2}(t+1) = \tilde{\sigma}_1^{\,3}(t) \tag{109}$$

$$\sigma_1^{\,2}(t+1) = \sigma_{1*}^{2}(t) \tag{110}$$

$$\sigma_2^{\,2}(t+1) = \sigma_{2*}^{2}(t) \tag{111}$$

The stresses σ_1^2, σ_2^1, σ_1^3, σ_2^2, $\tilde{\sigma}_1^2$, $\tilde{\sigma}_2^1$, $\tilde{\sigma}_1^3$ and $\tilde{\sigma}_2^2$ are shown in Fig. 7. The stresses σ_{1*}^{2} σ_{2*}^{1} etc. are introduced only in order to reduce the second order difference equations to an equivalent system of the first order [see Eqs. (98)-(102), (109) and (110)].

The system (97)-(111) can be presented in the matrix from (91) in which

$$\vec{\sigma} = \{\sigma_1^{\,2},\ \sigma_2^{\,1},\ \sigma_1^{\,3},\ \tilde{\sigma}_1^{\,2},\ \tilde{\sigma}_2^{\,1},\ \tilde{\sigma}_1^{\,3},\ \tilde{\sigma}_2^{\,2},\ \tilde{\sigma}_{1*}^{2},\ \sigma_{2*}^{1},\ \tilde{\sigma}_{2*}^{1},\ \sigma_{1*}^{3},\ \sigma_{1*}^{3},\ \sigma_{2*}^{2},\ \sigma_{1*}^{2},\ \sigma_{2*}^{2}\} \tag{112}$$

while

$$\vec{\sigma}_{t=0} = \{0,\ 0,\ -1,\ 0,\ 0,\ 1,\ 0,\ 0,\ 0,\ 0,\ 0,\ 0,\ 0,\ 0,\ 0\} \tag{113}$$

The matrix A in this particular case has the order m=15 (since the stress $\sigma_2^2(t)$ was eliminated from the system (46)-(60), while instead of it the only stress σ_{2*}^{2} is involved).

The non-zero elements of this matrix are:

$A_{1,2} = 1$, $A_{2,9} = 1$, $A_{3,11} = 1$, $A_{4,8} = 1$, $A_{5,10} = 1$, $A_{6,\,12} = 1$,

$$A_{7,13} = 1,\ A_{8,15} = 1,\ A_{5,4} = \xi_{T\ell},\ A_{9,7} = \zeta_{T\ell},\ A_{9,14} = \xi_{\ell\ell},$$

$$A_{10,15} = \zeta_{\ell\ell},\ A_{10,4} = \xi_{TT},\ A_{11,7} = \zeta_{TT},\ A_{10,14} = \xi_{\ell T},\ A_{10,15} = \zeta_{\ell T},$$
(114)

$$A_{11,4} = \zeta_{T\ell},\ A_{11,7} = \xi_{T\ell},\ A_{11,14} = \zeta_{\ell\ell},\ A_{11,15} = \xi_{\ell\ell},\ A_{12,4} = \zeta_{TT},$$

$$A_{12,7} = \xi_{TT},\ A_{13,8} = \xi_{TT},\ A_{12,14} = \zeta_{\ell T},\ A_{12,15} = \xi_{\ell T},\ A_{13,6} = 1,$$

$$A_{14,1} = 1,\ A_{15,3} = 1$$

Applying the method of successive matrix multiplications one finds the stress vector (112) at t=1,2,...etc.

$$\vec{\sigma}_1 = (0,\ 0,\ 0,\ 0,\ 0,\ 0,\ 0,\ 0,\ 0,\ 0,\ 0,\ 0,\ 1,\ 0,\ -1)$$

$$\vec{\sigma}_2 = (0,\ 0,\ 0,\ 0,\ 0,\ 0,\ 1,\ 0,\ \zeta_{\ell\ell},\ \zeta_{\ell T},\ \xi_{\ell\ell},\ \xi_{\ell T},\ 0,\ 0,\ 0)$$

$$\vec{\sigma}_3 = (0,\ \zeta_{\ell\ell},\ \xi_{\ell\ell},\ 0,\ \zeta_{\ell T},\ \xi_{\ell T},\ 0,\ 0,\ \zeta_{\ell T},\ \zeta_{TT},\ \xi_{T\ell},\ \xi_{TT},\ 0,\ 0,\ 0) \qquad (115)$$

$$\vec{\sigma}_4 = (\zeta_{\ell\ell},\ \zeta_{\ell T},\ \xi_{T\ell},\ 0,\ \zeta_{TT},\ \xi_{TT},\ 0,\ \zeta_{\ell T},\ 0,\ 0,\ 0,\ 0,\ \xi_{\ell T},\ 0,\ \xi_{\ell\ell})$$

$$\vec{\sigma}_5 = (\zeta_{\ell T},\ 0,\ 0,\ \zeta_{\ell T},\ 0,\ 0,\ \xi_{\ell T},\ \zeta_{TT},\ \zeta_{\ell\ell},\ \xi_{\ell T},$$
$$\zeta_{\ell T},\ \xi_{\ell\ell},\ \xi_{\ell\ell},\ \xi_{\ell T}\xi_{\ell\ell},\ \xi_{TT},\ \zeta_{\ell\ell},\ \xi_{T\ell}) \text{ etc.}$$

Thus, as follows from the solution (115) the magnitudes as well as the configurations of the stress vectors change by steps at t=1,2...., etc. It can be verified that these configurations (but not necessarily the magnitudes) repeat itself after each 75 time-steps, i.e., the motion is periodical.

For qualitative characteristics of the solution (115) (such as the rate of stability or damping) one has to investigate the characteristic roots of the matrix (114).

Conclusion

Thus, it has been demonstrated that the structural response to pulse excitations can be modelled by a system of equations with delay argument. The qualitative properties of the solutions to this system are defined by the fundamental matrix A which plays the same role as the stiffness matrix in finite element technique.

Besides of the Lyapunov - type stability depending upon the characteristic root modulii of the matrix A we have discussed another qualitative property of solutions which is associated with so called structural stability, i.e., the stability with respect to small changes in structural parameters such as reflection and transmission coefficients.

Special attention was paid to structural instability with respect to small changes in the lengths of structural members and in the characteristic speed of wave propagation. It has been shown that this instability may lead to transition to erogidicity i.e., to a loss of periodicity of pulse motions.

Thus, if a structure is subjected to impulsive concentrated loads, or to distributed, but discontinuous loads, its dynamical response will be affected by pulse propagation which cannot be modelled by modal methods. This means that structural response to pulse excitation is an additional structural characteristic which complement its modal properties. While within a framework of linear analysis, both results can be combined due to the superposition principle.

It is important to mention that prediction of structural response to pulse excitations allows oneto adjust a control law in order to eliminate the spillover effect.

Acknowledgement

The research described in this paper was carried out by the Jet Propulsion Laboratory, California Institute of Technology, under NASA Contract No. NAS7-918. This task was sponsored by Dr. Anthony K. Amos, Air Force Office of Scientific Research, Code NA.

References

1. Meirovitch, L., "Anaytical Methods in Vibrations," Macmillan Co., London, 1967.

2. Von Flotow, A., "A Traveling Wave Approach to the Dynamic Analysis of Large Space Structures," AIAA, 83-0964, pp. 509-519, 1983.

3. Balas, M. J., "Feedback Control of Feasible Systems," IEEE, Transactions on Automatic Control, Vol. AC-23, No. 4, pp. 673-679, 1978.

4. Flugge, W., Handbook of Engineering Mechanics, McGraw-Hill Co., New York, pg. 64-11, 1962.

5. Zukas, J. A., "Impact Dynamics," John Wiley and Sons, New York, pg. 17, 1982.

6. Hildebrand, F., "Finite-Difference Equations and Simulation," Prentice-Hall, Inc., pp. 26-27, 1968.

APPENDIX

The reflection and transmission coefficients ξ and ζ of joints between one - dimensional structural members introduced above were not specified. Strictly speaking, they depend on the type of joints design, and the most reliable way to define them is through experiments. Theoretical derivation of expressions for the reflection and transmission coefficients as functions of the parameters of joints is possible only for highly idealized mathematical models of these joints. We will introduce below such models in order to find at least qualitative dependence between the coefficients ξ and ζ, and the parameters characterizing joints such as material densities, cross-sectional areas, elastic and viscous properties, etc.

For the sake of simplicity we will consider only symmetric joints whose reflection and transmission coefficients are the same for waves coming from the right and from the left.

1. We will start with elastic joints and assume that their lengths are negligible in comparison with the lengths of the structural members. First of all, we will utilize the formulas for stress pulse propagations in structural members with discontinuous cross sections[5] assuming that joints and structural members are from the same material:

$$\frac{\sigma_R}{\sigma_I} = \frac{1-\eta}{1+\eta} \qquad \frac{\sigma_T}{\sigma_I} = \frac{2}{1+\eta}, \qquad \eta = \frac{S_1}{S_2} \tag{116}$$

in which S_1, S_2 are the respective cross-sectional areas, while σ_I, σ_R and σ_T are the incident, reflected and transmitted stresses.

Now turning to Fig. 9

one calculates that an incident wave σ_I coming to the joint from the left is reflected from its left side first as:

$$\sigma_R' = \frac{1-\eta}{1+\eta}\,\sigma_I \tag{117}$$

Then the transmitted wave

$$\sigma_T' = \frac{2}{1+\eta}$$

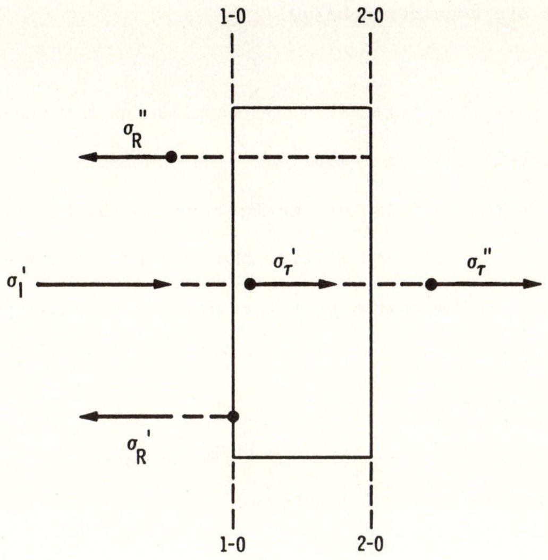

FIGURE 9

is reflected from the right side of the joint

$$\sigma_R'' = \frac{2\eta(\eta-1)}{(1+\eta)^2} \qquad (119)$$

Since the length of the joint is ignored, the total reflected stress will be:

$$\sigma_R = \sigma_R' + \sigma_R'' = \left(\frac{1-\eta}{1+\eta}\right)^2 \sigma_I \qquad (120)$$

and therefore, the reflection coefficient

$$\xi = \frac{1-\eta}{1+\eta} \leq 1 \qquad (121)$$

The wave σ_T^1 after second transmission through the right side of the joint will be:

$$\sigma_T = \eta \left(\frac{2}{1+\eta}\right)^2 \sigma_I \qquad (122)$$

and therefore, the transmission coefficient

$$\zeta = \eta \left(\frac{2}{1+\eta}\right)^2 \qquad (123)$$

The formulas (121) and (123) become more complicated if the material of the joints is different from the material of the structural members. More general models of joints are considered below.

2. Let us consider a two-bar joint (Fig. 9) modelled as a visco-elastic body which length is negligible in comparison with the lengths of the structural members. Assuming that a pulse at the left end of the bar 1 has caused an elastic compression σ_i, one can utilize the laws of conservation of momentum and kinetic energy in the course of reflection and transmission of the wave from the surface 1-0. These laws require that the following conditions to be satisfied at the interface:

a. The incident and reflected forces must be equal to the transmitted forces,

b. Particle velocities must be continuous.

Hence, at the interface 1-0:

$$S_1(\sigma_I + \sigma_R') = S_o\sigma_T' \tag{124}$$

$$\frac{\sigma_I}{\rho_1 C_1} - \frac{\sigma_R'}{\rho_1 C_1} = \frac{\sigma_T'}{\rho_o C_o + \nu_o} \tag{125}$$

in which S_1, S_o, ρ_1, ρ_o and C_1, C_o are the cross-sectional areas, linear densities and characteristics speeds of the structural member 1 and the joint 0, respectively, σ_I is the incident stress pulse, σ_R' and σ_T' are the reflected from/and transmitted through the interface 1-0 stress pulses, respectively, υ_o is the coefficient characterizing viscous properties of the joint so that the compression pulse σ_T' within the joint is presented as:

$$\sigma_T' = \rho_o C_o V + \nu_o V \tag{126}$$

in which V is the velocity pulse.

Here the first and second terms in the right hand side describe the elastic $\overset{*}{\sigma}_T'$ and viscous $\overset{**}{\sigma}_T'$ parts of the stress α_T'.

Since the viscous stress $\overset{**}{\sigma}_T'$ dissipates, the only elastic part of the stress will approach the interface 0-2, and therefore, for these interface, instead of Eqs. (124) and (125) one obtains:

$$S_o(\overset{*}{\sigma}_T' + \sigma_R'') = S_2\sigma_T'' \tag{127}$$

$$\frac{\overset{*}{\sigma}_T'}{\rho_o C_o + \nu_o} - \frac{\sigma_R''}{\rho_o C_o + \nu_o} = \frac{\sigma_T}{\rho_2 C_2} \tag{128}$$

in which S_2, ρ_2, C_2 are the cross-sectional area, density and the characteristic speed of the structural member 2, σ_R'' and σ_T'' are the stress pulses reflected from/and transmitted through the surface 0-2, respectively. Solving the system (124), (125), (127), and (128) with respect to σ_R', σ_R'', σ_T' and σ_T'' one obtains the reflection and transmission coefficients of the joint 0, respectively:

$$\xi = \frac{\sigma_R}{\sigma_I} = \frac{\sigma_R' + \sigma_R''}{\sigma_I} = \frac{1+\nu-\eta_1}{1+\nu+\eta_1} + 2\gamma_1 \frac{\eta_2 - 1 - \nu}{(\eta_2+1+\nu)(\eta_1+1+\nu)} \tag{129}$$

$$\zeta = \frac{\sigma_T''}{\sigma_I} = \frac{4\gamma_2\eta_2}{\gamma_1(\eta_2+1+\nu)(\eta_1+1+\nu)} \tag{130}$$

in which

$$\eta_1 = \frac{S_1}{S_o}\frac{\rho_1}{\rho_o}\frac{C_1}{C_o}, \quad \eta_2 = \frac{S_2}{S_o}\frac{\rho_2}{\rho_o}\frac{C_2}{C_o}, \quad \gamma_1 = \frac{S_1}{S_o}, \quad \gamma_2 = \frac{S_2}{S_o}, \quad \nu = \frac{\nu_o}{\rho_o C_o} \tag{131}$$

3. The interaction between a stress pulse and a joint becomes more complicated if two-structural members form some angle, Fig. 10.

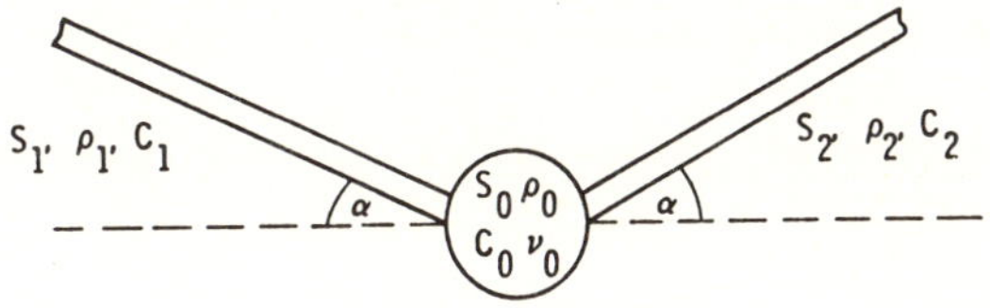

FIGURE 10

In this case we will introduce stress vectors

$$\vec{\sigma}_I = (\sigma_I, \tilde{\sigma}_I)^T, \quad \vec{\sigma} = (\sigma_R, \tilde{\sigma}_R)^T, \quad \sigma_T = (\sigma_T, \tilde{\sigma}_T)^T \text{ etc.} \tag{132}$$

in which σ and σ are the longitudinal and transverse stresses, respectively.

Following the same strategy as in the previous item one can rewrite Eqs. (124), (125), (127), and (128) in the following vector-matrix form:

$$S_1(\vec{\sigma}_I+\vec{\sigma}_R) = S_o B\vec{\sigma}_T' \tag{133}$$

$$Q_1^{-1}(\vec{\sigma}_I - \vec{\sigma}_R') = Q_o^{-1}B\vec{\sigma}_T' \tag{134}$$

$$S_o(\sigma_T' + \vec{\sigma}_R'') = S_2 B\vec{\sigma}_T'' \tag{135}$$

$$Q_o^{-1}(\vec{\sigma}_T^* + \vec{\sigma}_R'') = Q_2 B\vec{\sigma}_T'' \tag{136}$$

in which

$$B = \begin{pmatrix} \text{Cos } \alpha & \text{Sin } \alpha \\ \text{Sin } \alpha & \text{Cos } \alpha \end{pmatrix}, \quad Q_1 = \begin{pmatrix} \rho_1 C_1 & 0 \\ 0 & \tilde{\rho}_1\tilde{C}_1 \end{pmatrix}$$

$$Q_o = \begin{pmatrix} \rho_o C_o + \nu_o & 0 \\ 0 & \tilde{\rho}_o\tilde{C}_o + \tilde{\nu}_o \end{pmatrix} \quad Q_2 = \begin{pmatrix} \rho_2 C_2 & 0 \\ 0 & \tilde{\rho}_2\tilde{C}_2 \end{pmatrix} \tag{137}$$

Here 2α is the angle between the structural members 1 and 2, while ρ, C, ν and $\tilde{\rho}$, $\tilde{C}$, $\tilde{\nu}$ are related to longitudinal and transverse stresses, respectively. Solving the system (133)-(136) with respect to $\vec{\sigma}_R'$, $\vec{\sigma}_R''$, $\vec{\sigma}_T'$ and $\vec{\sigma}_T''$ one obtains the reflection and transmission coefficinets of the joint O:

$$\xi_{\ell\ell} = \frac{1+\nu-\eta_1}{1+\nu+\eta_1} + 2\gamma_1 \frac{\eta_2-1-\nu}{(\eta_2+1+\nu)(1+\nu+\eta_1)} \text{ Cos } \alpha \tag{138}$$

$$\xi_{\ell T} = -2\gamma_1 \frac{\eta_2-1-\nu}{(\eta_2+1+\nu)(1+\nu+\eta_1)} \text{ Sin } \alpha \tag{139}$$

$$\xi_{T\ell} = 2\gamma_1 \frac{\tilde{\eta}_2-1-\tilde{\nu}}{(\tilde{\eta}_2+1+\tilde{\nu})(1+\tilde{\nu}+\tilde{\eta}_1)} \text{ Sin } \alpha \tag{140}$$

$$\xi_{TT} = \frac{1+\tilde{\nu}-\tilde{\nu}_1}{1+\tilde{\nu}+\tilde{\eta}_1} + 2\gamma_1 \frac{\tilde{\eta}_2-1-\tilde{\nu}}{(\tilde{\eta}_2+1+\tilde{\nu})(1+\tilde{\nu}+\tilde{\eta}_1)} \text{ Cos } \alpha \tag{141}$$

$$\zeta_{\ell\ell} = \frac{4\gamma_1\eta_2}{\gamma_2(\eta_2+1+\nu)(\eta_1+1+\nu)} \text{ Cos } 2\alpha \tag{142}$$

$$\zeta_{\ell T} = \frac{4\gamma_1\eta_2}{\gamma_2(\eta_2+1+\nu)(\eta_1+1+\nu)} \text{ Sin } 2\alpha \tag{143}$$

$$\zeta_{T\ell} = \frac{4\gamma_1\tilde{\eta}_2}{\gamma_2(\tilde{\eta}_2+1+\tilde{\nu})(\tilde{\eta}_1+1+\tilde{\nu})} \text{ Sin } 2\alpha \tag{144}$$

$$\zeta_{TT} = \frac{4\gamma_1\tilde{\eta}_2}{\gamma_2(\tilde{\eta}_2+1+\tilde{\nu})(\tilde{\eta}_1+1+\tilde{\nu})} \text{ Cos } 2\alpha \tag{145}$$

while

$$\begin{pmatrix} \sigma_R \\ \tilde{\sigma} \end{pmatrix} = \begin{pmatrix} \xi_{\ell\ell} & \xi_{\ell T} \\ \xi_{T\ell} & \xi_{TT} \end{pmatrix} \begin{matrix} \sigma_I \\ \tilde{\sigma}_I \end{matrix} ; \quad \begin{pmatrix} \sigma_T'' \\ \tilde{\sigma}_T'' \end{pmatrix} = \begin{pmatrix} \zeta_{\ell\ell} & \zeta_{\ell T} \\ \zeta_{T\ell} & \zeta_{TT} \end{pmatrix} \begin{pmatrix} \sigma_I \\ \tilde{\sigma}_I \end{pmatrix} \tag{146}$$

in which, in addition to the notations (131)

$$\tilde{\eta}_1 = \frac{S_1}{S_o} \frac{\tilde{\rho}_1\tilde{C}_1}{\tilde{\rho}_o\tilde{C}_o}, \quad \tilde{\eta}_2 = \frac{S_2}{S_o} \frac{\tilde{\rho}_2}{\tilde{\rho}_o} \frac{\tilde{C}_2}{\tilde{C}_o}, \quad \tilde{\nu} = \frac{\tilde{\nu}_o}{\tilde{\rho}_o\tilde{C}_o} \tag{147}$$

A Review of Modelling Techniques for the Open and Closed-Loop Dynamics of Large Space Structures

Peter M. Bainum
Professor of Aerospace Engineering
Department of Mechanical Engineering
Howard University
Washington, D.C. 20059, USA

Abstract

This paper reviews the steps in the development of mathematical models that can be used to simulate the in orbit dynamic behavior of large flexible systems. A general continuum formulation is compared with the hybrid coordinate formulation and also a finite element representation of the total system. A review of structural analysis routines emphasizes the use of computer generated graphics to help understand the different modal elastic shape functions of complex systems. Numerical techniques employed to synthesize shape and attitude control laws are summarized. Finally, the modelling of environmental disturbance torques due to the interaction of solar radiation pressure on vibrating and thermally deflected systems is discussed.

I. Introduction

Large space systems or subsystems are under consideration for use in the space station, space platforms, communications, electronic orbital based mail systems, and in solar energy collection. Many of these systems will have characteristic dimensions of hundreds of meters and extremely small weight to area ratios. Thus, the effects of system flexibility must be incorporated into any mathematical model used to predict the dynamic behavior in orbit. For many applications both orientation and surface shape control will often be required.

Fig. 1 illustrates a plan of development of mathematical models for large space systems showing four different stages in the evolution of the open and closed loop over-all system model. The most basic component is that of modelling the open-loop unperturbed dynamics of the system in orbit.

II. Modelling of System Dynamics

Although modern space vehicles can be found in a wide variety of configurations, certain dynamic features are sufficiently common to be described as charac-

---

*Presented at the 15th International Symposium on Space Technology and Science, Tokyo, Japan, May 18-24, 1986.

Springer Series in Computational Mechanics
S. N. Atluri, A. K. Amos (Eds.)
Large Space Structures: Dynamics and Control

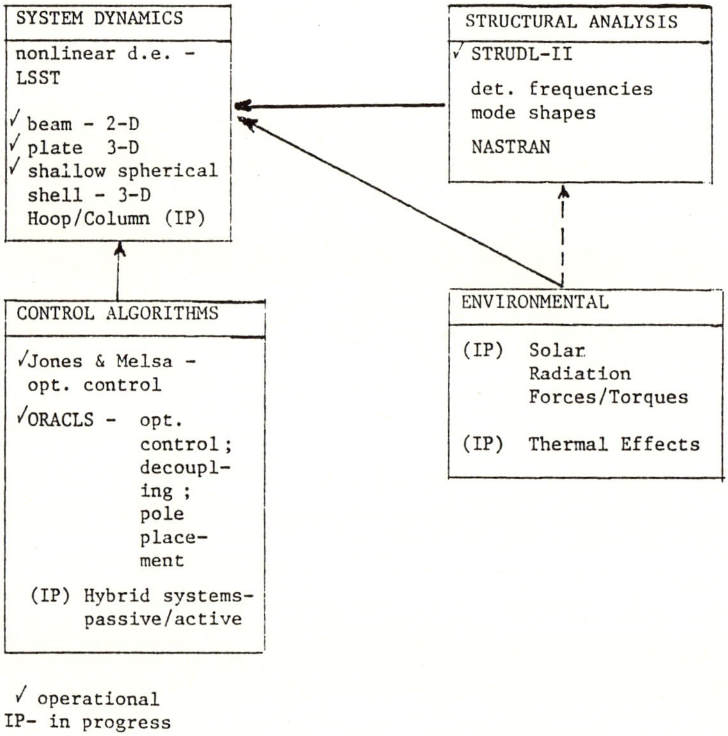

Fig. 1. Development of System Software for LSST Dynamics Analysis

teristic of space vehicles. Most current space vehicles can be modelled mathematically as a combination of one or more essentially rigid bodies with one or more flexible bodies. The structural subsystems of a spacecraft are often required to undergo substantial relative motions during mission performance, such as when large antennas, solar arrays, propulsion devices, etc., change their orientations relative to the main part (hub).

In the discrete-coordinate or discrete parameter method, the entire system is considered to be a collection of interconnected rigid bodies. Each rigid body of the system model is discrete, and the coordinates employed are coordinates of position and/or attitude of the individual bodies. In case one of the bodies is considered as the main part (such as the hub), then typical coordinates employed could be the position and Euler rotational angles of the main part, together with relative rotational (or position) coordinates of the auxiliary members relative to the main part (or to one another). The approach most frequently adopted (Refs. 1-4) involves direct application of the Newton-Euler equations of translation and rotation to various subsets of bodies in the assembly.

Flexibility can be introduced into this type of model by adjusting the stiffness of sets of springs assumed to be located at the connection points of pairs of the rigid bodies. The springs can be assumed to provide both translational and rotational flexibility. In addition, energy dissipation at the connecting joints can be modelled by equivalent viscous damping, at least within the linear range. For inherently flexible large systems each of the basic structural component systems (i.e., solar panels, appendages, antennas, etc.) can be subdivided into groups of spring connected rigid bodies. Since there are no limitations on the relative motions of the constituent parts of the spacecraft, the discrete coordinate approach is conceptually ideal for spacecraft undergoing large changes in orientation and configuration. The disadvantages of employing discrete coordinate equations of motion of a system of point-connected rigid bodies is that no degree of sophistication in mathematical modelling can provide proper representation of actual local deformations, as required for the calculation of stresses. In addition, although large amplitude displacements in the discrete coordinates can be treated, there may be difficulty in appropriately modelling local stiffness and damping properties outside of the linear range.

The hybrid coordinate method as introduced by Meirovitch and Nelson (Ref. 5) and further developed by Likins (Ref. 6) has been widely used to model modern spacecraft systems consisting of a primary rigid hub and elastic appendages which represent solar panels, antenna booms, instrumentation platforms, etc. In the Meirovitch approach flexibility is introduced into the model in the form of additional terms in the kinetic and potential energy expressions associated with a Lagrangian development. The elastic displacements and velocities contained in the energy expressions are related to a set of admissible (modal) shape functions consistent with the boundary conditions, and these admissible functions are approximated by a certain number of truncated terms for a finite number of modes. One of the difficulties of this method is in the development of a set of admissible functions for a complex, nonhomgeneous structure. A second difficulty is that linearization is accomplished only after expansion of complex algebraic expressions for systems requiring a large number of degrees of freedom. An advantage of this approach is that a (Liapunov) stability analysis about a desired equilibrium motion is facilitated since the energy expressions are available and can be reduced to the necessary quadratic forms (Refs. 7,8).

For the case when maneuvers associated with all or part of the vehicle subsystems permits one to assume small, linearly elastic deformations, Likins suggests an efficient simulation which combines discrete coordinates with distributed (modal) coordinates, retaining the validity of the discrete coordinates where necessary, but securing the computational advantages of the (truncated) modal coordinates, where possible. The result is a hybrid coordinate system (Refs. 9-12) that permits accurate simulation of complex modern space vehicles with a minimum number of coordinates. The hybrid coordinate equations of motion for a finite element model of a flexible appendage attached to a rigid base, undergoing unrestricted motions are developed in Ref. 9. In addition, some advantages of this technique are noted. In the finite element method, the structure is decomposed into geometrically simple elements for which the displacement field is described by the displacement vectors of some well-chosen nodal (grid) points. A proper representation of the actual local deformations at appropriate positions on the appendages can be obtained, within the linear range. However, a disadvantage of this approach, is that the formulation would have to be significantly modified and expanded to provide validity during large amplitude displacements which could occur e.g. - during a slewing maneuver of a flexible antenna (appendage). In addition, the connection between each appendage and the hub is idealized to be at a single point which may not be the most accurate model of how an antenna might be attached to the hub (for example).

When the size and weight to area ratio of such proposed future large space structural systems indicates that the entire system must be considered to be flexible, the hybrid coordinate formulation may not always be readily adoped by simply assuming the mass and inertias of the (previously) rigid central part tend to zero in the limit. Santini (Ref. 13) has developed a mathematical formulation for predicting the motion of a general orbiting flexible body using a continuum approach. The effects of higher harmonics in the Earth's gravitational potential are also included. Elastic deformations are considered small as compared with characteristic body dimensions. Equations are developed for both the rigid and elastic (generic) modes. This development is based on an a priori knowledge of the frequencies and modal shape functions of all modes to be included in the truncated system model. Kumar and Bainum (Ref. 14) have modified the development of

Ref. 13 based on vector calculus so that elastic modal shape functions expressed in arbitrary systems of coordinates may be accommodated. This method has been applied to the development of the equations of motion of relatively simple homogeneous and isotropic structures in orbit such as free-free beams, free plates, and shallow spherical shell structures. For more complicated, non-homogeneous, non-isotropic structures, the implementation of this method would involve the evaluation of many complex volume integrals dependent on both the spatial variation in the modal shape function(s) as well as the density. As an alternative, finite element modelling (FEM) techniques could be used to model the composite configuration. However, the coupling terms (induced by the orbital motion) between the rigid and flexible modes and between different flexible modes are usually neglected in the FEM. Potential modelling errors could result for large systems even within the linear range and are discussed in Ref. 15.

III. Structural Analysis

For simple isotropic structures, such as homogeneous beams and circular plates, closed form expressions are available for elastic modal frequencies and shape functions. For more complex and/or nonisotropic systems numerical methods must be employed to obtain this information. Commonly used routines include versions of STRUDL and NASTRAN (the latter more complex algorithm may also be useful in the simulation of thermo-elastic environmental effects). The use of computer generated graphics may also prove useful in understanding the elastic shape functions of complex systems when excited at different modal frequencies.

For systems requiring a large number of elastic modes to accurately represent the system dynamics, graph theoretic techniques may provide an alternative to the numerical problems involved in calculating the eigenvalues (modal frequencies) (Ref. 16). With this approach the system (stiffness) matrix can often be reduced to a system of lower ordered submatrices so that, under certain conditions, the eigenvalues of the original matrix are given by the union of the eigenvalues of the submatrices, including their multiplicity (Ref. 16). For large order systems this approach can substantially reduce the numerical effort involved with an improvement in accuracy. As an example, a free-free homogeneous square plate was considered in Ref. 16 where the original stiffness matrix contained a dimensionality of 16.

Before surface and orientation control systems can be designed, it is necessary to understand the dynamics and stability of the uncontrolled system. For large order systems an analytical approach to the stability problem is not feasible and numerical techniques must be employed to develop the system characteristic equation and the loci of its roots for different sets of system parametes (Ref. 17). As the number of modes retained in the truncated system model increases, expansion of the characteristic determinental equation becomes algebraically prohibitive. As an alternative an algorithm due to Leverrier (Refs. 18, 19) can be used to numerically determine the coefficients in the characteristic equation. In order to implement this algorithm the linearized equations must be written in standard state variable format.

IV. Control Modelling

At this point the modelling of the control actuators can be added to the previously developed open-loop system models (Refs. 20,21). In general, an actuator placed at an arbitrary location on a large space structure will affect both the rigid and flexible modes. The location of such an actuator has definite implications on the system controllability. For large order systems the reachability matrix and term rank concepts, also developed from graph theoretic techniques, may be used to verify controllability and can be computationally more effective than numerical rank tests of the system controllability matrix. The free-free square plate is again considered as an illustrative example in Ref. 16.

Three techniques are commonly used to develop control laws once the system controllability has been established. These include: (a) decoupling techniques; (b) pole placement (clustering); and (c) an application of the linear regulator problem from optimal control theory. These three techniques approach the controls problem from different points of view and each will be briefly discussed.

The decoupling technique can be applied in two distinct sub-cases: (1) where the linear state equations in the original coordinates are decoupled by using state variable feedback techniques; and (2) where the open loop linear equations are first transformed into a decoupled set in the modal coordinates and then control laws are developed independently for each mode. It then becomes necessary to transform the control laws as expressed in modal coordinates to the actual control in the original coordinates.

As an example of sub-case (1), we assume that the linearized equations can be expressed as:

$$\ddot{Z} = D\dot{Z} + EZ + BU \tag{1}$$

where $Z = (z_1, \ldots z_r, z_{r+1}, \ldots z_{r+n})^T$

describes the rigid body position displacements (1, 2 ... r) plus the n elastic (position) coordinates retained in any truncated model. After selecting $U = K_r\dot{Z} + K_pZ$, we can rewrite the controlled motion equations as:

$$\ddot{Z} = (D+BK_r)\dot{Z} + (E+BKp)Z \tag{2}$$

where K_r and K_p are evaluated such that $(D+BK_r)$ and $(E+BK_p)$ are diagonalized and, thus, yield the required damping and frequency of the controlled modes. The total number of rigid+elastic modes, (r+n), must be equal to the number of actuators here to avoid the necessity of using pseudo-inverse matrices.

For the second sub-case, as an example, let us consider a different form of Eq. (1) in terms of the mass (M) and stiffness (K) matrices, where D=0.

$$M\ddot{Z} + KZ = F = BU \tag{3}$$

With the following type of transformation: $Z = \phi q$, Eq. (3) may be recast in terms of the modal coordinates, q, and the transformation matrix (involving the eigenvectors), ϕ, as:

$$[\phi^T M \phi]\ \ddot{q} + [\phi^T K \phi]\ q = \phi^T F \tag{4}$$

such that $\phi^T M \phi$ and $\phi^T K \phi$ are diagonalized

$$[\ulcorner M_{i\lrcorner}]\ \ddot{q} + [\ulcorner K_{i\lrcorner}]\ q = \phi^T F = F' \tag{5}$$

It is then possible to design the control laws, F', in the modal space so that independent control of each of the modes can be achieved. A transformation is then required to obtain the control laws in the original coordinates, F, and, then, for a given location of actuators, the actual control from;

$$U = B^{-1}F = B^{-1}(\phi^T)^{-1}F'.$$

In the pole clustering method the overall transient requirements of the system are considered instead of concentrating on the behavior of the individual coordinates. The linearized system equations, Eq. (1), can be recast in the state space format as:

$$\dot{X} = AX + BU \tag{6}$$

where

$$X = [z_1,\ ,,\ z_r,\ z_{r+1}\ ..\ z_{r+n},\ \dot{z}_1,\ ..\ \dot{z}_r,\ \dot{z}_{r+1},\ ..\ \dot{z}_{r+n}]^T$$

The control, $U = -KX$, is then selected by using a digital computer algorithm such as ORACLS (Ref. 22) such that (A-BK) has the identical negative real part in each of its eigenvalues. Although the number of actuators can be less than the number of modes, a limitation of this particular algorithm is that the gains are selected such that all of the closed loop poles lie on a line parallel to the imaginary axis. The algorithm is useful, however, when it is important that each mode in the system satisfy some minimum damping characteristics.

The linear regulator theory allows one to set, a priori, distinct penalty weighting functions on the control effort as well as the state variables. The control law, $U = -KX$, is selected such that the following performance index is minimized

$$J = \int_0^\infty (X^T Q X + U^T R U)\ dt \tag{7}$$

where Q and R are positive semi-definite and positive definite matrices, respectively. The steady state solution of the matrix Riccati equation of dimension equal to the state has to be solved in order to obtain the gain matrices, K (which represent the positive definite solution to the algebraic Riccati equations).

A computer algorithm within the ORACLS (Ref. 22) software package can be used to obtain the gain matrices, K, for different combinations of the Q and R penalty matrices. This algorithm utilizes the Newton-Raphson method of solving the Riccati equation.

Both the linear regulator problem and the pole clustering methods can result in some of the closed loop frequencies being orders of magnitude greater than those of the uncontrolled system. These higher frequencies may also correspond to the frequencies of higher modes not included in the previously truncated system model. In order to completely consider such effects the order of the original system model would have to be increased in order to avoid the effects of control spillover. On the other hand, these methods have the advantage that they can be applied to situations where the number of actuators is less than the number of modes in the mathematical model, in contrast to the usual applications of the decoupling methods. Examples of the application of the various control algorithms are given in Ref. 21. As an example, a typical application of decoupling, using state variable feedback for the orientation and shape control of a free-free square plate nominally following the local vertical with its larger surface normal to the orbit normal is considered. The model contains three rigid rotational modes and the first three transverse flexible modes, with five actuators assumed to be located with their thrust axes normal to the plate (four at the corners and one at the middle edge), and a sixth actuator at one of the corners thrusting normal to the other five. The decoupling gains are selected in order to produce 20% of critical damping in each of the rigid modes and the fundamental elastic mode and 10% of critical damping in the second and third flexible modes. The corresponding time history of the required control forces is illustrated in Fig. 2.

V. Modelling of Environmental Disturbances

The principal disturbance forces and torques acting on a large flexible system in orbit are the gyroscopic and gravity-gradient torques associated with the orbital motion, the control torques, and those torques due to the environment. In the formulation of Refs. 13 and 14 the gyroscopic and gravity-gradient torques are included in the model of the system dynamics. If other formulations are employed, such as general finite element methods, which do not account for the orbital dynamics, the effects of the gyroscopic and gravity-gradient torques should be carefully considered before deleting them from the dynamic model.

Environmental disturbances can be attributed mainly to the effects of solar radiation pressure, except in very low earth orbit where the aerodynamic drag forces predominate. Moments due to solar radiation pressure are induced if the center of solar radiation pressure is not co-located with the system center of mass. The location of the center of pressure is dependent on the surface characteristics as well as the geometrical shape of the structure. In addition, due to

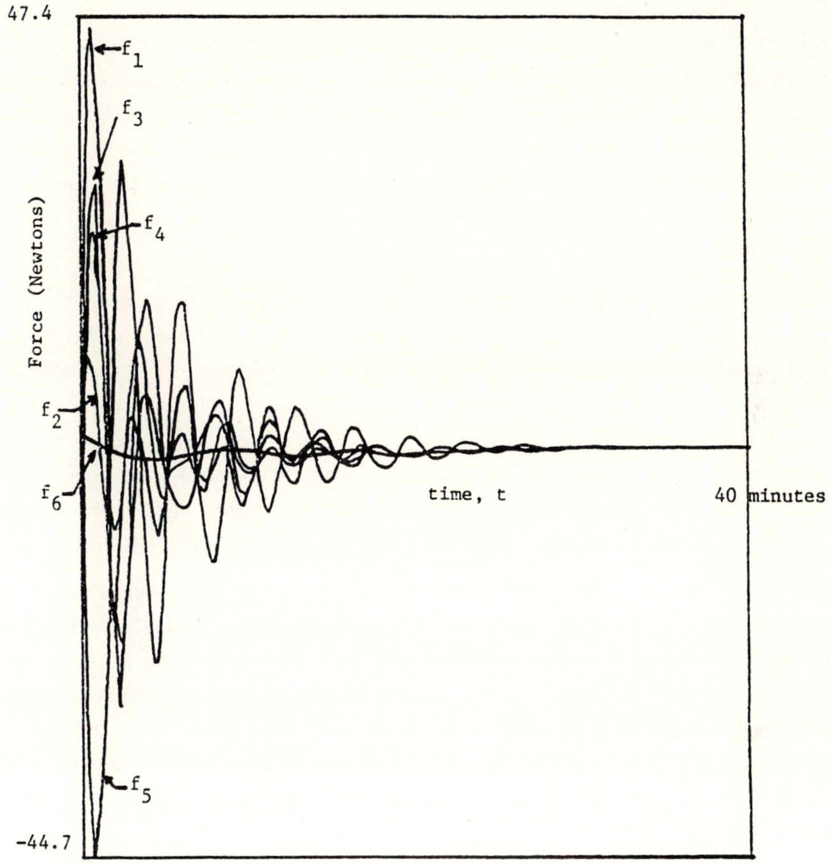

Fig. 2. Control Force Time History for Case (iii)-II.

solar heating, thermal gradients can be induced in the structure which may result in appreciable thermal strains. As a result, the structure will undergo deformations which will further contribute to the forces and torques caused by the solar pressure.

Several investigators have considered the effect of solar radiation pressure on the dynamics of spacecraft. The majority of the spacecraft modelled consisted of a smaller rigid central satellite to which flat plate appendages, also treated as rigid, were assumed to be attached. A few authors showed how the solar pressure moments generated could be used for satellite attitude control by controlling the orientation of plates and/or vanes which could rotate at the ends of the appendages (Refs. 23, 24).

Recently the uncontrolled dynamics of an orbiting flexible beam in the presence of solar radiation forces was considered (Ref. 25). For the special case of a completely absorbing surface analytical expressions for the force and moments induced on the vibrating structure were obtained (Fig. 3). For the more general surfaces, numerical techniques were employed. It was found that moments result only from the flexible symmetric modal deformations. For small pitch amplitudes and at geosynchronous altitudes, the solar radiation moments due to the deformations of the beam can be greater than those due to the gravity-gradient.

Subsequently, the closed loop dynamics of flexible orbiting beams and plates exposed to solar radiation has been studied (Refs. 26, 27). In Ref. 27 the effect of thermally induced deflections are also included. The resulting rigid modal oscillations induced are found to be an order of magnitude larger than for those cases previously considered in which only the solar radiation pressure effect on the vibrating structure was treated. Modifications of the control laws are considered in order to improve transient response characteristics from the effects of thermal induction.

VI. Concluding Comments

This paper has attempted to review the steps required for the modelling and simulation of the dynamics and control of future proposed large flexible orbiting systems which will require, in general, both shape as well as orientation (attitude) control. Problem areas, mainly associated with the large order of such system models, are emphasized. Considerable insight into the behavior of more complex non-homogeneous, non-isotropic systems can be gained by first modelling simple systems such as flexible beams and plates in orbit.

Acknowledgment

Research supported by: NASA/Howard U. Large Space Structures Institute; NASA Grant NSG-1414; and Nippon Telegraph and Telephone Corp.

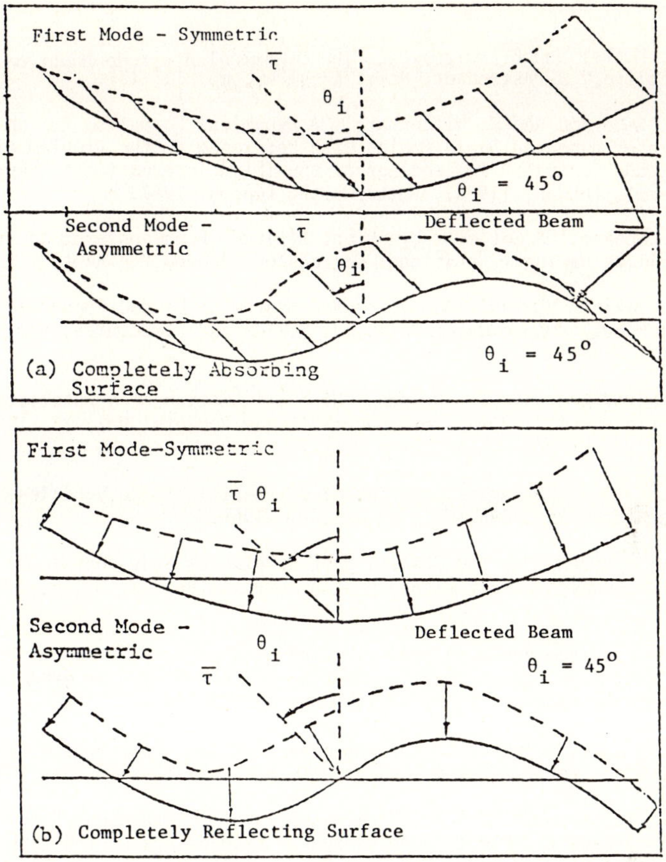

Fig. 3. Solar Radiation Force Distribution on the First Two Modes-Free-Free Beam

References

1. W.W. Hooker and G. Margulies, "The Dynamical Attitude Equations for an N-Body Satellite," J. Astronaut. Sci., 12, 1965, pp. 123-128.

2. R.E. Roberson and J. Wittenburg, "A Dynamical Formalism for an Arbitrary Number of Interconnected Rigid Bodies with Reference to the Problem of Satellite Attitude Control," Proceedings of the Third International Congress of Automatic Control, London, (1966), Butterworth, London, 1967.

3. W.W. Hooker, "A Set of r Dynamical Attitude Equations for an Arbitrary N-Body Satellite Having r Rotational Degrees of Freedom," AIAA J., 8, 1970, pp. 1205-1207.

4. P.W. Likins, "Dynamic Analysis of a System of Hinged-Connected Rigid Bodies with Nonrigid Appendages," Int. J. Solids and Structures, Vol. 9, 1973, pp. 1473-1487.

5. L. Meirovitch and H.D. Nelson, "High Spin Motion of a Satellite Containing Elastic Parts," Journal of Spacecraft and Rockets, Vol. 13, 1966, pp. 1597-1602.

6. P.W. Likins, "Dynamics and Control of Flexible Space Vehicles," Jet Propulsion Laboratory, Pasadena, Ca., TR 32-1329, 1969.

7. L. Meirovitch, "On the Stability of a Spinning Body Containing Elastic Parts via Lyapunov's Direct Method," AIAA Journal, Vol. 8, No. 7, 1970, pp. 1193-1200.

8. L. Meitrovitch and R. A. Calico, "The Stability of Motion of Force Free Spinning Satellites with Flexible Appendages," AAS/AIAA Astrodynamics Specialists Conference, August 17-19, 1971, Paper No. 71-345.

9. P.W. Likins, "Finite Element Appendage Equations for Hybrid Coordinate Dynamic Analysis," Int. J. Solids and Structures, Vol. 8, 1972, pp. 709-731.

10. P.W. Likins and A.H. Gale, "The Analysis of Interactions Between Attitude Control Systems and Flexible Appendages," Proceedings of the 19th International Astronautical Congress, Pergamon Press, 1970, Vol. 2, pp. 67-90.

11. G.E. Fleischer and L.F. McGlinchy, "Viking Thrust Vector Control Dynamics Using Hybrid Coordinates to Model Vehicle Flexibility and Propellent Slosh," AAS/AIAA Astrodynamics Specialists Conference, August 17-19, 1971, Paper No. 71-348.

12. P.W. Likins, "Large-Deformation Modal Coordinates for Non-rigid Vehicle Dynamics," Jet Propulsion Laboratory, TR 32-1565, 1972.

13. P. Santini, "Stability of Flexible Spacecrafts," Acta Astronautica, Vol. 3, 1976, pp. 685-713.

14. V.K. Kumar and P.M. Bainum, "Dynamics of a Flexible Body in Orbit," AIAA/AAS Astrodynamics Conference, Palo Alto, Calif., 1978, Paper No. 78-1418; also J. of Guidance and Control, Vol. 3, 1980, pp. 90-92.

15. C.M. Diarra and P.M. Bainum, "On the Accuracy of Modelling the Dynamics of Large Space Structures," 36th International Astronautical Congress, Stockholm, Sweden, Oct. 7-12, 1985, Paper No. IAF85-228; to appear, Acta Astronautica.

16. A.S.S.R. Reddy and P.M. Bainum, "Graph Theory Approach to the Eigenvalue Problem of Large Space Structures," Proc. of the Third VPI&SU/AIAA Symposium on Dynamics and Control of Large Flexible Spacecraft, Blacksburg, Va., June 15-17, 1981 (VPI&SU Press, Ed: L. Meirovitch, 1982), pp. 175-189.

17. P.M. Bainum and V.K. Kumar, "On the Dynamics of Large Orbiting Flexible Beams and Platforms Oriented along the Local Horizontal," Acta Astronautica, Vol. 9, 1982, pp. 119-127.

18. L.A. Zadeh and C.A. Desoer, Linear System Theory, McGraw Hill Book Co., New York, 1963, pp. 303-305.

19. J.L. Melsa and S.K. Jones, Computer Programs for Computational Assistance in the Study of Linear Control Theory, McGraw Hill Book Co., New York, 1970, pp. 6-11.

20. P.M. Bainum and A.S.S.R. Reddy, "On the Controllability of a Long Flexible Beam in Orbit," Proceedings of the Second AIAA Symposium on Dynamics and Control of Large Flexible Spacecraft, June 21-23, 1979, (VPI&SU Press, Ed.: L. Meirovitch, 1980), pp. 145-159.

21. A.S.S.R. Reddy, P.M. Bainum, H.A. Hamer and R. Krishna, "Control of a Large Flexible Platform in Orbit," Journal of Guidance and Control, Vol. 4, 1981, pp. 642-648.

22. E.S. Armstrong, "ORACLS - A System for Linear-Quadratic-Gaussian Control Law Design," NASA Technical Paper 1106, April 1978.

23. V.J. Modi and K.C. Pande, "Solar Pressure Control of a Dual Spin Satellite," Journal of Spacecraft and Rockets, Vol. 10, 1973, pp. 355-361.

24. K.C. Pande, M.S. Davis and V.J. Modi, "Time Optimal Pitch Control of Satellite using Solar Radiation Pressure," Journal of Spacecraft and Rockets, Vol. 11, 1974, pp. 601-603.

25. R. Krishna and P.M. Bainum, "Effect of Solar Radiation Disturbance on a Flexible Beam in Orbit," AIAA Journal, Vol. 22, May 1984, pp. 677-682.

26. Peter M. Bainum and R. Krishna, "Control of an Orbiting Flexible Square Platform in the Presence of Solar Radiation," Acta Astronautica, Vol. 12, No. 9, 1985, pp. 699-704

27. R. Krishna and P.M. Bainum, "Dynamics and Control of Orbiting Flexible Structures Exposed to Solar Radiation," Journal of Guidance, Control and Dynamics, Vol. 8, Sept. - October 1985, pp. 591-596.

Dynamic Friction

A. V. Srinivasan
United Technologies Research Center
East Hartford, Connecticut 06108

1. INTRODUCTION

Friction forces manifesting at structural joints and interfaces are an important source of energy dissipation. When materials with low inherent damping are used in environments offering essentially no aerodynamic damping, the ability of the structure to withstand vibratory stress due to sudden maneuvers etc. depends almost entirely on the extent of friction damping. Therefore, it is important to understand the parameters that influence this phenomenon so that guidelines can be established for the design of structural components.

Friction between contacting interfaces which undergo relative vibratory motion is known to dissipate the energy of vibration resulting in damped oscillations. The phenomenon of friction between contacting surfaces is probably the most elusive physical mechanism that defies clear comprehension. The complexity of the phenomenon of friction damping arises from the variations in the type of time dependent motions developing at an interface. These variations in the relative motion at the contacting surfaces span the extremes between microslip and gross motion and include local slip, stick-slip motion, chatter, etc. The parameters that control the resulting motion include the normal forces holding the surfaces together and their distribution, properties of materials in contact, surface treatments, temperature, frequency of vibration, level of vacuum and coefficient of dynamic friction. The phenomenon is clearly nonlinear and the feasibility of linearization needs to be established in each application.

In the context of vibration engineering, the basic requirements are to (a) quantify the nature and magnitude of friction forces that manifest between contacting interfaces of vibrating components, (b) quantify the nature and magnitude of vibratory motion at these interfaces, and (c) predict the extent of damping that may be present. Within the present context, friction forces are considered to be useful, i.e., they control vibratory amplitudes which otherwise may escalate. On the other hand, any consideration of friction forces cannot ignore the influence of these forces on wear of the components resulting in loss of useful life of machines. The phenomena of friction and wear are thus inseparable. The emphasis in this study, however, is in outlining those aspects of friction forces that pertain to vibration

Springer Series in Computational Mechanics
S. N. Atluri, A. K. Amos (Eds.)
Large Space Structures: Dynamics and Control

damping. This chapter will present recent efforts to model dynamic friction and discuss results of laboratory testing aimed at understanding this difficult phenomenon.

2. BACKGROUND

An excellent treatment of the science of friction and wear is found in a book Friction and Wear: Calculation Methods (Ref. 1) by the Russian authors Kragelsky, Dobychin and Kombalov. The chapter on "Dry and Boundary Friction" should be of special interest to researchers in this field and has an excellent treatment of the subject beginning as far back as 1508 when Leonardo da Vinci established the simple law of friction which assumed the friction force to be proportional to the normal load. Researchers should also find the three volumes of the book on Friction, Wear, Lubrication (Ref. 2) edited by Kragelsky and Alisin interesting. Another book, also by Russian authors Panovko and Gubanova entitled Stability and Oscillation of Elastic Systems (Ref. 3) treats the subject of self-induced oscillations with dry friction. In addition, the proceedings volume of a NASA-sponsored symposium on Interdisciplinary Approach to Friction and Wear (Ref. 4) edited by Ku contains material which covers the several aspects of surface topography, friction, adhesion, wear, etc. Another excellent source of reference on this interdisciplinary topic is a recent report by Oden and Martins (Ref. 5). Reference 5 provides a thorough discussion of the physical aspects of dynamic friction and points to the need to include the deformability of interfaces in the normal direction. Both continuum mechanics and finite element models are examined in Ref. 5. These books and reports, along with the papers that will be referred to in the sections below, have served as the basis for the viewpoints presented here. Because of the interdisciplinary nature of the subject matter, no claim will be made that all the relevant literature has been surveyed here.

2.1 Characteristics of Friction Forces

A great variety of phenomena occur in rubbing contact between surfaces of components. This is because of the dual nature of friction by which surface friction overcomes molecular interaction forces between surfaces and mechanical resistances associated with changes in the profile of the surface layer. Kragelsky et al. (Ref. 1) point to data that indicate molecular attraction at an interface could contribute substantially to the observed frictional forces. As can be seen from Fig. 1, the contribution to friction from molecular forces can dominate particularly for lighter load conditions. It is assumed that a sublayer of decreased resistance is formed at the interface between rubbing components. One speaks of dry friction forces when the sublayer is in a solid phase.

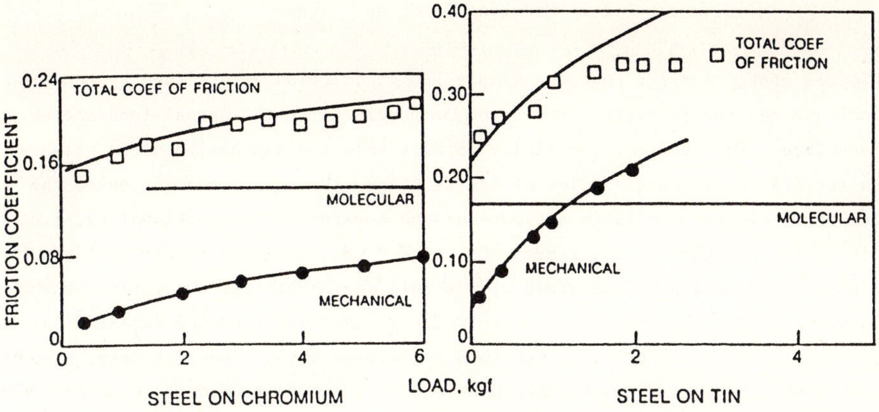

Figure 1. Variation of Friction Coefficient Components with Normal Load

(Ref. #1, Page 192)

Mechanical deformation may occur due to either penetration of surfaces with ploughing or continuous formation and rupture of welds. All these processes are influenced by the environment. As observed by Kragelsky et al. (Ref. 1) "in spite of the complex nature of mechanical, physical, and chemical processes involved, it is still possible to pinpoint certain principles which are common to all friction and wear processes". They are outlined below:

1. The contact between solids occurs at discrete points because of surface roughness.

2. Increase in load leads to increase in contact area due to increase in number of contact points. Therefore, the real pressure at any point increases only slightly.

3. Tangential resistances are additive. Dissipation of energy in the formation and rupture of an individual frictional bond is determined by the resistance in overcoming molecular interaction at points of contact, and the mechanical ploughing effects.

4. Three interconnected processes occur simultaneously in sliding contact

   a) Interaction of the surfaces
   b) Changes in the surface layers and films
   c) Rupture of the surface layer.

5. Surface sliding requires formation and rupture of a thin layer (weakened layer of base material or films) whose shear strength must be less than the shear strength of the substrate.

2.2 Modeling Friction Forces

It was stated earlier that the simple law of friction established by Leonardo de Vinci assumed the friction force to be proportional to the normal load acting on the interface. The apparent simplicity of this relationship has been the primary reason for its use in most studies of dry friction. However, in addition to the problem of obtaining a reliable estimate of the constant of proportionality, i.e. the coefficient of friction μ, the representation $F = \mu N$ leads to computational complications even when applied to the study of a single degree-of-freedom system, as shown by Den Hartog (Ref. 6). Oden and Pires (Ref. 7) have discussed the mathematical and physical difficulties associated with such a representation. Nevertheless, a substantial number of attempts have been made and continue to be made to use modifications of this basic model (Refs. 8 to 24). Such modifications are adaptations of the basic model to the analysis of particular components under study. Adding more degrees-of-freedom to the basic model leads to additional complexity (Ref. 11) which can be avoided by the use of linearized versions of the model. As shown by Jacobsen (Ref. 12), such linearization can provide a good representation of system behavior if the friction force is small so that it does not distort the motion appreciably. Subsequent analyses, many of which have been devised to represent specific vibratory components, have generally relied upon the basic model in its linearized version. These developments have mainly involved more degrees-of-freedom and different configurations (Refs. 13, 14, 15, 16). Improved friction models have been tried, for example, making the friction force dependent on the vibration amplitude (Ref. 17) and by analyzing the interfacial slip in a vibrating beam (Ref. 18). However, there have been few studies which incorporate a detailed model of interface behavior into a model of a vibrating system.

In 1778, a two term formula was proposed by Coulomb according to which the friction force F was written as $F = AN+B$, where A can be identified as the familiar coefficient of friction. With such a representation, the coefficient of friction is not a property depending only on the materials in contact but also on the normal load. Interface behavior is a difficult area of study which is why it is often avoided by invoking a constant overall coefficient of friction. Any discussion of friction in the context of vibratory rub must recognize the need to distinguish between the tangential forces of external static friction and those of external dynamic friction. The former exist in a region of small partially reversible displacements and peak at the boundary of such a region. Dynamic friction forces are not dependent on the magnitude of displacement. The fully developed static friction force is characterized by a coefficient which is different from that corresponding to the dynamic friction force. In the solution of vibration problems, prior to the initiation of any sliding motion, the static coefficient of friction should govern. The coefficient of friction is either chosen on the basis of practical experience or used as a disposable parameter to obtain the best agreement between theory and experiment. However, a deeper understanding of the effects of interface damping on vibrating systems can only be obtained by studying the friction processes in more detail.

Introducing the concepts of a local coefficient of friction and combining it with a study of stress distribution in the contact leads to a microslip solution in which only a part of the contact undergoes relative motion. Goodman (Ref. 19) has reviewed the basis of this approach which has since been applied to practical systems through the use of finite element techniques (Refs. 20, 21, 22).

Postulating an overall friction force, but allowing it to vary throughout each vibration cycle leads to the study of stick-slip behavior. Antoniou et al. (Ref. 23) have reviewed this approach and show the degree of complication which this can add to a one degree of freedom system.

The classical mathematical model of dry friction, i.e.

$$F = \begin{cases} +F_o \; ; \; \dot{x} > 0 \\ -F_o \; ; \; \dot{x} < 0 \end{cases} \qquad (1)$$

(where F_o is the force of slipping friction, $\dot{x}$ is the velocity of slipping) leads to a force-displacement relationship for alternating input displacement illustrated by two fixed horizontal lines and two vertical lines as shown in Fig. 2. Experimental observations, however, suggest that the transition from one equation of slip ($-F_o$) to the other ($+F_o$) is smoother than indicated by Eq. (1), and more like the characteristics shown in Fig. 3 (see Ref. 25). This important observation allows the introduction of hysteresis characteristics of contact interaction and may serve as a basis of the analysis of friction damping. Accordingly, if x and F represent displacement and friction force respectively and x_o, F_o their maximum values, the hysteresis representation may be represented as

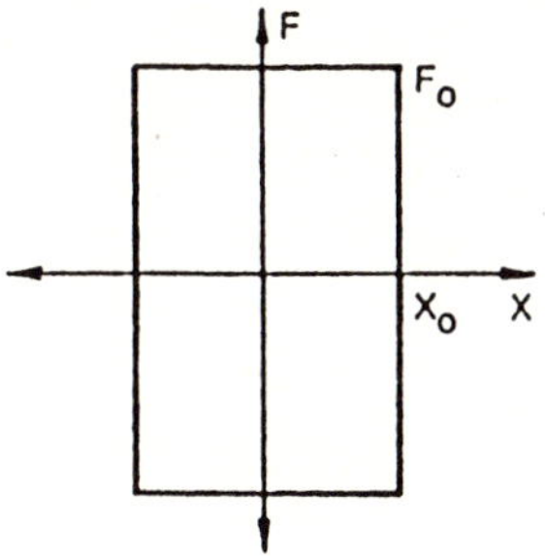

Figure 2

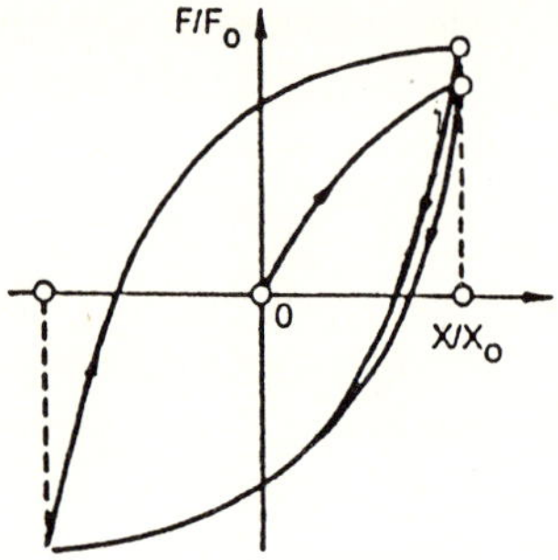

Figure 3. Hysteresis Characteristic of Dry Friction

$$\alpha = \begin{cases} 1 - e^{-\beta(\xi - C_1)} & ; \dot{\xi} > 0 \\ -1 + e^{\beta(\xi - C_2)} & ; \dot{\xi} < 0 \end{cases} \tag{2}$$

where

$\alpha = F/F_o$

$\xi = x/x_o$

$\beta = bx_o$

b = is a material constant

The constants C_1, C_2 can be shown to be equal and opposite to each other, i.e., $C_1 = -C_2$

$$\therefore \frac{d\alpha}{d\xi} = \begin{cases} \beta(1-\alpha) & ; \dot{\xi} > 0 \\ \beta(1+\alpha) & ; \dot{\xi} < 0 \end{cases} \tag{3}$$

Given initial values of ξ, α, the constants C_1, C_2 can be generated for the ascending and descending parts of the loop. The procedure can be repeated for several vibration cycles.

An amplitude dependent model of friction can be derived as shown in Ref. 25 for any prescribed value of β. For example, for $\xi = A$, from Eq. (2)

$$\begin{aligned} \alpha_A &= 1 - e^{-\beta(A - C_o)} \\ \alpha_{-A} &= -1 + e^{\beta(-A + C_o)} \end{aligned} \tag{4}$$

which leads to $$C_o = \frac{-1}{\beta} \ln (\cosh \beta A) \tag{5}$$

using which, α can be shown to be

$$\alpha = \begin{cases} 1 - \dfrac{e^{-\beta\xi}}{\cosh \beta A} & \dot{\xi} > 0 \\ -1 + \dfrac{e^{\beta\xi}}{\cosh \beta A} & \dot{\xi} < 0 \end{cases} \tag{6}$$

Equation (6) is an amplitude dependent friction model. The area of the hysteresis loop due to friction can be shown to be equal to

$$4(A - \frac{1}{\beta} \tanh \beta A) \tag{7}$$

from which β can be determined from tests. It may be noted that 4A in the above expression is the area of the loop consistent with the friction characterization shown in Eq. (1).

Another approach was presented by Srinivasan and Cassenti (Ref. 26) and is based on the concept of nonlocal nature of deformations introduced by Oden and Pires (Ref. 7). A brief discussion of the concept is included here for the sake of completeness and is followed by the formulation of a modified law.

Recognizing the continuum nature of bodies in contact, the friction forces at a point $\bar{x}$ is represented as the sum total of the influence of normal stresses over a neighborhood as shown schematically in Fig. 4. Thus, unlike the local law

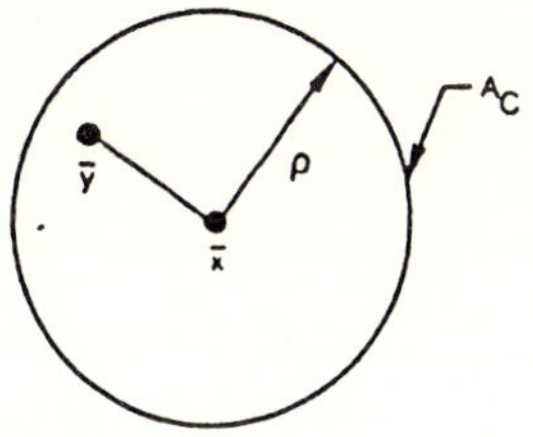

Figure 4. Concept of Nonlocal Laws of Friction (Oden & Pires)

$$\tau(\bar{x}) = \mu\sigma_n(x) \tag{8}$$

a nonlocal law is represented by

$$\tau(\bar{x}) = \mu \int_{A_c} W_\rho(\bar{x}-\bar{y})\ \sigma_n(\bar{y})dA_c \tag{9}$$

where W_ρ is a weighting function that distributes the influence of σ_n in a circle of characteristic radius ρ. Further, as the friction (shear stress) vector is always opposed to the direction of relative motion

$$\bar{\tau}(\bar{x}) = -\mu\bar{n}_f \int_{A_c} W_\rho(\bar{x}-\bar{y})\sigma_n(\bar{y})dA_c$$

$$\bar{n}_f = \frac{\bar{u}_f}{|\bar{u}_f|} \tag{10}$$

where $\bar{u}_f$ is the vector of relative displacement along the surface. The requirement that the friction force be dependent on the motion (displacement or velocity) leads to

$$\bar{\tau}(\bar{x}) = -\mu \bar{n}_f \Phi_e(u_f) \int_{A_c} W_\rho(\bar{x}-\bar{y})\sigma_n(\bar{y})dA_c \qquad (11)$$

An extension to the above law (Ref. 26) includes velocity dependence for the forces at the interface and an integration along the path followed by the points at the interfaces undergoing relative motion. The velocity dependence accounts for the dynamics of relative motion at interfaces and the type of integration proposed recognizes a nonisotropic weighting function. The nature of this extension is presented below. The details may be obtained from Ref. 26.

The modification proposed recognizes that bonds between asperities may form, deform together, and break continuously at an interface. The adhesive strength of this bond will depend upon many factors including the extent of time any two points remain in contact. A mathematical representation of these concepts is summarized by

$$\tau(X) = -\mu_s \int_{-\infty}^{\infty} \int_{-\infty}^{\infty} \phi(R/\varepsilon)W(\eta/\rho)\sigma_n(X_1) \frac{u}{|u|} \frac{dX_1 dY_1}{\varepsilon_o \rho} \qquad (12)$$

The total friction force at the interface is

$$f = \int_A \tau(X)dA \qquad (13)$$

where A is the total area of the interface. The four-fold integration evident in Eq. (13) averages the local effects and therefore there is no need to model individual asperities on the surface.

The formulation presented above is of sufficient generality to permit its use in specific engineering problems in which rubbing at interfaces influences the dynamics of components. Computing time may become an important factor in solving problems of practical interest in view of the need to compute four-fold integrals for total forces.

2.2.1 Coefficient of Friction

An accurate calculation of friction forces induced at interfaces in relative vibratory motion depends on the accuracy with which the coefficient of sliding friction can be estimated. As observed earlier, there appears to be no experimental data obtained in a vibratory environment from which the nature of the coefficient of friction can be obtained. The following is a summary of such an experiment performed at the United Technologies Research Center under contract to NASA LeRC (Ref. 15).

The approach that was chosen in that investigation was to provide sinusoidal relative motion between two test pieces which were held together with a constant normal force and to measure the resulting frictional forces directly. The test pieces were a pair of titanium alloy (8-1-1) bars with their rubbing interfaces flame sprayed with tungsten carbide. The size of the rubbing surface was 22.35 x 6.35mm. The test assembly, shown in Figs. 5 and 6 uses an electrodynamic shaker to induce sinusoidal excitation of the lower test piece. The upper piece which rubs on the lower piece during test was held in such a way that longitudinal forces on the piece could be measured using strain gaged load bolts without any appreciable movement. The normal load on the joint was applied by setting the required weights on the loading platform. Testing consisted of setting the required frequency and input acceleration and recording the frictional force-slip loop on an oscilloscope for a range of normal loads.

The nominal condition for testing was defined as a displacement of 0.127mm double amplitude at a frequency of 280Hz with normal loads ranging from about 250 to 700N. Three loads were used, based on the weights available and the dead load of the upper arm; these were 271, 492, and 672N. The displacement was varied from 0.056 to 0.132mm DA and frequencies of 80, 140, 280 and 420Hz were input. Two typical friction force-slip loops are shown in Fig. 7.

The equivalent friction coefficient was determined from the measured area of the recorded (photograph of oscilloscope picture) loop and a knowledge of the input frequency and displacement. The resulting values for friction coefficient are shown plotted against normal load, frequency and velocity, in Fig. 8. A least squares fit of an equation of the form: coef = (A/Load)+B was attempted using all the data. The resulting curve is shown superimposed on the test data points in Fig. 8. An examination of these data indicates that there is no obvious correlation with either input frequency or maximum velocity.

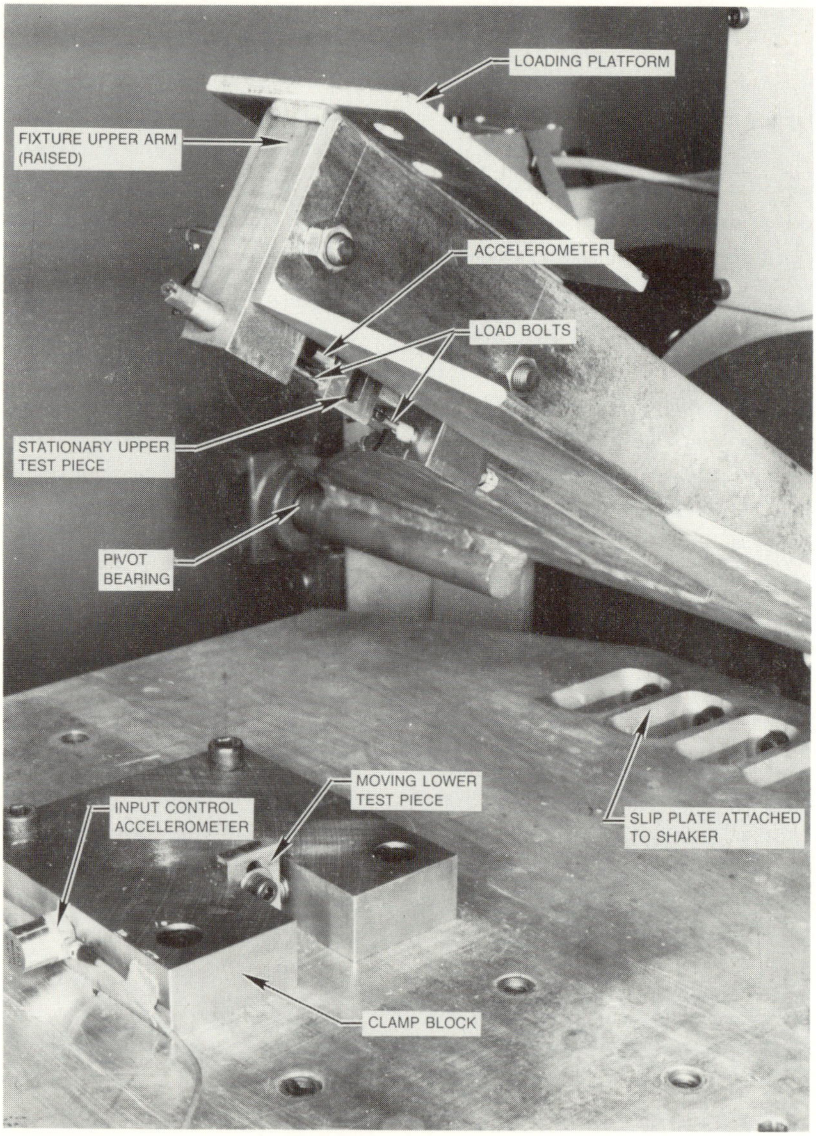

Figure 5. Friction Test Assembly (Upper Arm Raised)

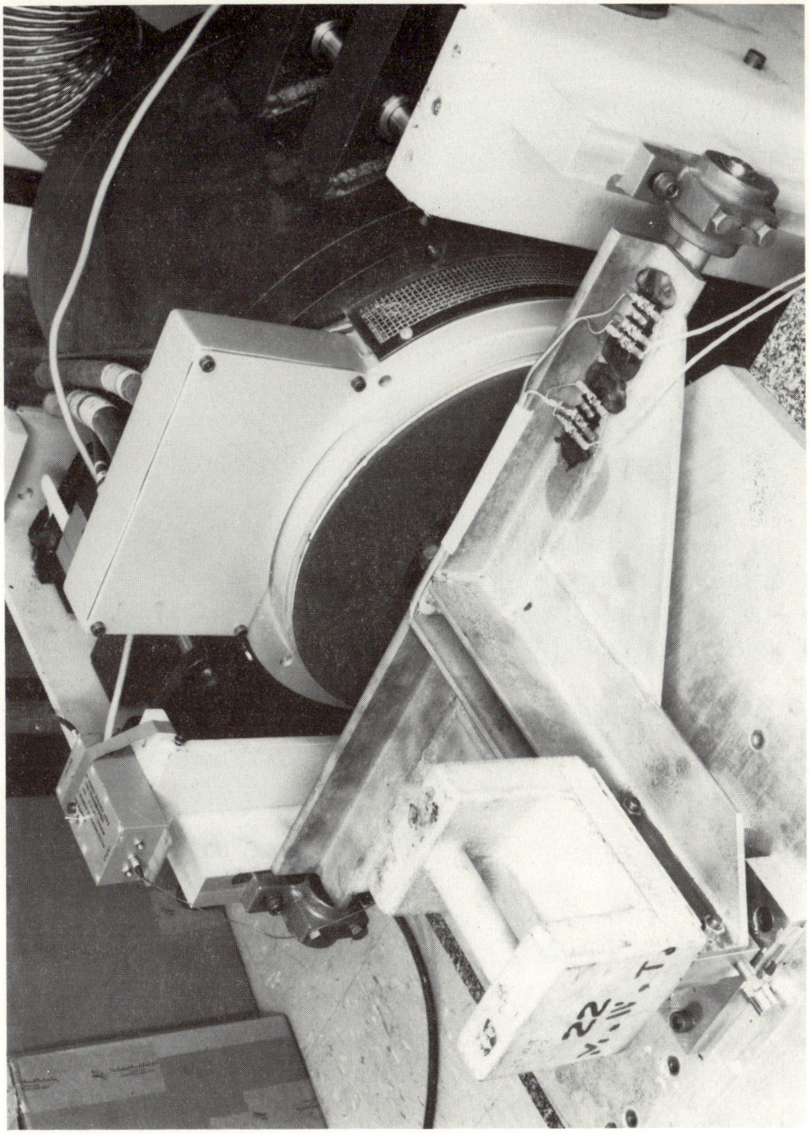

Figure 6. Friction Test Assembly

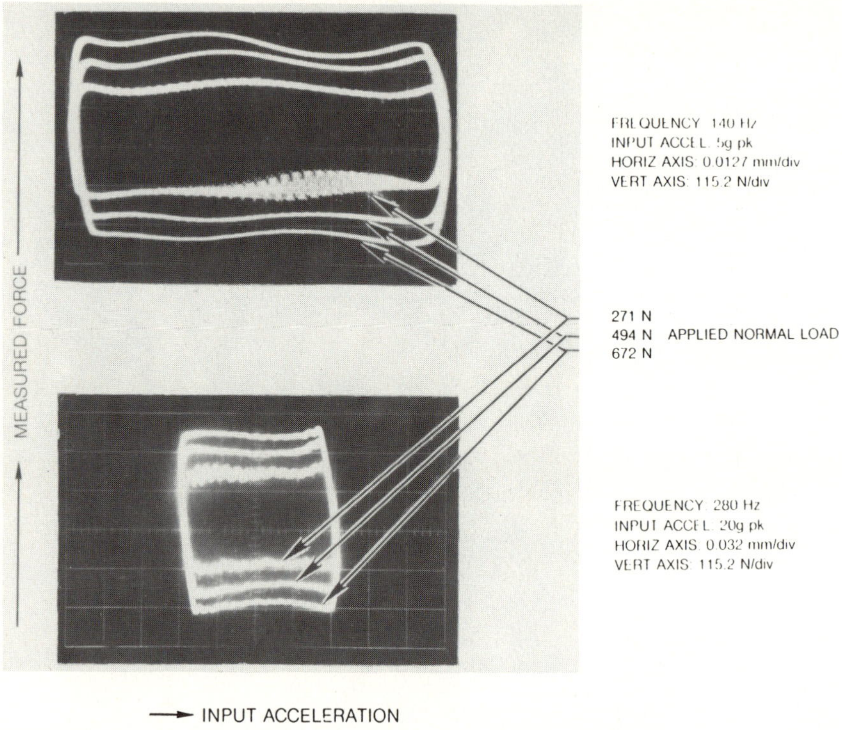

Figure 7. Typical Measured Friction Force - Slip Loops

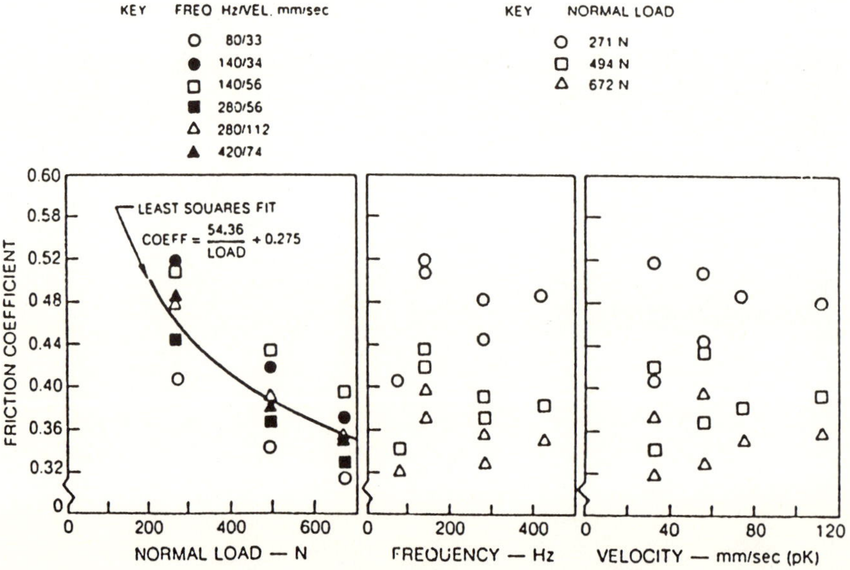

Figure 8. Variation of Equivalent Friction Coefficient During Sinusoidal Motion with Normal Load, Frequency and Maximum Relative Velocity

2.3 Dry Friction Damping Technology

Control of resonant vibration in light, flexible structures has been found to be largely the result of damping available at mating surfaces. The potentials of such damping depends, as has been observed earlier, on a host of parameters that include the magnitude of normal force holding the surfaces in close contact and the characteristics of the surfaces. These and other parameters determine the nature and extent of slip at the interfaces so that an optimum in energy dissipation can be expected with an appropriate combination of normal load and relative motion (Ref. 24). Therefore, an understanding of the processes involved in relative vibratory motion at joints becomes a basic requirement that may influence the design of structures where friction becomes an important source of damping.

The analytical approaches for estimating friction damping can be broadly classified as follows: (1) macroslip approach and (2) microslip approach. In the macroslip approach, the entire interface is assumed to be either slipping or stuck. The friction mechanism is either replaced by an equivalent linear viscous model or assumed to be governed by some form of Coulomb's law of dry friction. The effect of friction damping is obtained by determining the forced response of the component. The analysis involved is relatively straightforward and the justification for widespread use of this approach is its effectiveness at predicting the actual response. However, there is some question as to the validity of the macroslip approach when the interface has a large area, and is subject to a nonuniform load distribution. In the microslip approach, a relatively detailed analysis of the stress distribution at the interface is carried out, typically via a finite-element procedure. The extent of local slip, not necessarily throughout the interface, between pairs of contacting points is determined by applying Coulomb's law of friction to the normal and tangential stresses. A detailed knowledge of interface slip dynamics can be obtained from this approach. Although much useful information can be gained in a linearized analysis, a fully nonlinear time history solution may be necessary and feasible with modern computational methods and equipment.

Calculation of energy dissipation over a period of vibration requires an accurate assessment of both the forces and motion of an interface. Much effort will be needed before satisfactory analytical models become available to calculate these parameters. In the meantime, the process of developing and calibrating analytical models requires advances in instrumentation that can measure extremely small vibratory motion at interfaces.

3. CONCLUDING REMARKS

An accurate assessment of damping due to friction will obviously depend upon the accuracy with which the interfaces and the environment are characterized. On the basis of observations made earlier, five important interdisciplinary aspects may now be outlined.

1. Surface Science: involving a definition of the interacting surfaces in terms of (a) geometry, (b) bulk material properties such as moduli of elasticity, hardness, etc., (c) surface topography and contact area, and (d) surface chemistry.

2. Environment: involving (a) definition of mechanical and thermal loading, (b) contaminants, gases and liquids present, (c) humidity, and (d) electrical and magnetic fields.

3. Laws of friction that relate 1 and 2 above.

4. Calculation of response of a machine element using 1, 2 and 3 above along with a knowledge of structural properties such as mass, stiffness, etc.

5. Calculation of energy dissipation using 2 and 4 above.

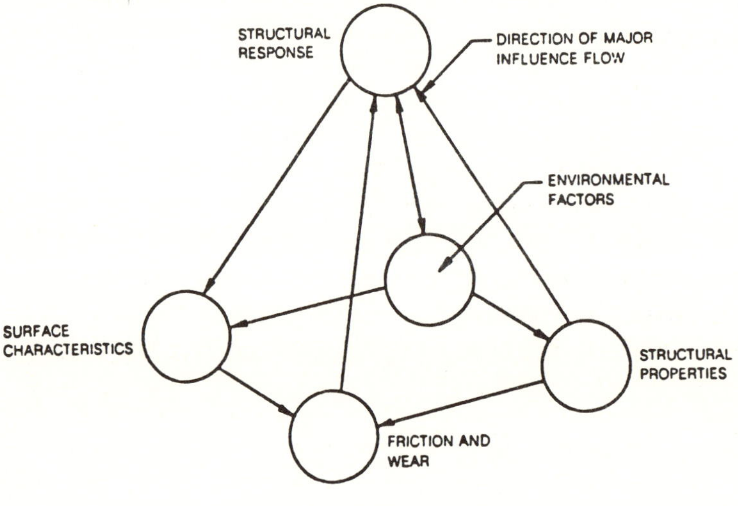

Figure 9. Factors Influencing Structural Response When Frictional Forces are Significant

These interdependent and influential factors may be represented in the form of a friction damping pyramid as shown in Fig. 9. Determination of those parameters that are most influential in a given application is an important task, although certain aspects of research in the technology of friction damping will be of a basic nature. For example, the general nature of the coefficient of friction may be established although its magnitude will vary from one pair of materials to another. Analytical modeling for both single degree of freedom and multiple degrees of freedom must be developed from first principles and calibrated on the basis of carefully planned experimental investigations.

As noted earlier, an important aspect of friction analysis is the determination of the law of friction that is valid and appropriate for application to a given vibrating component. The representations in the form of a nonlocal law afford the flexibility needed in this complex problem area. Also this concept provides a rational basis for deriving more general empirical descriptions which are appropriate to a vibratory environment.

REFERENCES

1. Kragelski, I. V., M. N. Dobychin and V. S. Kombalov: Friction and Wear; Calculation Methods. Pergamon Press, Inc. 1982.
2. Kragelski, I. V., and V. V. Alisin (ed.): Friction Wear Lubrication; Tribology Handbook. Mir Publishers, 1981.
3. Panovko, Y. G., and I. I. Gubanovo: Stability and Oscillations of Elastic Systems. Consultants Bureau, New York, 1965.
4. Ku, P. M. (ed.): Interdisciplinary Approach to Friction and Wear: NASA SP-181, 1968.
5. Oden, J. T. and Martins, J. A. C.: Computational Methods for Dynamic Friction Phenomena. To appear in Computer Methods in Applied Mechanics and Engineering, North Holland, Amsterdam.
6. Den Hartog, J. P.: Forced Vibrations with Combined Coulomb and Viscous Friction. Trans. ASME, APM-53-9, pp. 107-115, 1931.
7. Oden, J. J. and E. B. Pires: Nonlocal and Nonlinear Friction Laws and Variational Principles for Contact Problems in Elasticity. J. App. Mech., Vol. 50, 1983.
8. Bowden, F. P. and D. Tabor: Friction and Lubrication of Solids: Oxford Univ. Press (London), Pt. I, 1954 and Pt. II, 1964.
9. Zhuravlev, V. A.: On the Physical Basis of the Amontons-Coulomb Law of Friction. J. Tech. Phys. (USSR), Vol. 10, 1940, p. 1447.
10. Greenwood, J. A. and J. B. P. Williamson: The Contact of Nominally Flat Surfaces. Proc. 2nd Int. Conf. on Electric Contacts (Graz, Austria), 1964.
11. Yeh, G. C. K.: Forced Vibrations of a Two-Degree-of-Freedom System with Combined Coulomb and Viscous Damping. J. Acoustical Soc. of America, Vol. 39, (1966), pp. 14-24.
12. Jacobsen, L. S.: Steady Forced Vibration as Influenced by Damping. Trans. ASME, Vol. 52 (1930), pp. 169-181.

13. Pratt, T. K. and R. Williams: Non-Linear Analysis of Stick/Slip Motion. J. Sound and Vibration, Vol. 74, No. 4 (1981), pp. 531-542.

14. Griffin, J. H.: Friction Damping of Resonant Stresses in Gas Turbine Engine Airfoils. ASME Paper #79-GT-109, March 1979.

15. Srinivasan, A. V., D. G. Cutts, and S. Sridhar: Turbojet Engine Blade Damping. NASA CR 165406, July 1981.

16. Mayer, R. L. and N. A. Mowbray: The Effect of Coulomb Damping on Multidegree of Freedom Elastic Structures. Earthquake Engineering and Structural Dynamics, Vol. 3 (1975), pp. 275-286.

17. Earles, S. W. E. and E. J. Williams: A Linearized Analysis for Frictionally Damped Systems. Jour. Sound and Vibration 24(4), pp. 445-458, 1972.

18. Pian, T. H. H.: Structural Damping of a Simple Built-up Beam with Riveted Joints in Bending. ASME J. Appl. Mechanics, Vol. 24 (1957), pp. 35-38.

19. Goodman, L. E.: A Review of Progress in Analysis of Interfacial Slip Damping (in Structural Damping papers from a colloquium - ASME annual meeting, December 1959), J. E. Ruzicka (ed.), Pergamon (1960).

20. Bielawa, R. L.: An Analytic Study of the Energy Dissipation of Turbomachinery Bladed Disk Assemblies Due to Inter-Shroud Segment Rubbing. ASME Paper No. 77-DET-73 (1977).

21. Annigeri, B. S.: Finite Element Analysis of Planar Elastic Contact with Friction. ME Thesis, Graduate School of Illinois Institute of Technology (August 1976).

22. Rimkunas, D. A. and H. M. Frye: Investigation of Fan Blade Shroud Mechanical Damping. Wright-Patterson Air Force Base, Aero Propulsion Lab., Report No. FR-11065, 1979.

23. Antoniou, S. S., A. Cameron, and C. R. Gentle: The Friction Speed Relation from Stick-Slip Data. Wear, Vol. 36 (1976), pp. 235-254.

24. Beards, C. F.: The Damping of Structural Vibration by Controlled Interfacial Slip in Joints. ASME Paper 81-DET-86 (1981).

25. Busarov, Y. P. and M. S. Ostrovskii: Mathematical Model of Hysteresis of Surface Friction. Mashinovedenic 1976, No. 5, pp. 82-87.

26. Srinivasan, A. V., Cassenti, B. S.: A Nonlinear Theory of Dynamic Systems with Dry Friction Forces, Vol. 108, Journal of Engineering for Gas Turbines and Power, July 1986.

Control of Distributed Structures

Leonard Meirovitch
Department of Engineering Science and Mechanics
Virginia Polytechnic Institute and State University
Blacksburg, VA 24061 USA

The motion of distributed structures is governed by partial differential equations to be satisfied inside a given domain defining the structure and by boundary conditions to be satisfied at points bounding this domain. In essence, distributed structures are infinite-dimensional systems, so that control of distributed structures presents problems not encountered in lumped-parameter systems. Indeed, for the most part, the control theory was developed for lumped-parameter systems, and many of the concepts are not applicable to distributed systems. The situation is considerable better in using modal control, which amounts to controlling a structure by controlling its modes. In this case, many of the concepts developed for lumped-parameter structures do carry over to distributed-parameter structures, as both types of structures can be described in terms of modal coordinates. The main difficulty arises in computing the control gains, as this implies infinite-dimensional gain matrices. This question can be obviated by using the independent modal-space control method, but this requires a distributed control force, which can be difficult to implement. Implementation by point actuators is possible, but this implies control of a reduced number of modes, i.e., modal truncation. Another approach to the control of distributed structures is direct feedback control, whereby the sensors are collocated with the actuators, and the actuator force at a given point depends only on the sensor signal at the same point. Here the difficulty is in deciding on suitable control gains. Damping tends to complicate matters, except for the case of proportional damping. Indeed, in this case modal controls can be designed with the same ease as for undamped structures.

1. Equation of Motion of a Distributed Structure

We are concerned with the problem of controlling a distributed structure whose behavior is governed by the partial differential equation (pde) (see [1])

$$\mathcal{L}w(P,t) + m(P)\ddot{w}(P,t) = f(P,t), \quad P\varepsilon D \tag{1}$$

where $w(P,t)$ is the displacement of a typical point P inside the domain D and at time t, $\mathcal{L}$ is a homogeneous differential operator of order 2p, and known as a stiffness operator, m(P) is the mass density and f(P,t) is a distributed control force. The solution w(P,t) of Eq. (1) is subject to the boundary conditions

$$B_i w(P,t) = 0, \quad P\varepsilon S, \quad i = 1,2,\ldots,p \tag{2}$$

Springer Series in Computational Mechanics
S. N. Atluri, A. K. Amos (Eds.)
Large Space Structures: Dynamics and Control

where S is the set of all points belonging to the boundary of D.

If f(P,t) depends explicitly on P and t, as indicated in Eq. (1), then the control is said to be open-loop. In this case, Eqs. (1) and (2) describe an ordinary response problem, which can be solved by modal analysis [1]. However, if the objective is to drive the motion of the structure to zero, it is unlikely that a function f(P,t) capable of achieving this objective can be found, so that open-loop control must be ruled out as an effective approach to the control of structures.

In view of the above, we wish to consider feedback control. By analogy with lumped-parameter systems, we can consider distributed control of the form

$$f(P,t) = f(w(P,t),\dot{w}(P,t)), \quad P \in D \tag{3}$$

where f is in general a nonlinear function of the displacement w(P,t) and velocity $\dot{w}(P,t)$. Inserting Eq. (3) into Eq. (1), we obtain the closed-loop pde for nonlinear control

$$\mathcal{L}w(P,t) + m(P)\ddot{w}(P,t) - f(w(P,t), \dot{w}(P,t)) = 0, \quad P \in D \tag{4}$$

It turns out that nonlinear distributed control is not feasible, so that we consider linear control in the form of proportional and rate feedback control, or

$$f(P,t) = - \mathfrak{G}(P)w(P,t) - \mathcal{H}(P)\dot{w}(P,t) \tag{5}$$

where $\mathfrak{G}(P)$ and $\mathcal{H}(P)$ are control gain operators. Inserting Eq. (5) into Eq. (1), we obtain the closed-loop pde for linear control

$$\mathcal{L}^*w(P,t) + \mathcal{H}\dot{w}(P,t) + m\ddot{w}(P,t) = 0, \quad P \in D \tag{6}$$

where

$$\mathcal{L}^* = \mathcal{L} + \mathfrak{G} \tag{7}$$

is a closed-loop stiffness operator. Hence, the problem of linear control of distributed structures reduces to the determination of operators $\mathfrak{G}$ and $\mathcal{H}$ such that the motion of the structures approaches zero asymptotically. Unfortunately, there are no algorithms capable of producing operators $\mathfrak{G}$ and $\mathcal{H}$. Later in this chapter, we consider an approach achieving the same result without requiring explicit expressions for $\mathfrak{G}$ and $\mathcal{H}$.

2. Modal Equations

Control of a structure using the formulation based on a pde is not feasible, so that a different approach is desirable. To this end, we propose to replace the pde, Eq. (1), by a set of ordinary differential equations (ode's) known as modal equations. This in turn, requires the solution of the eigenvalue problem.

The open-looop eigenvalue problem has the form

$$\mathcal{L}\phi(P) = \lambda m\phi(P), \quad P\varepsilon D \tag{8}$$

where $\phi(P)$ is subject to the boundary conditions

$$B_i\phi(P) = 0, \quad P\varepsilon S, \quad i = 1,2,\ldots,p \tag{9}$$

The solution of the eigenvalue problem consists of a denumerable infinite set of eigenvalues λ_r and associated eigenfunctions $\phi_r(P)$ $(r = 1,2,\ldots)$. We assume that the operator $\mathcal{L}$ is self-adjoint [1] and nonnegative definite, so that the eigenvalues are real and nonnegative. We denote them by $\lambda_r = \omega_r^2$ $(r = 1,2,\ldots)$, where ω_r are the natural frequencies. Moreover, the eigenfunctions are orthogonal and can be normalized so as to satisfy the orthonormality relations [1]

$$(\phi_s, m\phi_r) = \delta_{rs}, \quad (\phi_s, \mathcal{L}\phi_r) = \omega_r^2\delta_{rs}, \quad r,s = 1,2,\ldots \tag{10a,b}$$

where

$$(g,h) = \int_D gh\, dD \tag{11}$$

represents an inner product of the two functions g and h and δ_{rs} is the Kronecker delta.

Using the expansion theorem [1], the displacement w(P,t) can be expressed as the linear combination

$$w(P,t) = \sum_{r=1}^{\infty} \phi_r(P)q_r(t) \tag{12}$$

where $q_r(t)$ $(r = 1,2,\ldots)$ are generalized coordinates ordinarily known as <u>modal coordinates</u>. Similarly, we can expand the distributed force f(P,t) in the series

$$f(P,t) = \sum_{r=1}^{\infty} m(P)\phi_r(P)f_r(t) \tag{13}$$

where

$$f_r(t) = (\phi_r(P), f(P,t)), \quad r = 1,2,\ldots \tag{14}$$

are referred to as <u>modal forces</u>, or <u>modal controls</u>. Then, inserting Eqs. (12) and (13) into Eq. (1), multiplying through by $\phi_s(P)$, integrating over the domain D and considering Eqs. (10), we obtain

$$\ddot{q}_r(t) + \omega_r^2 q_r(t) = f_r(t), \quad r = 1,2,\ldots \tag{15}$$

which are the desired <u>modal equations</u>.

In control, it is customary to work with state equations instead of configuration equations such as Eqs. (15). To this end, we adjoin the identities $\dot{q}_r(t) = \dot{q}_r(t)$ to Eqs. (15), so that, introducing the rth modal state vector $\underset{\sim}{w}_r(t) = [q_r(t) \;\; \dot{q}_r(t)]^T$, Eqs. (15) can be replaced by

$$\dot{\underset{\sim}{w}}_r(t) = \Lambda_r \underset{\sim}{w}_r(t) + \underset{\sim}{B}_r f_r(t), \quad r = 1,2,\ldots \tag{16}$$

where

$$\Lambda_r = \begin{bmatrix} 0 & 1 \\ -\omega_r^2 & 0 \end{bmatrix}, \quad \underset{\sim}{B}_r = \begin{bmatrix} 0 \\ 1 \end{bmatrix}, \quad r = 1,2,... \tag{17}$$

are coefficient matrices. Equations (16) are known as modal state equations.

In state feedback control, it is necessary to measure the system output and to estimate the state from the output. Consistent with the modal state equations, we assume that the modal states are related to the modal measurements $y_r(t)$ by

$$y_r(t) = \underset{\sim}{C}_r^T \underset{\sim}{w}_r(t), \quad r = 1,2,... \tag{18}$$

where in the case of displacement measurements

$$\underset{\sim}{C}_r = [1\ 0]^T, \quad r = 1,2,... \tag{19a}$$

and in the case of velocity measurements

$$\underset{\sim}{C}_r = [0\ 1]^T, \quad r = 1,2,... \tag{19b}$$

The connection between modal measurements and actual measurements is discussed in Sec. 3.

3. Mode Controllability and Observability

The concept of state controllability was developed in conjunction with lumped-parameter systems [2]. In this section, we wish to examine the concept in the context of distributed systems. To this end, we refer to developments in [2] and introduce the modal controllability matrix

$$\mathbf{C}_r = [\underset{\sim}{B}_r \quad \Lambda_r \underset{\sim}{B}_r] = \begin{bmatrix} 0 & 1 \\ 1 & 0 \end{bmatrix}, \quad r = 1,2,... \tag{20}$$

in which we considered Eqs. (17). Equations (20) permit us to state that the distributed system is modal-state controllable if and only if each and every controllability matrix $\mathbf{C}_r$ is of full rank 2; this is clearly the case. The preceding statement implies, of course, that each and every modal control $f_r(t)$ is nonzero, in which case the application of the controllability criterion is a trivial formality. Note that an infinity of modal controls $f_r(t)$ is tantamount to an actual distributed control function $f(P,t)$, as indicated by Eq. (13).

Next, we consider the concept of observability [2]. Referring to developments in [2] and considering the modal output equations, Eqs. (18), we can introduce the modal observability matrix, defined as

$$\mathbf{O}_r = [\underset{\sim}{C}_r \quad \Lambda_r^T \underset{\sim}{C}_r], \quad r = 1,2,... \tag{21}$$

and state that the distributed system is modal-state observable if and only if each and every observability matrix $\mathbf{O}_r$ is of full rank 2. For dislacement measurements

$$O_r = \begin{bmatrix} 1 & 0 \\ 0 & 1 \end{bmatrix}, \quad r = 1,2,\ldots \tag{22a}$$

and for velocity measurements

$$O_r = \begin{bmatrix} 0 & -\omega_r^2 \\ 1 & 0 \end{bmatrix}, \quad r = 1,2,\ldots \tag{22b}$$

so that the system is in general observable with either displacement measurements or velocity measurements. Notable exceptions are semidefinite systems [1], which admit rigid-body modes with zero eigenvalues. Indeed, semidefinite systems are not observable with velocity measurements alone, as in this case the modal observability matrices corresponding to the rigid-body modes do not have rank 2.

At this point, let us return to the relation between the modal outputs and the actual outputs. Multiplying Eq. (12) by $m(P)\phi_s(P)$, integrating over the domain D and considering the orthonormality conditions (10a), we can write for displacement measurements

$$y_r(t) = q_r(t) = \int_D m(P)\phi_r(P)w(P,t)\, dD, \quad r = 1,2,\ldots \tag{23a}$$

and for velocity measurements

$$y_r(t) = \dot{q}_r(t) = \int_D m(P)\phi_r(P)\dot{w}(P,t)\, dD, \quad r = 1,2,\ldots \tag{23b}$$

From Eqs. (23) we conclude that an infinity of modal displacement observations implies a distributed displacement measurement $w(P,t)$ and an infinity of modal velocity observations implies a distributed velocity measurement $\dot{w}(P,t)$.

4. Closed-Loop Modal Equations

Let us return now to the closed-loop pde, Eq. (6). Retracing the steps leading from Eq. (1) to the modal equations, Eqs. (15), we obtain

$$\ddot{q}_s(t) + \sum_{r=1}^{\infty} h_{sr}\dot{q}_r(t) + \sum_{r=1}^{\infty} (\omega_s^2\delta_{sr} + g_{sr})q_r(t) = 0, \quad s = 1,2,\ldots \tag{24}$$

where

$$g_{sr} = (\phi_s, \mathfrak{G}\phi_r), \; h_{sr} = (\phi_s, \mathcal{H}\phi_r), \quad r,s = 1,2,\ldots \tag{25a,b}$$

Equations (24) represent the closed-loop modal equations. In contrast, Eqs. (15) in which the modal forces $f_r(t)$ depend explicitly on the time t and not on $q_r(t)$ and/or $\dot{q}_r(t)$ are referred to as open-loop modal equations. The open-loop modal equations, Eqs. (15) are independent. In general, $g_{sr} \neq 0$ and $h_{sr} \neq 0$ for $s \neq r$, so that Eqs. (24) represent an infinite set of simultaneous second-order ode's. Hence, the effect of feedback control is to couple the open-loop modal equations. Physically, the term $g_{sr}q_r(t)$ implies a generalized spring force and the term $h_{sr}\dot{q}_r(t)$ a generalized viscous damper force. The fact that the cross-product terms are not zero implies that the feedback control provides nonproportional augmenting stiffness and nonproportional impressed viscous damping [1], respectively. Because feedback control in which $g_{sr} \neq 0$ and $h_{sr} \neq 0$ for $s \neq r$ destroys the independence of the open-loop modal equations, we refer to this case as coupled control.

The closed-loop modal equations can be written in the compact form

$$\ddot{\underset{\sim}{q}}(t) + H\dot{\underset{\sim}{q}}(t) + (\Omega^2 + G)\underset{\sim}{q}(t) = \underset{\sim}{0} \tag{26}$$

where $\underset{\sim}{q}(t) = [q_1(t)\ q_2(t)\ \ldots]^T$ is the infinite-dimensional configuration vector, $\Omega = \text{diag}\ [\omega_r]$ is the infinite-order diagonal matrix of natural frequencies and $G = [g_{sr}]$ and $H = [h_{sr}]$ are square control gain matrices of infinite order. Note that in the general case the matrices G and H are not only nondiagonal but they are unlikely to be even symmetric.

Before the behavior of the closed-loop system can be established, it is necessary to determine the control gain operators $\mathcal{G}$ and $\mathcal{H}$ or the associated gain matrices G and H. However, there are no algorithms capable of producing the operators $\mathcal{G}$ and $\mathcal{H}$ or the infinite-order matrices G and H. Hence, distributed feedback control realized through coupled control is not possible.

For future reference, we would like to cast Eq. (26) in state form. To this end, we introduce the 2∞-dimensional modal state vector $\underset{\sim}{w}(t) = [\underset{\sim}{q}^T(t) \mid \dot{\underset{\sim}{q}}^T(t)]^T$, so that, adjoining that identity $\dot{\underset{\sim}{q}}(t) = \dot{\underset{\sim}{q}}(t)$, Eq. (26) can be rewritten in the state form

$$\dot{\underset{\sim}{w}}(t) = A\underset{\sim}{w}(t) \tag{27}$$

where

$$A = \left[\begin{array}{c|c} 0 & I \\ \hline -(\Omega^2 + G) & -H \end{array}\right] \tag{28}$$

is the coefficient matrix of order 2∞, in which I is the identity matrix of infinite order. The problem of determining the control gain matrices G and H remains. In this regard, we could consider pole allocation or optimal control. As shown in [2], in the pole allocation method the problem reduces to the solution of a set of nonlinear algebraic equations, which is not feasible for infinite-dimensional systems. Similarly, for optimal control using a quadratic performance index, one is faced with the solution of a matrix Riccati equation of order 2∞, which is not possible.

5. Independent Modal-Space Control

There is one case in which distributed feedback control is possible, namely the one in which the operators $\mathcal{G}$ and $\mathcal{H}$ satisfy the eigenvalue problems

$$\mathcal{G}\phi_r(P) = g_r m(P)\phi_r(P),\quad \mathcal{H}\phi_r(P) = h_r m(P)\phi_r(P),\quad r = 1,2,\ldots \tag{29a,b}$$

which imply that $\mathfrak{S}$ and $\mathcal{H}$ are such that

$$(\phi_s, \mathfrak{S}\phi_r) = g_r\delta_{rs}, \quad (\phi_s, \mathcal{H}\phi_r) = h_r\delta_{rs}, \quad r,s = 1,2,... \tag{30a,b}$$

In this case the closed-loop modal equations reduce to the independent set

$$\ddot{q}_s(t) + h_s\dot{q}_s(t) + (\omega_s^2 + g_s)q_s(t) = 0, \quad s = 1,2,... \tag{31}$$

Because of the independence of the closed-loop modal equations, this type of control is called independent modal-space control (IMSC). It is characterized by modal control forces of the form

$$f_s(t) = - g_s q_s(t) - h_s\dot{q}_s(t), \quad s = 1,2,... \tag{32}$$

In open-loop response problems, the coordinates $q_s(t)$ corresponding to independent equations of motion are called natural [1]. Because IMSC guarantees the independence of the closed-loop equations, we refer to IMSC as natural control.

The fact that both the open-loop and closed-loop modal equations are independent has very important implications. Indeed, this implies that the open-loop eigenfunctions ϕ_s are closed-loop eigenfunctions as well. Hence, in natural control, the control effort is directed entirely to altering the eigenvalues, leaving the eigenfunctions unaltered. In this regard, it should be recalled that the stability of a linear system is determined by the system eigenvalues, with the eigenfunctions playing no role, so that in natural control no control effort is used unnecessarily.

One question remaining is how to determine the modal gains g_s and h_s (s = 1,2,...). Two of the most widely used techniques are pole allocation and optimal control:

i. Pole allocation

In the pole allocation method, the closed-loop poles are selected in advance and the gains are determined so as to produce these poles. In the IMSC, the procedure is exceedingly simple. Denoting the closed-loop eigenvalue associated with the sth mode by $-\alpha_s + i\beta_s$, the solution of Eqs. (31) can be written as

$$q_s(t) = c_s e^{(-\alpha_s + i\beta_s)t}, \quad s = 1,2,... \tag{33}$$

Inserting Eqs. (33) into Eqs. (31) and separating the real and imaginary parts, we obtain the modal gains

$$g_s = \alpha_s^2 + \beta_s^2 - \omega_s^2, \quad h_s = 2\alpha_s, \quad s = 1,2,... \tag{34}$$

To guarantee asymptotic stability, however, it is only necessary to impart the open-loop eigenvalues some negative real part and it is not necessary to alter the frequencies. This can be achieved by letting $\beta_s = \sqrt{\lambda_s} = \omega_s$ (s = 1,2,...), where ω_s is the sth natural frequency of the open-loop system. Hence, the frequency-preserving control gains are

$$g_s = \alpha_s^2, \; h_s = 2\alpha_s, \quad s = 1,2,... \tag{35}$$

ii. Optimal control

In optimal control, the closed-loop poles are determined by minimizing a given performance index. Consistent with previous developments, we are interested in constant gains and, to this end, we consider the performance functional [3]

$$J = \int_0^\infty [(\dot{w},m\dot{w}) + (w,\mathcal{L}w) + (f,rf)]dt \tag{36}$$

where the various quantities are as defined in Eq. (1), except for $r = r(P)$ which is a weighting function assumed to satisfy [3]

$$(f,rf) = \sum_{r=1}^{\infty} R_r f_r^2 \tag{37}$$

where R_r are modal weights. Inserting Eqs. (12) and (37) into Eq. (36) and recalling Eqs. (10), we obtain

$$J = \sum_{r=1}^{\infty} J_r \tag{38}$$

where

$$J_r = \int_0^\infty (\dot{q}_r^2 + \omega_r^2 q_r^2 + R_r f_r^2)dt, \quad r = 1,2,... \tag{39}$$

are modal performance indices. Because in IMSC the modal control f_r is independent of any other modal control, it follows that

$$\min J = \min \sum_{r=1}^{\infty} J_r = \sum_{r=1}^{\infty} \min J_r \tag{40}$$

so that the minimization can be carried out independently for each mode.

The minimization of J_r leads to a 2×2 matrix Riccati equation that in the steady-state case can be solved in closed form, yielding the modal control gains [3]

$$\begin{aligned} g_r &= -\omega_r^2 + \omega_r(\omega_r^2 + R_r^{-1})^{1/2} \\ h_r &= [R_r^{-1} - 2\omega_r^2 + 2\omega_r(\omega_r^2 + R_r^{-1})^{1/2}]^{1/2} \end{aligned} \quad , \; r = 1,2,... \tag{41}$$

Because no constraint has been imposed on the control function $f = f(P,t)$, the solution defined by Eqs. (13), (32) and (41) is globally optimal, and is unique because the solution to the linear optimal control problem is unique [3].

It should be pointed out that the solution presented above requires distributed sensors and actuators. Indeed, inserting Eqs. (32) into Eq. (13), we obtain the distributed feedback control force

$$f(P,t) = -\sum_{r=1}^{\infty} m(P)\phi_r(P)[g_r q_r(t) + h_r \dot{q}_r(t)] \tag{42}$$

Equation (42) indicates that control implementation requires the entire infinity of modal displacements $q_r(t)$ and modal velocities $\dot{q}_r(t)$ $(r = 1,2,...)$ for feedback. This, in turn, implies a distributed sensor, as can be concluded from Eqs. (23). Note that, inserting Eq. (12) into Eq. (5) and comparing the results with Eq. (42), we can verify Eqs. (29). At this point, we observe that the gain operators $\mathcal{G}$ and $\mathcal{H}$ are never determined explicitly, nor is it necessary to do so, as the determination of the modal gains g_r and h_r $(r = 1,2,...)$ is sufficient to produce the feedback control density function $f(P,t)$.

6. Control by Point Actuators

As pointed out in Sec. 5, globally optimal control of a distributed structure requires a distributed actuator. On the assumption that distributed actuation is not feasible, we seek control by means of a finite number m of discrete actuators acting at the points $P = P_i$ $(i = 1,2,...,m)$ of the structure.

Discrete actuators can be treated as distributed by writing

$$f(P,t) = \sum_{i=1}^{m} F_i(t)\delta(P - P_i), \quad P\varepsilon D \tag{43}$$

where $F_i(t)$ are force amplitudes and $\delta(P - P_i)$ are spatial Dirac delta functions. Introducing Eq. (43) into Eq. (1), we obtain

$$\mathcal{L}w(P,t) + m(P)\ddot{w}(P,t) = \sum_{i=1}^{m} F_i(t)\delta(P - P_i), \quad P\varepsilon D \tag{44}$$

Control of the structure in terms of pde's is not feasible, however, so that we once again transform the pde, Eq. (44) in this case, into a set of modal equations. Inserting Eq. (12) into Eq. (44), multiplying by $\phi_s(P)$, integrating over the domain D and considering the orthonormality relations, Eqs. (10), we have

$$\ddot{q}_r(t) + \omega_r^2 q_r(t) = \sum_{i=1}^{m} \phi_r(P_i)F_i(t), \quad r = 1,2,... \tag{45}$$

Equations (45) can be written in the matrix form

$$\ddot{\underset{\sim}{q}}(t) + \Omega^2 \underset{\sim}{q}(t) = \Phi \underset{\sim}{F}(t) \tag{46}$$

where

$$\Phi = [\phi_{ri}] = [\phi_r(P_i)] \tag{47}$$

is an ∞xm matrix known as the modal participation matrix. As in Sec. 4, Eq. (46) can be written in the modal state form

$$\dot{\underset{\sim}{w}}(t) = A\underset{\sim}{w}(t) + B\underset{\sim}{F}(t) \tag{48}$$

where

$$A = \begin{bmatrix} 0 & I \\ -\Omega^2 & 0 \end{bmatrix}, \quad B = \begin{bmatrix} 0 \\ \Phi \end{bmatrix} \tag{49a,b}$$

are $2\infty \times 2\infty$ and $2\infty \times m$ coefficient matrices and $\underset{\sim}{F}(t)$ is the m-dimensional vector of actuator forces. For linear feedback control, the control vector is related to the modal state vector according to

$$\underset{\sim}{F}(t) = -G\underset{\sim}{w}(t) \tag{50}$$

where G is an $m \times 2\infty$ control gain matrix. No confusion should arise from using the same notation for the $\infty \times \infty$ gain matrix defined in Sec. 4. Determination of infinite-dimensional gain matrices is not possible, so that control of the entire infinity of modes is not feasible, nor is it necessary. Indeed, higher modes have only minimal participation in the motion, as they are difficult to excite. Moreover, various assumptions made in deriving the equation of motion of a structure limit the validity of the theory to lower modes only. Hence, we propose to control only a limited number of lower modes, where the number is sufficiently large that the accuracy will not suffer very much.

In view of the above, we propose to control n modes only. Ordinarily, the lower modes are the ones in need of control. Hence, we assume that the displacement can be aproximated by

$$w(P,t) \cong \sum_{r=1}^{n} \phi_r(P)q_r(t) \tag{51}$$

so that, retracing the steps leading to Eq. (48), we obtain

$$\dot{\underset{\sim}{w}}_c(t) = A_c\underset{\sim}{w}_c(t) + B_c\underset{\sim}{F}(t) \tag{52}$$

where

$$A_c = \begin{bmatrix} 0 & I_c \\ -\Omega_c^2 & 0 \end{bmatrix}, \quad B_c = \begin{bmatrix} 0 \\ \Phi_c \end{bmatrix} \tag{53a,b}$$

are $2n \times 2n$ and $2n \times m$ matrices, respectively, in which the notation is obvious. Equation (52) represents a 2n-order discrete system, and control of the system can be carried out by one of the methods for the control of lumped systems [2]. Of course, in this case Eq. (50) must be replaced by

$$\underset{\sim}{F}(t) = -G_c\underset{\sim}{w}_c(t) \tag{54}$$

where G_c is an $m \times 2n$ control gain matrix.

There remains the question of modal state estimation. To this end, we consider point sensors and let

$$y_i(t) = w(P_i,t) = \sum_{r=1}^{n} \phi_r(P_i)q_r(t), \quad i = 1,2,\ldots,s \tag{55a}$$

be the outputs from s sensors at points $P = P_i$ $(i = 1,2,\ldots,s)$ if displacement sensors are used and

$$y_i(t) = \dot{w}(P_i,t) = \sum_{r=1}^{n} \phi_r(P_i)\dot{q}_r(t), \quad i = 1,2,\ldots,s \tag{55b}$$

if velocity sensors are used. Introducing the notation

$$\Phi_e = [\phi_{ir}] = [\phi_r(P_i)], \quad i = 1,2,\ldots,s; \; r = 1,2,\ldots,n \tag{56}$$

where we note that the entries in Φ_e are transposed relative to the entries in Φ_c (disregarding the dimensions of the matrices, which are in general different), we can write Eqs. (55) in the compact form

$$\underset{\sim}{y}(t) = C_e\underset{\sim}{w}_c(t) \tag{57}$$

where in the case of displacement sensors

$$C_e = [\Phi_e \,\vdots\, 0] \tag{58a}$$

and in the case of velocity sensors

$$C_e = [0 \,\vdots\, \Phi_e] \tag{58b}$$

To estimate the full controlled modal state $\underset{\sim}{w}_c(t)$ from the output $\underset{\sim}{y}(t)$, we can consider a Luenberger observer or a Kalman-Bucy filter [2]. In either case, we can write the modal observer in the form

$$\dot{\hat{\underset{\sim}{w}}}_c(t) = A_c\hat{\underset{\sim}{w}}_c(t) + B_c\underset{\sim}{F}(t) + K(t)[\underset{\sim}{y}(t) - C_e\hat{\underset{\sim}{w}}_c(t)] \tag{59}$$

where $\hat{\underset{\sim}{w}}_c(t)$ is the estimated controlled modal state. In the case of the Luenberger observer, the observer gain matrix K(t) can be obtained by a pole allocation technique. In the case of the Kalman-Bucy filter, K(t) can be determined optimally by solving a matrix Riccati equation, which assumes that the noise intensities associated with the actuators and sensors are known [2]. Upon obtaining the estimated controlled modal state from Eq. (59), we compute the feedback control forces by writing

$$\underset{\sim}{F}(t) = -G_c\hat{\underset{\sim}{w}}_c(t) \tag{60}$$

In the above, we treated the distributed system as if it were discrete. In the process, we ignored the uncontrolled modes. At this point, we wish to examine the effect of the control forces on the uncontrolled modes. To this end, we refer to the uncontrolled modes as residual and denote them by the subscript R. Hence, we define

$$\underset{\sim}{w}(t) = [\underset{\sim}{w}_C^T(t) \mid \underset{\sim}{w}_R^T(t)]^T, \quad A = \left[\begin{array}{c|c} A_C & 0 \\ \hline 0 & A_R \end{array}\right], \quad B = \left[\begin{array}{c} B_C \\ \hline B_R \end{array}\right] \tag{61a,b,c}$$

so that, inserting Eqs. (61) into Eq. (48) and considering Eqs. (60), we obtain

$$\dot{\underset{\sim}{w}}_C(t) = A_C\underset{\sim}{w}_C(t) - B_C G_C \hat{\underset{\sim}{w}}_C(t) \tag{62a}$$

$$\dot{\underset{\sim}{w}}_R(t) = A_R\underset{\sim}{w}_R(t) - B_R G_C \hat{\underset{\sim}{w}}_C(t) \tag{62b}$$

Moreover, inserting Eqs. (57) and (60) into Eq. (59), we can write the observer equation in the form

$$\dot{\hat{\underset{\sim}{w}}}_C(t) = (A_C - B_C G_C)\hat{\underset{\sim}{w}}_C(t) + K(t)C_e[\underset{\sim}{w}_C(t) - \hat{\underset{\sim}{w}}_C(t)] \tag{63}$$

Next, we introduce the error vector

$$\underset{\sim}{e}_C(t) = \hat{\underset{\sim}{w}}_C(t) - \underset{\sim}{w}_C(t) \tag{64}$$

so that Eqs. (62) and (63) can be rearranged as

$$\dot{\underset{\sim}{w}}_C(t) = (A_C - B_C G_C)\underset{\sim}{w}_C(t) - B_C G_C \underset{\sim}{e}_C(t) \tag{65a}$$

$$\dot{\underset{\sim}{w}}_R(t) = -B_R G_C \underset{\sim}{w}_C(t) + A_R \underset{\sim}{w}_R(t) - B_R G_C \underset{\sim}{e}_C(t) \tag{65b}$$

$$\dot{\underset{\sim}{e}}_C(t) = (A_C - KC_e)\underset{\sim}{e}_C(t) \tag{65c}$$

where K(t) was assumed to be constant. Equations (65) can be written in the matrix form

$$\left[\begin{array}{c} \dot{\underset{\sim}{w}}_C(t) \\ \hline \dot{\underset{\sim}{w}}_R(t) \\ \hline \dot{\underset{\sim}{e}}_C(t) \end{array}\right] = \left[\begin{array}{c|c|c} A_C - B_C G_C & 0 & -B_C G_C \\ \hline -B_R G_C & A_R & -B_R G_C \\ \hline 0 & 0 & A_C - KC_e \end{array}\right] \left[\begin{array}{c} \underset{\sim}{w}_C(t) \\ \hline \underset{\sim}{w}_R(t) \\ \hline \underset{\sim}{e}_C(t) \end{array}\right] \tag{66}$$

Because of the block-triangular nature of the coefficient matrix, the eigenvalues of the closed-loop system are determined by the submatrices $A_C - B_C G_C$, A_R and $A_C - KC_e$. The term $-B_R G_C$ is responsible for the excitation of the residual modes by the control forces and is known as <u>control spillover</u>. Because the term has no effect on the eigenvalues of the closed-loop system, we conclude that <u>control spillover cannot destablize the system</u>, although it can cause some degradation in the system performance. As in the case of lumped systems, we observe that the <u>separation principle</u> is valid here as well [2].

Equation (57) implies that the sensors measure only the contribution of the controlled modes to the motion of the structure. In reality, however, the sensor signals will include contributions from all the modes, so that the proper expression for the output vector is

$$\underset{\sim}{y}(t) = C\underset{\sim}{w}(t) = C_e\underset{\sim}{w}_C(t) + C_R\underset{\sim}{w}_R(t) \tag{67}$$

where the notation is obvious. In this case, the observer equation, Eq. (63), must be replaced by

$$\dot{\hat{\underset{\sim}{w}}}_c(t) = (A_c - B_cG_c)\hat{\underset{\sim}{w}}_c(t) + KC_e[\underset{\sim}{w}_c(t) - \hat{\underset{\sim}{w}}_c(t)] + KC_R\underset{\sim}{w}_R(t) \tag{68}$$

so that the error equation, Eq. (65c), becomes

$$\dot{\underset{\sim}{e}}_c(t) = (A_c - KC_e)\underset{\sim}{e}_c(t) + KC_R\underset{\sim}{w}_R(t) \tag{69}$$

Combining Eq. (69) with Eqs. (65a) and (65b), we can write

$$\begin{bmatrix} \dot{\underset{\sim}{w}}_c(t) \\ \hline \dot{\underset{\sim}{w}}_R(t) \\ \hline \dot{\underset{\sim}{e}}_c(t) \end{bmatrix} = \left[\begin{array}{c|c|c} A_c - B_cG_c & 0 & -B_cG_c \\ \hline -B_RG_c & A_R & -B_RG_c \\ \hline 0 & KC_R & A_c - KC_e \end{array}\right] \begin{bmatrix} \underset{\sim}{w}_c(t) \\ \hline \underset{\sim}{w}_R(t) \\ \hline \underset{\sim}{e}_c(t) \end{bmatrix} \tag{70}$$

This time, however, the separation principle is no longer valid, as the term KC_R causes the closed-loop system eigenvalues to be affected by the observer. This effect is known as observation spillover and can produce instability in the residual modes [4]. This is particularly true if the actuators and sensors are not collocated. The residual modes are particularly vulnerable because, in the absence of observation spillover, the eigenvalues associated with these modes lie on the imaginary axis and have no stability margin. Note, however, that a small amount of damping inherent in the structure is often sufficient to overcome the observation spillover effect [5]. At any rate, observation spillover can be eliminated if the sensor signals are prefiltered so as to screen out the contribution of the uncontrolled modes.

The effect of the observation spillover can be greatly diminished if a large number of sensors is used. Indeed, if the displacement $w(P_i,t)$ is measured at a large number of points P_i, it is possible to reconstruct an approximate displacement profile $\hat{w}(P,t)$ through spatial interpolation. Then, using Eqs. (23a), we can estimate the controlled modal coordinates as follows:

$$\hat{q}_r(t) = \int_D m(P)\phi_r(P)\hat{w}(P,t)\, dD, \quad r = 1,2,\ldots,c \tag{71}$$

Equations (64) are referred to as modal filters [6]. Modal velocity estimates can be produced by measuring velocities $\dot{w}(P_i,t)$ and generating the velocity profile $\dot{w}(P,t)$ through interpolation and inserting this velocity profile into the modal fiters, Eqs. (71). Alternatively, it is possible to estimate modal velocities from modal displacements by using modal Luenberger observers or modal Kalman-Bucy filters [2].

In the above discussion, we treated the vector $\underset{\sim}{w}_R(t)$ of residual modal coordinates as if it had finite dimension when in fact it is infinite-dimensional. In practice, however, only a finite number of modes can be excited, so that $\underset{\sim}{w}_R(t)$ can be treated as if it were finite-dimensional without incurring much error.

7. Direct Feedback Control

One problem that can prove troublesome in modal control of distributed structures is the estimation of the modal states for feedback. As pointed out in Sec. 6, the problem of observation spillover is potentially more serious than the problem of control spillover, as it can lead to instability. Hence, a procedure not requiring modal state estimation appears desirable.

One approach not requiring modal state estimation is direct feedback control, whereby the sensors are collocated with the actuators and a given actuator force is a function of the sensor output at the same point. For simplicity, we assume that the control law is linear, although nonlinear control laws can be used. We consider m discrete actuators acting at the points $P = P_i$ $(i = 1,2,\dots,m)$, where the force amplitudes are

$$F_i(t) = -g_i w(P_i,t) - h_i \dot{w}(P_i,t), \quad i = 1,2,\dots,m \tag{72}$$

in which g_i and h_i $(i = 1,2,\dots,m)$ are actual control gains. Clearly, the gains must be positive. As in Sec. 6, we can treat the actuators as distributed by writing

$$f(P,t) = -\sum_{i=1}^{m} [g_i w(P,t) + h_i \dot{w}(P,t)]\delta(P - P_i), \quad P\varepsilon D \tag{73}$$

To make the connection with the approach of Sec. 1, we can regard the operators $\mathfrak{G}$ and $\mathcal{H}$ defined in Eq. (5) as having the expressions

$$\mathfrak{G}(P) = \sum_{i=1}^{m} g_i \delta(P - P_i), \mathcal{H}(P) = \sum_{i=1}^{m} h_i \delta(P - P_i), \quad P\varepsilon D \tag{74a,b}$$

The closed-loop pde can be obtained by inserting Eqs. (74) into Eqs. (6) and (7). We shall not pursue this approach, but turn our attention to the modal equations.

We showed in Sec. 4 that the closed-loop modal equations have the form

$$\ddot{q}_s(t) + \sum_{r=1}^{\infty} h_{sr}\dot{q}_r(t) + \sum_{r=1}^{\infty} (\omega_s^2\delta_{sr} + g_{sr})q_r(t) = 0, \quad s = 1,2,\dots \tag{75}$$

where

$$g_{sr} = (\phi_s, \mathfrak{G}\phi_r), \; h_{sr} = (\phi_s, \mathcal{H}\phi_r), \quad r,s = 1,2,\dots \tag{76a,b}$$

Introducing Eqs. (74) into Eqs. (76), we obtain

$$g_{sr} = \sum_{i=1}^{m} g_i \phi_r(P_i)\phi_s(P_i), \; h_{sr} = \sum_{i=1}^{m} h_i \phi_r(P_i)\phi_s(P_i), \quad r,s = 1,2,\dots \tag{77a,b}$$

The modal equations can be written in the compact form

$$\ddot{\underset{\sim}{q}}(t) + H\dot{\underset{\sim}{q}}(t) + (\Omega^2 + G)\underset{\sim}{q}(t) = \underset{\sim}{0} \tag{78}$$

where $\underset{\sim}{q}(t)$ is the infinite-dimensional configuration vector, $\Omega = \text{diag}\,[\omega_r]$ is the infinite-order diagonal matrix of natural frequencies and

$$G = \sum_{i=1}^{m} g_i \underset{\sim}{\phi}(P_i)\underset{\sim}{\phi}^T(P_i), \; H = \sum_{i=1}^{m} h_i \underset{\sim}{\phi}(P_i)\underset{\sim}{\phi}^T(P_i) \tag{79a,b}$$

are square control gain matrices of infinite order, where $\underline{\phi}(P_i) = [\phi_1(P_i)\ \ \phi_2(P_i)\ \ldots]^T$ is an infinite-dimensional vector of eigenfunctions evaluated at $P = P_i$. Because g_i and h_i are positive, it follows that the matrices G and H are positive semidefinite. In fact, because the matrices $\underline{\phi}(P_i)\underline{\phi}^T(P_i)$ represent outer products of given vectors, the matrices have rank one.

From Eq. (78), we conclude that the effect of displacement feedback is to increase the structure stiffness, thus increasing the natural frequencies. On the other hand, the effect of velocity feedback is to provide viscous damping. Assuming that the damping is pervasive, i.e., it couples all the modal equations of motion, Eq. (78), the net effect is to impart negative real part to the open-loop eigenvalues, which implies the assumption that the damping is sufficiently small that the closed-loop eigenvalues are complex. Hence, if the object is to ensure asymptotic stability, then velocity feedback is sufficient. On the other hand, if the object is to alter the structure natural frequencies, then displacement feedback is all that is necessary. Ordinarily, the object in control is asymptotic stability, so that displacement feedback is not really necessary. Notable exceptions are unrestrained structures, for which the operator is only positive semidefinite. Such structures are referred to as semidefinite and admit rigid-body modes characterized by zero natural frequencies. Semidefinite structures cannot be stabilized with velocity feedback alone and displacement feedback is essential.

Once again the problem is that of determining the control gains. The problem is different here because there is only a finite number of gains g_i and h_i ($i = 1,2,\ldots,p$) and the system is infinite-dimensional. There is no computational algorithm permitting the computation of the control gains in conjunction with either pole allocation or optimal control, so that one must consider modal truncation. Even for the truncated model, the situation remains questionable. The reason for this is that pole allocation and optimal control most likely will require gain matrices with entries independent of each other while direct feedback control implies that the entries of G and H are not independent, as can be seen from Eqs. (79). In fact, there is some question whether arbitrary pole placement is possible for direct feedback control. This statement can be explained by the fact that, for given preselected closed-loop poles, it is not possible to guarantee that all the control gains h_i ($i = 1,2,\ldots,m$) are positive, thus causing some of the uncontrolled modes to become unstable due to control spillover [7]. Moreover, because the entries of G and H are not independent, there is some question whether optimal control is possible in the presence of constraints on the control gains.

8. Systems with Proportional Damping

In the preceding discussion the assumption that the system possessed no inherent damping was made implicitly. At this point we wish to relax this assumption and include the damping effect. To this end, we reconsider Eq. (1) and assume that the motion of the distributed structure is described by the pde

$$\mathcal{L}w(p,t) + \mathcal{C}\dot{w}(P,t) + m(P)\ddot{w}(P,t) = f(P,t), \quad P \varepsilon D \tag{80}$$

where $\mathcal{C}$ is a damping operator. The reamining quantities are as defined in Sec. 1. Then, following the approach of Sec. 1, we conclude that the closed-loop pde for damped systems has the form

$$\mathcal{L}^* w(P,t) + \mathcal{H}^* \dot{w}(P,t) + m\ddot{w}(P,t) = 0, \quad P \varepsilon D \tag{81}$$

where

$$\mathcal{H}^* = \mathcal{C} + \mathcal{H} \tag{82}$$

is a closed-loop damping operator.

As pointed out in Sec. 2, designing controls for a structure based on a pde is not feasible, so that once again we propose to investigate the use of modal equations. An attempt to use the pattern of Sec. 2 will reveal difficulties from the onset. Indeed, the open-loop eigenvalue problem no longer has the simple form given by Eqs. (8) and (9), so that in general the solution is no longer real. In fact, a closed-form solution of the differential eigenvalue problem for damped system is generally not feasible. A notable exception is the case in which damping is of the proportional type. This implies that the damping operator $\mathcal{C}$ is a linear combination of the stiffness operator $\mathcal{L}$ and the mass density m, or

$$\mathcal{C} = am + b\mathcal{L} \tag{83}$$

where a and b are proportionality constants of appropriate dimensions. In this case, the eigenfunctions of the associated undamped system can still be used to decouple the open-loop equations. Indeed, using the expansion theorem, Eq. (12), and retracing the steps of Sec. 2, we obtain the modal equations

$$\ddot{q}_r(t) + 2\zeta_r\omega_r\dot{q}_r(t) + \omega_r^2 q_r(t) = f_r(t), \quad r = 1,2,... \tag{84}$$

where

$$\zeta_r = \frac{1}{2\omega_r}(a + b\omega_r^2), \quad r = 1,2,... \tag{85}$$

are modal damping factors. The remaining quantities in Eqs. (84) are as defined in Sec. 2. The modal state equations retain the general form

$$\dot{\underset{\sim}{w}}_r(t) = \Lambda_r \underset{\sim}{w}_r(t) + \underset{\sim}{B}_r f_r(t), \quad r = 1,2,... \tag{86}$$

but here the coefficient matrix Λ_r has the form

$$\Lambda_r = \begin{bmatrix} 0 & 1 \\ -\omega_r^2 & -2\zeta_r\omega_r \end{bmatrix}, \quad r = 1,2,... \tag{87}$$

From Sec. 3, we conclude that the controllability matrix becomes

$$\mathbf{C}_r = [\underset{\sim}{B}_r \quad \Lambda_r \underset{\sim}{B}_r] = \begin{bmatrix} 0 & 1 \\ 1 & -2\zeta_r\omega_r \end{bmatrix}, \quad r = 1,2,\ldots \tag{88}$$

Moreover, the observability matrix for displacement measurements retains the form (22a), but for velocity measurements it becomes

$$\mathbf{O}_r = \begin{bmatrix} 0 & \omega_r^2 \\ 1 & -2\zeta_r\omega_r \end{bmatrix} \tag{89}$$

Hence, all the conclusion reached in Sec. 3 concerning modal-state controllability and observability remain valid here.

Following the pattern of Sec. 4, the closed-loop modal state equations can be shown to retain the form (27), but the coefficient matrix now is

$$A = \left[\begin{array}{c|c} 0 & I \\ \hline -(\Omega^2 + G) & -(2\zeta\Omega + H) \end{array}\right] \tag{90}$$

where $\zeta\Omega = \text{diag}\,(\zeta_r\omega_r)$.

If the control gain operators and satisfy Eqs. (29a,b), then the closed-loop modal equations are still independent, but this time they have the form

$$\ddot{q}_s(t) + (2\zeta_s\omega_s + h_s)\dot{q}_s(t) + (\omega_s^2 + g_s)q_s(t) = 0, \quad s = 1,2,\ldots \tag{91}$$

In using IMSC, the poles can be placed with the same ease as in Sec. 5 and the modal control gains corresponding to the closed-loop pole $-\alpha_s + i\beta_s$ can be shown to have the expressions

$$g_2 = \alpha_s^2 + \beta_s^2 - \omega_s^2, \; h_s = 2\alpha_s - 2\zeta_s\omega_s, \quad s = 1,2,\ldots \tag{92a,b}$$

from which we conclude that damping affects only the rate feedback. Consistent with physical intuition, Eq. (92b) shows that internal damping reduces the need for rate feedback. Similarly, using the same performance index as in Sec. 5, it is not difficult to show that optimal IMSC is characterized by the modal control gains

$$\begin{aligned} g_r &= -\omega_r^2 + \omega_r(\omega_r^2 + R_r^{-1})^{1/2}, \\ h_r &= -2\zeta_r\omega_r + [4\zeta_r^2\omega_r^2 + R_r^{-1} - 2\omega_r^2 + 2\omega_r(\omega_r^2 + R_r^{-1})^{1/2}]^{1/2}, \end{aligned} \quad r = 1,2,\ldots \tag{93}$$

Once again, we observe that damping affects the rate feedback only and, morever, it reduces the need for rate feedback.

Damping has another beneficial effect, in addition to reducing the need for rate feedback control. Indeed, we recall from Sec. 6 that damping increases the stability margin to the extent that the possibility of observation spillover instability is virtually eliminated.

When damping is not of the proportional type, but it is relatively small, it is possible to treat it as if it were proportional by simply ignoring the coupling terms in the modal equations [1].

9. Conclusions

Control of structures can be carried out conveniently by controlling the natural modes of the structure. Mode controllability and observability dictate that control be carried out by distributed actuators and state estimation by distributed sensors. Practical considerations, however, demand that control be carried out by point actuators and state estimation by point sensors. This, in turn, requires modal truncation, giving rise to phenomena known as control and observation spillover. A method not requiring state estimation is direct feedback control. Although the control gains can be chosen so that most modes are controlled, and no mode is destabilized, problems with an optimal choice of gains remain in direct feedbck control. Damping tends to affect the rate feedback only, and it reduces the need for control.

References

1. Meirovitch, L., Computational Methods in Structural Dynamics, Sijthoff & Noordhoff, The Netherlands, 1980.

2. Meirovitch, L., Dynamics and Control of Structures, Wiley-Interscience, New York, 1988.

3. Meirovitch, L. and Silverberg, L. M., "Globally Optimal Control of Self-Adjoint Distributed Systems," Optimal Control and Applications (Special issue dedicated to Professor Angelo Miele), Vol. 4, No. 4, 1983, pp. 365-386.

4. Balas, M. J., "Active Control of Flexible Systems," Journal of Optimization Theroy and Applications, Vol. 25, No. 3, 1978, pp. 415-436.

5. Meirovitch, L. and Baruh, H., "On the Problem of Observation Spillover in Distributed-Parameter Systems," Journal of Optimization Theory and Applications, Vol. 39, No. 2, 1983, pp. 269-291.

6. Meirovitch, L. and Baruh, H., "On the Implementation of Modal Filters for Control of Structures," Journal of Guidance, Control, and Dynamics, Vol. 8, No. 6, 1985, pp. 707-716.

7. Meirovitch, L., "Some Problems Associated with the Control of Distributed Structures," Journal of Optimization Theory and Applications, Vol. 54, No. 1, 1987.

The Acoustic Limit of Control of Structural Dynamics

A. H. von Flotow
Department of Aeronautics and Astronautics
Massachusetts Institute of Technology
Cambridge, MA, 02139, USA

ABSTRACT

This paper investigates the acoustic limit of active control of structural dynamics; the limit as the control bandwidth includes a very large number of natural modes of the structure. The point is made that in this limit modal analysis cannot provide reasonably accurate models of the structural dynamics, and that control design with respect to modal models is then of questionable value. Alternative modeling approaches are reviewed. A particular wave propagation formalism, applicable to modeling the acoustic response of networks of slender structural members, is described in some detail. Control options designed with reference to this formalism are reviewed, and speculations as to future developments of such control are offered.

1. INTRODUCTION

Modal analysis is a powerful analysis technique, central to the discipline of structural dynamics since the publication of Rayleigh's[1] book. Nevertheless, practical limitations to the applicability of modal analysis do exist. This paper attempts a definition of one such limitation, the high-frequency acoustic limit, and points out the relevance of this limitation to the technology of active control of structural dynamics.

It is difficult to define the boundary between structural dynamics and structural acoustics, indeed, one might even insist that the former includes the latter. The boundary between analysis techniques is somewhat more clear; modal analysis relies upon a global description of an entire structure, while structural acoustic response is typically described in terms of the scattering properties of local components. Exceptions exist. It may be convenient to describe some portions of a structure in terms of acoustic parameters and other portions via modal analysis. Structures which are coupled to a fluid or elastic body[2] of infinite extent are examples of such exceptions, since it is then convenient to employ modal analysis for the structural response, and acoustic techniques to define the effect of the infinite medium.

Springer Series in Computational Mechanics
S. N. Atluri, A. K. Amos (Eds.)
Large Space Structures: Dynamics and Control

The discussion of this paper will be confined to structures of finite extent. Even in such situations techniques of structural acoustics may be preferable to modal analysis. A structural component, though finite, may be effectively infinite. This limit is approached if the component is much larger than disturbance wavelengths or if damping levels are high enough to attenuate a disturbance before many reflections can occur. Both situations tend to occur when excitation frequencies include many of the structure's natural modes. The frequency boundary between structural acoustics and structural dynamics thus depends upon the structure under consideration. A reasonable division for aircraft might be a few tens of Hz. Ultrasonic devices are well described by modal analysis at frequencies of several hundred kHz. Large flexible spacecraft, with fundamental frequencies below one Hz, would enter the acoustic response regime at frequencies as low as a few Hz. Figure 1 attempts to provide a graphical version of these arguments.

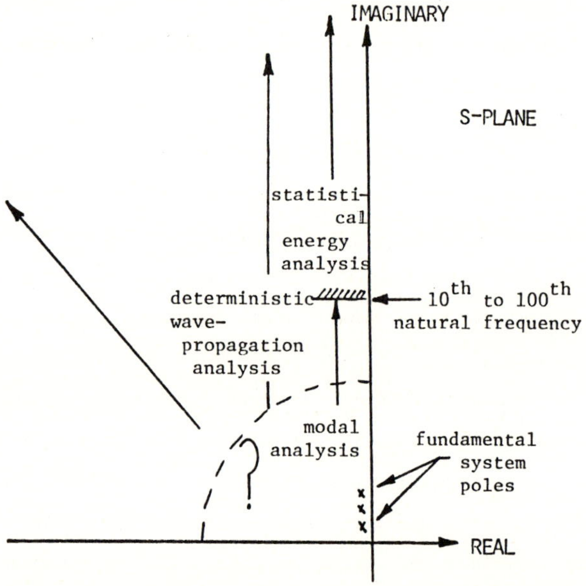

FIGURE 1 Approximate boundaries between the regions of applicability of modal analysis and local acoustic analysis are sketched in the *s – plane*. The high-fequency limit of modal analysis is due to sensitivity to parameter uncertainty. The low frequency limit of acoustic modeling is less well defined, and depends on geometrical complexity of the structure.

A relatively strong argument can be made for a high-frequency limit to the applicability of modal analysis. It is more difficult to define a low-frequency limit to the applicability of acoustic techniques. Local descriptions of component scattering behaviour can, in principle, be evaluated at any frequency, including zero, and can be linked into a global description, either in the frequency or the time domain. The convenience of such an approach will depend upon the geometric complexity of the structure, since this will govern the number of components and interconnections that must be independently modeled. The low-frequency limit of structural acoustic analysis is thus set by questions of convenience.

The advent of active control of flexible structures has underscored the limitations of structural modal analysis; high-performance active control depends upon a design model of high fidelity. Analyses of the response of proposed large flexible spacecraft suggest that hundreds of modes[3,4] will contribute significantly to the performance metric, often defined as a line-of-sight error or other measure of image quality. One is thus interested in the response at frequencies well within the acoustic regime. Unfortunately modal analysis is unable to predict details of such responses reliably, and control design with reference to such a model is then of questionable value.

An increase in the level of passive damping can make a major contribution towards an engineering solution of active control of structural dynamics and structural acoustics. Passive damping treatments tend to result in modal damping ratios which increase with mode number. Strongly damped modes can often be safely ignored, since they will not contribute strongly to degradation of the performance metric, nor couple with the control system. The virtues of passive damping can be quantified with respect to the impact on control design.[5] Mass penalties associated with passive damping treatments need not be enormous. Recent estimates[6] suggest that damping ratios of five percent can be obtained in the fundamental mode with a mass penalty of five percent. Higher modes can be damped passively with a much lower mass penalty, and the bandwidth of such damping can be effectively infinite.

If passive damping treatment is excluded or insufficient, and active control of structural dynamics to acoustic frequencies must be accomplished, then it is appropriate to base the control design upon acoustic models. Since these models are local, acoustic control will also be local. Such control theory is not well developed; only a few studies have recently been published[7,8,9,10].

2. THE ACOUSTIC LIMIT OF STRUCTURAL DYNAMIC MODELING

Mathematical modeling of an elastic structure invariably introduces a sequence of mathematical idealisations. One of these is the constitutive law assumed to apply, others are the introduction of simplifying kinematic assumptions leading to beam, plate, shell, membrane and other idealized models of components and boundary conditions for their connection. Assumptions of linear elasticity and infinitesimal deformations lead to a linear model of the structural dymanics of each component. If the components are linked by linear boundary conditions, they can be assembled into a global linear model which describes the dynamics of the entire structure.

2.1 A LIMITATION OF MODAL ANALYSIS

Modal analysis is a further manipulation of this global model. For relatively simple structural idealisations, spatial discretisation may be avoided and the modal parameters (natural frequencies and mode shapes) may be calculated as exact solutions of the global model. For most structures of practical interest, discretisation must be introduced and the calculated modal parameters are then only approximate solutions of the global model. The difference is perhaps only of academic interest, since any level of the model is merely an approximate description of the structure. Although one alternative produces modal parameters which are exact solutions of the lowest level of the model, neither approach yields the exact modal parameters of the structure. These exact modal parameters may not even exist, since even the slightest non-linearity or temporal variation of parameters excludes rigorous modal analysis.

It is often stated that a structure is infinite dimensional, and that discretisation of the mathematical model obscures this property by the enforcement of a finite number of degrees of freedom. Several analysts have questioned the validity of this claim,[11] pointing out that the origin of the infinite dimensionality of the structural model can be traced to the introduction of the idealisation of continuum constitutive laws, and that the infinite-dimensional viewpoint certainly fails when the model dimension exceeds the number of atoms in the structure. This paper refrains from contributing to this debate, and rather points out that a practical limit of modal analysis is reached long before the number of modeled natural modes approaches the number of atoms in the structure. The origin of this limit is the extreme sensitivity of modal analysis to modeling errors.

Several perturbation analyses have been published which define this sensitivity analytically. These analyses depend upon assumptions of linearity and time invariance of the model and its perturbation. Courant and Hilbert[12] offer an analysis in terms of operator notation, specific to self-adjoint operators:

Let the eigenvalue problem be defined by

$$L(u_n) + \lambda_n u_n = 0 \tag{1}$$

where L is a linear self-adjoint operator assumed to describe the structural dynamics and λ_n, u_n, $n = 1,2,3,\ldots$, are pairs of eigenvalues and eigen-functions of L. If the structure is actually described by another linear operator, slightly perturbed from L;

$$L(\bar{u}_n) - \epsilon r \bar{u}_n + \bar{\lambda}_n \bar{u}_n = 0 \tag{2}$$

where the function r defines this perturbation and ϵ is a small parameter, then the eigenvalues $\bar{\lambda}_n$ and eigen-functions $\bar{u}_n$ of this perturbed operator can be related to those of L by a classic perturbation analysis. For the case of non-repeated eigenvalues λ_n the analysis goes as follows:

Expand the perturbed eigenvalues and eigen-functions in terms of the small parameter ϵ;

$$\begin{aligned} \bar{u}_n &= u_n + \epsilon v_n + \epsilon^2 w_n + \ldots \\ \bar{\lambda}_n &= \lambda_n + \epsilon \mu_n + \epsilon^2 \nu_n + \ldots \end{aligned} \tag{3}$$

The first-order perturbation of the nth eigenvalue is then the inner product $\mu_n = < r u_n, u_n >$, and the first order perturbation of the nth eigen-function, v_n, is given by a sum of the contributions of the other unperturbed eigen-functions u_j, $j \neq n$;

$$v_n = \sum_{j \neq n} \frac{< r u_n, u_j >}{< L(u_n), u_n >} \frac{\lambda_n}{\lambda_n - \lambda_j} u_j \tag{4}$$

A similar argument in terms of matrix notation, valid for non-self-adjoint systems, has been presented in reference [13].

Modal density is invariably an increasing function of frequency; natural frequencies become ever more closely spaced as the mode number increases. Inspection of equation (4) reveals that this results in high-frequency eigen-functions with extreme sensitivity to small modeling errors. Modal analysis is thus limited to frequency regimes where relative spacing of natural frequencies remains large compared to the relative parameter uncertainty;

$$\frac{< r u_n, u_j >}{< L(u_n), u_n >} \ll \frac{\lambda_n - \lambda_j}{\lambda_n} \tag{5}$$

Experience suggests that this limitation will not include hundreds or even tens of modes of any structure. Many modes can be calculated, but the information, though detailed, will be useless.

Alternatives to modal analysis of linear structural dynamics, applicable to the high-frequency regime, have been developed by acousticians. These analysis techniques can be classified according to whether a stochastic or deterministic approach is taken. Hodges and Woodhouse[14] recent review paper is a reasonable starting point for study of the stochastic approaches. These approaches include asymptotic modal analysis and statistical energy analysis, and predict mean levels of response to broadband excitation. Response to narrow band excitation is not available, nor is deterministic response to any form of deterministic excitation. Such stochastic approaches are not useful for the design of active control, but can be used for performance analyses of open or closed-loop systems of very high order.

2.2 WAVE PROPAGATION ANALYSIS

A wave propagation analysis of structural acoustic response yields a deterministic model upon which active control can be based. A complex structure is modeled by an assemblage of local component models. Each component is described in the frequency domain by frequency-dependent scattering or propagation coefficients and by the equivalent impulse responses in the time domain.

Only a few books[2,15,16,17,18] have treated the subject of structural acoustics. The focus of these books has varied, this variation reflecting the wide variation of approaches to the problem. Lyon's[15] book on statistical energy analysis does not contribute to deterministic solution techniques. Junger and Feit[2] are primarily concerned with the coupling of structural response to the acoustic response of a surounding fluid. Auld[16] treats problems arising in ultrasonics and response of crystals and di-electric materials. Cremer and Heckl[17] and Graff[18] treat situations relevant to this paper; wave propagation and scattering in structural components which can be idealized as beams, plates, shells, membranes and rods.

The treatments of both references [17] and [18] tend to be very example-oriented; indeed it is difficult to develop a generic treatment of wave propagation in arbitrary structures. Too often each new example considered introduces new types of behaviour. If the scope of the analysis is restricted to structural components consisting of slender one-dimensional members and their interconnections, quite a general treatment is possible. Such a formalism was developed in a recent dissertation[19] and in two derivative publications[20,21].

2.2.1 WAVE PROPAGATION ON SLENDER ONE-DIMENSIONAL MEMBERS

Modeling of a slender one-dimensional member begins with the introduction of kinematic assumptions. Each cross section is assumed to deform from its reference condition according to a number of deflection variables. These variables are a function of only one spatial coordinate, the axial location of the section, hence the member is termed "one-dimensional". If the measures of cross-sectional deformation are continuous functions of the axial coordinate, introduction of a constituitive relation leads to a set of partial differential equations in time and in one spatial dimension. If the member is spatially periodic (an important subset, since this includes periodic truss beams), the cross-sectional deflections are defined at a set of discrete locations.

2.2.1.1 DISPERSION AND WAVE MODES IN CONTINUUM MODELS

A continuum model of a structural member is traditionally formulated as a system of coupled partial differential equations. Fourier transformation yields a system of coupled ordinary differential equations. For the purpose of this analysis, it is convenient to transform into a system of first-order, ordinary differential equations;

$$\frac{d\mathbf{y}}{dx} = \mathbf{A}(\omega)\,\mathbf{y} \tag{6}$$

in terms of the "cross-sectional state vector" $\mathbf{y}$ of physical cross-sectional variables. The dimension of $\mathbf{y}$ is equal to twice the number of deflection variables assigned to the cross section. The choice of the additional variables in $\mathbf{y}$ is not unique; they may represent internal forces, or spatial derivatives of the deflection variables.

Diagonalization of (6) may be interpreted in terms of wave propagation along the member. The eigenvalues of matrix $\mathbf{A}(\omega)$ are "propagation coefficients" $\gamma_j(\omega) = \alpha_j(\omega) + ik_j(\omega)$, $(i = \sqrt{-1})$, of traveling wave modes. The wave modes appear in forward and backward traveling pairs, thus the eigenvalues of $\mathbf{A}(\omega)$ appear in pairs $(\gamma_j, -\gamma_j)$. For non-dissipative models, $\mathbf{A}(\omega)$ is real. Its eigenvalues are not arbitrary complex pairs, but are restricted by the principle of conservation of energy to the first and third quadrants of the complex γ-plane. Thus they are either real (near fields), or pure imaginary conjugate pairs (traveling wave trains). For dissipative models $\mathbf{A}(\omega)$ becomes complex; the eigenvalues now appear anywhere in the first and third quadrants of the complex γ-plane.

The cross-sectional state vector $\mathbf{w}$ of the diagonalized system

$$\mathbf{\Gamma}(\omega) = diag(-\gamma_1 \ldots - \gamma_n, \gamma_1 \ldots \gamma_n)$$

$$\frac{d\mathbf{w}}{dx} = \mathbf{\Gamma}(\omega)\;\mathbf{w} \tag{7}$$

is related to $\mathbf{y}$ by a frequency dependent matrix of eigenvectors

$$\mathbf{y} = \mathbf{Y}(\omega)\,\mathbf{w} \tag{8}$$

Each element of $\mathbf{w}$ represents the amplitude of a wave mode, with the corresponding eigenvector occupying a column of $\mathbf{Y}(\omega)$. These wave modes travel independently of one another within the member; each has the form $w_j(\omega)\,\mathbf{y}_j(\omega)\,e^{-\gamma_j(\omega)\,x}$.

The polynomial equation $det[\mathbf{A}(\omega) - \gamma_j \mathbf{I}] = 0$ defines the dispersion relation between frequency and propagation coefficient. The phase speed c_p is defined by $c_p = \frac{\omega}{k}$. The group velocity c_g is defined by $c_g = \frac{\partial \omega}{\partial k}$. A medium for which these speeds are frequency dependent is called dispersive. In such media, the signal distorts as it propagates. Most structural models of interest, with the exception of the simple wave equation describing torsion and compression of a rod, and lateral deflection of a cable, are strongly dispersive.

The Timoshenko Beam Continuum Model

Bending and shear deformation of symmetric, solid section, slender beams is well described by the Timoshenko beam model, even to frequencies where the wavelength approaches the beam thickness. This model is also often used as an equivalent continuum description of truss-work beams in bending since the shear flexibility of such beams becomes important at much lower wave numbers than for solid sections. The governing partial differential equations of this model are typically given as a second order pair in terms of the deflection variables ψ (face rotation), and w (face translation)[18];

$$GA_s(\frac{\partial \psi}{\partial x} + \frac{\partial^2 w}{\partial x^2}) - \rho A \frac{\partial^2 w}{\partial t^2} = 0 \tag{9}$$

$$GA_s(\psi + \frac{\partial w}{\partial x}) - EI \frac{\partial^2 \psi}{\partial x^2} + \rho I \frac{\partial^2 \psi}{\partial t^2} = 0 \tag{10}$$

The cross-sectional state vector can be chosen to contain only deflections and the associated internal forces; $\mathbf{y} = (-w, \psi, M, V)^T$, where the additional variables $M = EI\frac{\partial \psi}{\partial x}$ (bending moment), and $V = GA_s(\frac{\partial w}{\partial x} + \psi)$ (shear force), have been used. With this choice, the equivalent system of ordinary differential equations is

$$\frac{d\mathbf{y}}{dx} = \begin{bmatrix} 0 & 1 & 0 & \frac{-1}{GA_s} \\ 0 & 0 & \frac{1}{EI} & 0 \\ 0 & -\rho I \omega^2 & 0 & 1 \\ \rho A \omega^2 & 0 & 0 & 0 \end{bmatrix} \mathbf{y} \tag{11}$$

To make this example specific, four beam parameters of a continuum model[22] of a lattice beam were chosen. These correspond to a proposed space lattice beam with very slender members, overall width and thickness of 5 m, and bay length of 7.5 m. The values are; $\rho A = 2.39\ kg/m$, $\rho I = 11.8\ kg - m$, $EI = 1.77 \times 10^8\ N - m^2$, $GA_s = 2.94 \times 10^6\ N$. Structural damping of $\eta = 0.01$ is assumed.

The resulting dispersion curves are displayed in Figures 2a and 2b. This member supports two wave modes in each direction; traditionally they have been termed bending and shear modes[18], according to the dominant entry of the corresponding eigenvector. A key point to note in these dispersion curves is that both the attenuation coefficient $\alpha(\omega)$, and the wave number $k(\omega)$ become proportional to the frequency ω for large frequencies. Both modes are dispersive, but have non-dispersive asymptotes. This non-dispersive high-frequency asymptotic behaviour can be exploited in calculating transient response by wave propagation.[21]

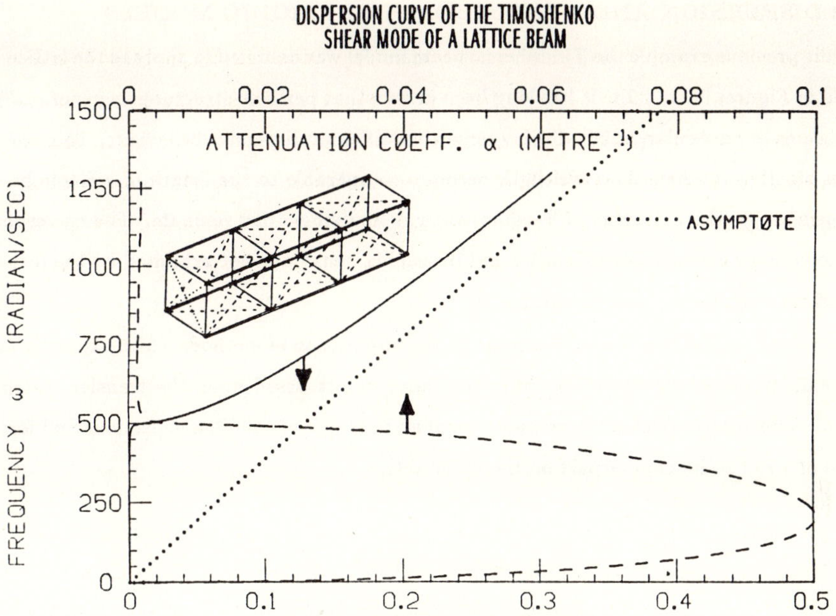

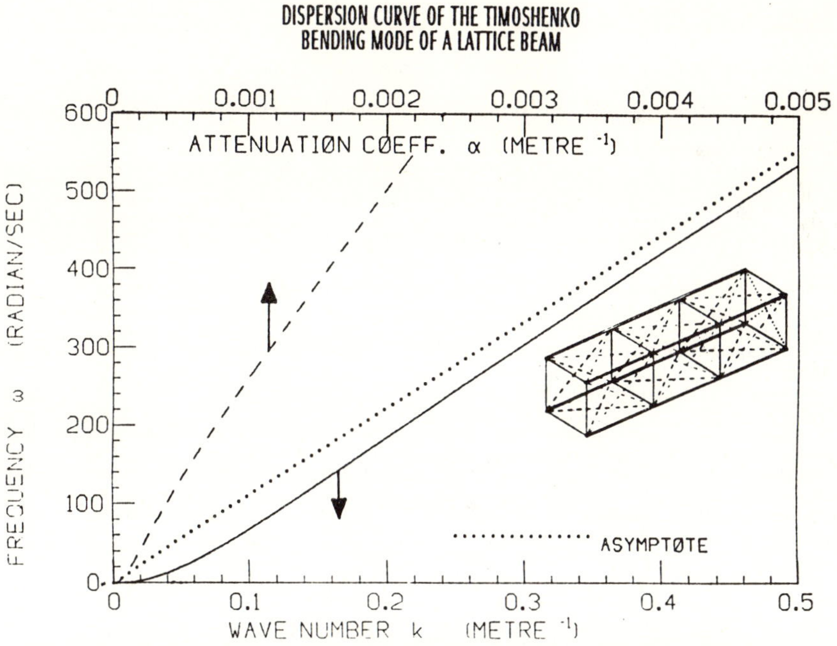

FIGURE 2 Dispersion curves for the two wave modes of the Timoshenko beam model. The beam parameters were taken from reference [22] where they were derived as an equivalent continuum model for the pictured lattice beam.

2.2.1.2 DISPERSION AND WAVE MODES IN PERIODIC MODELS

In the previous example the Timoshenko beam model was assumed to apply to the lattice beam sketched in Figures 2a and 2b. It has long been known that periodic structures in general[23], and lattice beams in particular[24,25], display somewhat different dispersive behaviour. This difference becomes significant when the wavelength becomes comparable to the length of a single bay, and at frequencies at which degrees of freedom internal to a single bay resonate. The corresponding dispersion curves display discontinuities and branching behaviour not exhibited by the dispersion curves of the "equivalent" continuum model.

Both these limitations can be overcome by the application of methods which explicitly exploit the periodicity of the structure. A convenient approach is based upon the transfer matrix of a basic cell. This matrix relates the cross-sectional state vector of coupling deflections and forces on one side of a cell to its counterpart on the other side;

$$\mathbf{y}_{i+1} = \mathbf{T}_y(\omega)\, \mathbf{y}_i \tag{12}$$

One method of calculating the transfer matrix, $\mathbf{T}_y(\omega)$, of a single bay employs finite-element derived mass and stiffness matrices[25].

The defining characteristic of a wave mode propagating along a periodic member is the fact that the entire cross-sectional state vector is multiplied by a complex factor, say ξ, as the wave passes through each cell;

$$\mathbf{y}_{i+1} = \xi\, \mathbf{y}_i \tag{13}$$

Equations (12) and (13) form an eigenvalue problem for ξ. These eigenvalues appear in pairs $(\xi_j, \frac{1}{\xi_j})$, corresponding to similar waves traveling in opposite directions. Equivalent propagation coefficients $\gamma_j(\omega)$ can be obtained from solution of the equation $\xi_j(\omega) = e^{\pm\gamma_j(\omega)\, l_{cell}}$ (where l_{cell} is the cell length). Care must be exercised to choose the correct branch of the complex logarithm. The corresponding eigenvectors of $\mathbf{T}_y(\omega)$ have an interpretation identical to that of the wave-mode eigenvectors of continuum members (equation(8)), but are meaningful only at cell interfaces.

The important effect of periodicity, from the point of view of wave propagation, is to introduce discontinuities into the dispersion curves. Two types of discontinuity may appear. Excitation of an internal degree of freedom results in an additional branch in the dispersion curves, at the resonant frequency of the internal degree of freedom. Other discontinuities occur when $k = \frac{n\pi}{l_{cell}}$ (when the cell length is an integral multiple of the spatial half-wavelength). This is well beyond the range where an equivalent continuum model may be expected to be valid.

A Periodic Member in Torsion

Perhaps the simplest model of a periodic truss-work member is an equivalent continuum model which has been made periodic by the addition(at regular intervals) of masses, springs, or arbitrary dynamic systems. Torsion represents the simplest of this class of problems. We choose Noor's[22] equivalent continuum model for torsion of the member treated in the previous example. Periodicity is introduced by mounting five percent of the inertia of the rod on torsional springs, fastened to the rod at intervals equal to the bay length of 7.5 m. These springs are chosen to resonate at $\omega_R = 40\ rad/sec$, and are meant to represent an arbitrary internal degree of freedom. A structural loss factor of $\eta = 0.01$ is assumed for this internal degree of freedom, and for the continuum model. The continuum model has the parameter values $GJ = 3.67 \times 10^7 N - m^2$, $\rho J = 23.6 kg - m$.

The cell transfer matrix for such a model may be calculated exactly. The transfer matrix of a single cell is given by the product $\mathbf{T}_{CELL} = \mathbf{T}_{FIELD}\,\mathbf{T}_{POINT}\,\mathbf{T}_{FIELD}$ where the "field" transfer matrix, $\mathbf{T}_{FIELD}$, is an exact solution of the governing partial differential equation, and relates the cross-sectional state vector at two points of a continuous rod, separated by a distance $\frac{l_{CELL}}{2}$;

$$\begin{pmatrix} \theta \\ GJ\frac{d\theta}{dx} \end{pmatrix}_{x=\frac{l_{CELL}}{2}} = \begin{bmatrix} cos(\bar{\omega}) & \frac{l_{CELL}}{2\bar{\omega}GJ} sin(\bar{\omega}) \\ -\frac{2\bar{\omega}GJ}{l_{CELL}} sin(\bar{\omega}) & cos(\bar{\omega}) \end{bmatrix} \begin{pmatrix} \theta \\ GJ\frac{d\theta}{dx} \end{pmatrix}_{x=0} \tag{14}$$

where $\bar{\omega} = \frac{l_{CELL}}{2}\omega\sqrt{\frac{\rho J}{GJ}}$. The "point" transfer matrix of a locally applied external torque $\Pi_{EXT} =$

$H(\omega)\theta$ is

$$\begin{pmatrix} \theta \\ GJ\frac{d\theta}{dx} \end{pmatrix}_{RIGHT} = \begin{bmatrix} 1 & 0 \\ H(\omega) & 1 \end{bmatrix} \begin{pmatrix} \theta \\ GJ\frac{d\theta}{dx} \end{pmatrix}_{LEFT} \tag{15}$$

The local degree of freedom is modeled as a simple oscillator, $H(\omega) = -\omega^2 I_R/(1 - (\frac{\omega}{\omega_R})^2)$.

The dispersion curves of this model are given in Figure 3. The internal degree of freedom introduces the discontinuity and the additional branch at its resonant frequency, $\omega_R = 40 rad/sec$. Figure 3 shows that the resonant frequency of the internal degree of freedom becomes an upper limit for the applicability of an equivalent continuum model.

A real truss-work member will have many such internal resonances. Each of these resonances creates its own discontinuity, and its own additional branch in the dispersion curves. Reference [25] reports a computational investigation into the wave propagation behaviour of a particular truss beam, with each bay modeled via finite elements. Each of the wave modes supported by this model exhibits many discontinuities in its dispersion curve. A new type of traveling wave mode is reported in this work; a "complex wave mode" which both travels and is spatially attenuated.

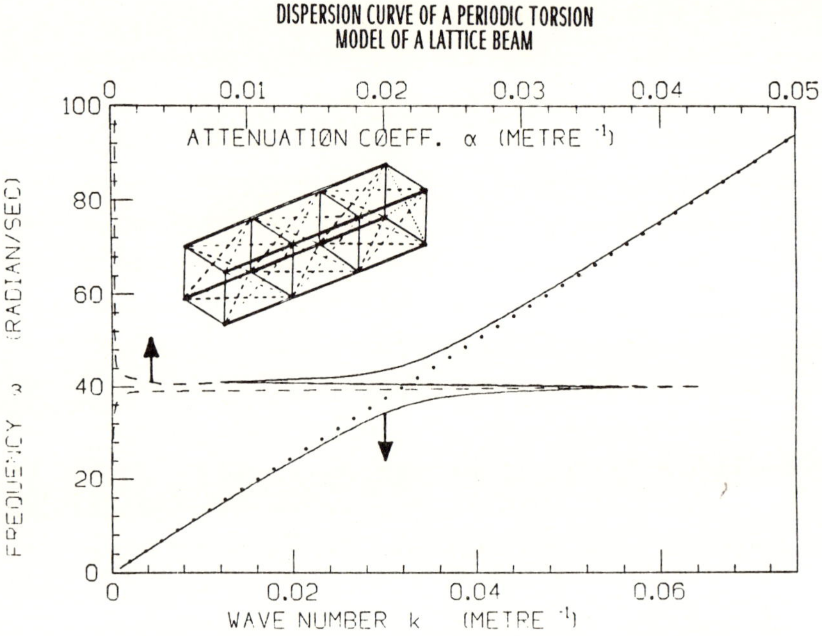

FIGURE 3 Dispersion curves of a periodic torsion model based upon an equivalent continuum model of the pictured lattice beam. The continuum model was developed in reference [22]. The discontinuity at $\omega = 40 rad/sec$ is due to resonance of local degrees of freedom.

2.2.2 WAVE-MODE TRANSIENTS ON MEMBER SEGMENTS

The transient response of disturbance propagation along elastic members is conveniently calculated in terms of the traveling wave modes. Indeed, it is this convenience which prompted their introduction in the frequency domain description of the previous section. This transient calculation has historically been the focus of much work[18]. A computational approach based on extensive use of the discrete Fourier transform is reported in reference [21]. These calculations are not central to the remainder of this paper, particularly to the control design of section 3; their description is omitted here in the interest of brevity.

2.2.3 SCATTERING AT JUNCTIONS AND DISCONTINUITIES

The following discussion is based upon the "generic" junction of Figure 4. This sketch, and the notation (with the exception of the external forces, $_{\beta}\mathbf{f}$) can be found in most basic texts on microwave circuits.[26]

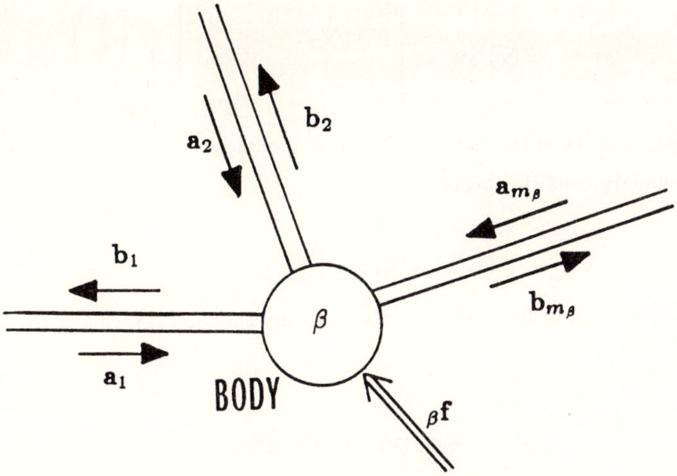

FIGURE 4 The generic junction. The junction can include a flexible body and can be connected to many members. Each member, j, transmits incoming wave modes, $\mathbf{a}_j$, and outgoing wave modes, $\mathbf{b}_j$. External forces are grouped in the vector ${}_\beta\mathbf{f}$. The notation is standard in microwave circuit analysis[26].

Associated with each member j, is a cross-sectional state vector $\mathbf{y}_j$, of size n_j, defined at the (arbitrary) interface between the junction and the member. This state vector may be transformed into wave-mode coordinates according to the transformation derived in the previous section. Each wave-mode state vector $\mathbf{w}_j$ can be grouped into incoming $\mathbf{a}_j$ and outgoing $\mathbf{b}_j$ wave modes.

The junction boundary conditions are conveniently defined in terms of the composite junction state vector, ${}_\beta\mathbf{y} = (\mathbf{y}_1^T, \mathbf{y}_2^T \cdots \mathbf{y}_{m_\beta}^T)^T$, having dimension ${}_\beta n = \sum_{j=1}^{m_\beta} n_j$. The boundary conditions can be expressed

$$ {}_\beta\mathbf{B}(\omega)\, {}_\beta\mathbf{y} = {}_\beta\mathbf{f}(\omega) \tag{16} $$

where ${}_\beta\mathbf{B}(\omega)$ is a (possibly lively) function of frequency and has ${}_\beta n$ columns and $\frac{{}_\beta n}{2}$ rows, and ${}_\beta\mathbf{f}(\omega)$ is a vector of size $\frac{{}_\beta n}{2}$, of externally applied forces and deflections. If the junction contains a flexible body, and is described by a system of ordinary differential equations, its description may be reduced to the form of equation (16). Such reduction is described in reference [19] and is a relatively standard procedure in structural dynamics.

The junction boundary conditions can be transformed into wave-mode coordinates by use of a block-diagonal matrix of member transformation matrices;

$$
{}_{\beta}\mathbf{y} = {}_{\beta}\mathbf{Y}(\omega)\ {}_{\beta}\mathbf{w} = \begin{bmatrix} \mathbf{Y}_1 & & & \\ & \mathbf{Y}_2 & & \\ & & \ddots & \\ & & & \mathbf{Y}_{m_\beta} \end{bmatrix} \begin{pmatrix} \mathbf{w}_1 \\ \mathbf{w}_2 \\ \vdots \\ \mathbf{w}_{m_\beta} \end{pmatrix} \tag{17}
$$

After partitioning into incoming and outgoing wave modes, ${}_{\beta}\mathbf{w} = ({}_{\beta}\mathbf{a}^T, {}_{\beta}\mathbf{b}^T)^T$, and re-ordering, the junction boundary condition becomes;

$$
[{}_{\beta}\mathbf{B}_a(\omega)\ {}_{\beta}\mathbf{B}_b(\omega)] \begin{pmatrix} {}_{\beta}\mathbf{a} \\ {}_{\beta}\mathbf{b} \end{pmatrix} = {}_{\beta}\mathbf{f}(\omega) \tag{18}
$$

If the sub-matrix ${}_{\beta}\mathbf{B}_b$ can be inverted, then the boundary equation (18) can be written in "causal" form;

$$
{}_{\beta}\mathbf{b}(\omega) = {}_{\beta}\mathbf{S}(\omega)\ {}_{\beta}\mathbf{a}(\omega) + {}_{\beta}\mathbf{\Psi}(\omega)\ {}_{\beta}\mathbf{f}(\omega) \tag{19}
$$

where the scattering matrix, ${}_{\beta}\mathbf{S}(\omega) = -{}_{\beta}\mathbf{B}_b^{-1}(\omega)\ {}_{\beta}\mathbf{B}_a(\omega)$, is a matrix of frequency-dependent transmission and reflection coefficients and ${}_{\beta}\mathbf{\Psi}(\omega) = {}_{\beta}\mathbf{B}^{-1}(\omega)$ is a matrix describing the generation of outgoing wave modes, ${}_{\beta}\mathbf{b}$, by the external forcing ${}_{\beta}\mathbf{f}$.

Junction of the Timoshenko Beam and Periodic Torsion Models

Numerous examples illustrating the application of the above derivation have been presented in reference [19]. The example presented here describes the perpendicular junction of the two members treated above. For this junction, the boundary conditions are;

$$
\begin{bmatrix} 1 & 0 & 0 & -1 & 0 & 0 & 0 & 0 \\ 1 & 0 & 0 & 0 & 0 & 0 & -1 & 0 \\ 0 & 0 & 0 & 0 & 0 & 1 & 0 & 0 \\ 0 & -1 & 0 & 0 & 1 & 0 & 0 & 1 \end{bmatrix} \begin{pmatrix} \mathbf{y}_{ROD_A} \\ \mathbf{y}_{BEAM} \\ \mathbf{y}_{ROD_B} \end{pmatrix} = 0 \tag{20}
$$

Transformation into wave-mode coordinates and derivation of the scattering matrix is done numerically. The scattering matrix is presented in Figure 5, over the frequency range $0 \leq \omega \leq 1200 rad/sec$. Periodicity of the rod member has been suppressed by setting $H(\omega) = 0$ in equation (15). Most of the activity in the frequency dependence of these scattering coefficients is due to the very dispersive nature of the Timoshenko shear mode. This mode exhibits a cut-off frequency at $\omega \approx 500 rad/sec$ (see Figure 2).

2.2.4 TRANSIENTS IN JUNCTIONS

The calculation of transient wave scattering through a junction described by equation (19) has been the topic of published work[21]. This transient behaviour is not relevant to the control design techniques proposed in the next section, and so is ommited here for brevity.

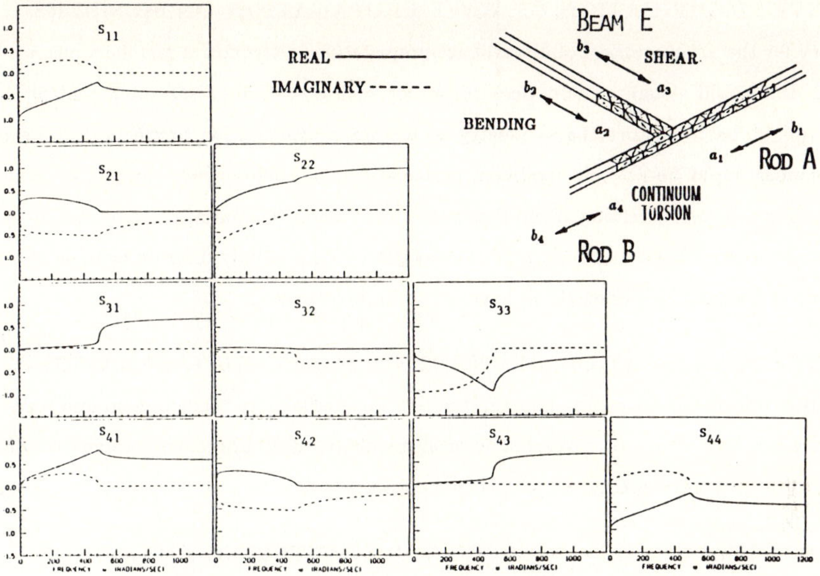

FIGURE 5 The scattering behaviour of the junction of a beam in bending modeled with Timoshenko beam theory and two rods in torsion. The periodicity of the torsion model has been suppressed by choice of $H(\omega) = 0$ in equation (15). The cut-off behaviour of the Timoshenko shear mode (at $\omega \approx 500 rad/sec$) has a strong effect on these curves.

3. ACTIVE CONTROL OF STRUCTURAL ACOUSTICS

Active control of structural acoustic response, with reference to an explicit structural acoustic model, is an almost untouched field of investigation with only four known publications[7,8,9,10]. Conversely, control design for structural dynamics is the subject of a large literature[27]. Control design techniques developed for systems of low order are being extended to models with many (tens or even hundreds) of the structure's natural modes of vibration. Such extension is a de facto attempt to achieve active control of structural acoustics. Unfortunately a modal description of structural dynamics to these frequencies is essentially useless, as has been discussed in a previous section.

Since high fidelity deterministic models of structural acoustics are spatially local, control based upon such models will share this property. Each local controller will be based upon the local dynamics of a component. Supervisory coordination between these local controllers is possible, but the theory for such an architecture has not yet been developed. This section reviews approaches to component control design based upon wave propagation models, and draws upon previously published work for specific examples.

3.1 ACTIVE MODIFICATION OF WAVE PROPAGATION ON MEMBERS

Theory for the active control of disturbance propagation in structural members has not been developed. One could visualize many possibilities for such a spatially distributed control; design objectives might be to distort the symmetry of the wave propagation properties of the member. A member might be actively modified, perhaps in selected frequency ranges, to propagate disturbances in only one direction. Such member control might be spatially discretized for implementation with discrete actuators, or might be designed for a spatially discrete member such as a truss beam. Many possibilities exist, all have yet to be investigated.

3.2 ACTIVE MODIFICATION OF JUNCTION SCATTERING BEHAVIOUR

Equation (19) is the frequency domain description of reflection, transmission and wave generation at a junction of one or several structural members. Two approaches to control of such scattering have been developed, both build upon equation(19).

3.2.1 SPECIFICATION OF CLOOSED-LOOP SCATTERING BEHAVIOUR

One might, based upon intuitive reasoning or analysis, wish to specify values for selected reflection and transmission coefficients of the closed-loop junction behaviour. Obvious choices for such a specification would be zero for some coefficients, although the specification might be more general, perhaps frequency dependent. This would result in selective absorption of incoming wave modes, or shunting of some incoming disturbances into selected outgoing directions.

If the controller exerts influences (forces and relative deflections) upon the junction which are grouped in the vector ${}_{\beta}\mathbf{f}$, and this control effort depends linearly upon the incoming wave modes $\mathbf{a}_{\beta}$, then the control law

$$ {}_{\beta}\mathbf{f}(\omega) = {}_{\beta}\mathbf{\Psi}^{-1}(\omega)[{}_{\beta}\mathbf{S}_{CL}(\omega) - {}_{\beta}\mathbf{S}(\omega)]{}_{\beta}\mathbf{a}(\omega) = \mathbf{C}_a(\omega){}_{\beta}\mathbf{a}(\omega) \tag{21} $$

leads to the desired closed-loop behaviour;

$$ {}_{\beta}\mathbf{b}(\omega) = {}_{\beta}\mathbf{S}_{CL}(\omega){}_{\beta}\mathbf{a}(\omega) \tag{22} $$

Measurement of the incoming wave modes ${}_{\beta}\mathbf{a}(\omega)$ may not be practical since they are related to physical variables through a frequency-dependent transformation (equation (17)). Use of this transformation permits manipulation of equation (21) into the form

$$ {}_{\beta}\mathbf{f}(\omega) = {}_{\beta}\mathbf{S}_{CL}(\omega){}_{\beta}\mathbf{a}(\omega) = \mathbf{C}_y(\omega){}_{\beta}\mathbf{y}(\omega) \tag{23} $$

where the control forces are now given in terms of the physical variables ${}_{\beta}\mathbf{y}$.

The preceeding discussion has not considered the possibility that only a few actuators are available at a given junction to effect control. In this situation, the number of independent entries in $_{\beta}\mathbf{f}(\omega)$ would be less than the dimension of $_{\beta}\mathbf{a}(\omega)$; more wave modes depart the junction than one has actuators available. Several options have been developed for this situation[9,19]. One might attempt to minimize a sum of squares of departing wave-mode amplitudes. One might set selected departing wave modes to zero, while letting the others behave as they will, or one might influence only subsets of the scattering coefficients. Each approach leads to a control of the form of equation (23).

Examples of Wave-Shunting Control Design

Two examples, taken from prior published work, are offered here to clarify the above discussion. Figure (5), from reference [9], gives the open-loop scattering behaviour of the junction of three members. One is modeled in bending with Timoshenko beam theory, the other two are modeled in torsion with simple rod theory. The member models thus support four incoming and four outgoing wave modes. As an arbitrary design exercise, a compensator has been calculated which prevents waves from departing the junction along one of the members, that is, with reference to Figure (5), $b_1 = 0$. The control force to accomplish this was (arbitrarily) chosen to be an external moment applied to the junction.

An external torque, M_{EXT} can be included in the boundary conditions of equation (20) by introduction of an external forcing vector $_{\beta}\mathbf{f} = (0\,0\,0\,1)^T M_{EXT}$. The boundary conditions are then readily manipulated into the form of equation (19), where $_{\beta}\mathbf{S}(\omega)$ is given in Figure 5. A bit of algebra, done numerically at each frequency, yields a compensator of the form

$$\begin{aligned} M_{EXT} &= C_{\theta_A}\theta_A + C_{\tau_A}\tau_A + C_{w_E}w_E + C_{\psi_E}\psi_E \\ &\quad + C_{M_E}M_E + C_{V_E}V_E + C_{\theta_B}\theta_B + C_{\tau_B}\tau_B \end{aligned} \tag{24}$$

where θ (rotation) and τ (torque) are the cross-sectional state variables of the two torsion members at the junction, and w (lateral deflection), ψ (face rotation), M (bending moment) and V (shear force) are the four cross-sectional state variables of the bending member at the junction. Subscripts denote which member the variable corresponds to. Note that all eight local cross-sectional state variables are used by this compensator. The open loop scattering matrix of this junction (Figure 5) shows that all arriving wave modes must be countered. One of the eight compensators of equation (24) is displayed in Figure 6 (the other seven are available in reference [19]). Note that this compensator is both infinite dimensional and infinite in bandwidth.

Reference [9] takes this computational example somewhat further. The junction is imbedded into a structure, and transcendental transfer functions are calculated with and without this wave-absorbing control. The change in the structural response is dramatic.

A second example is taken from reference [10] and treats reflection cancellation for the free end of a beam in bending. A beam, if modeled with Bernoulli-Euler beam theory, supports one

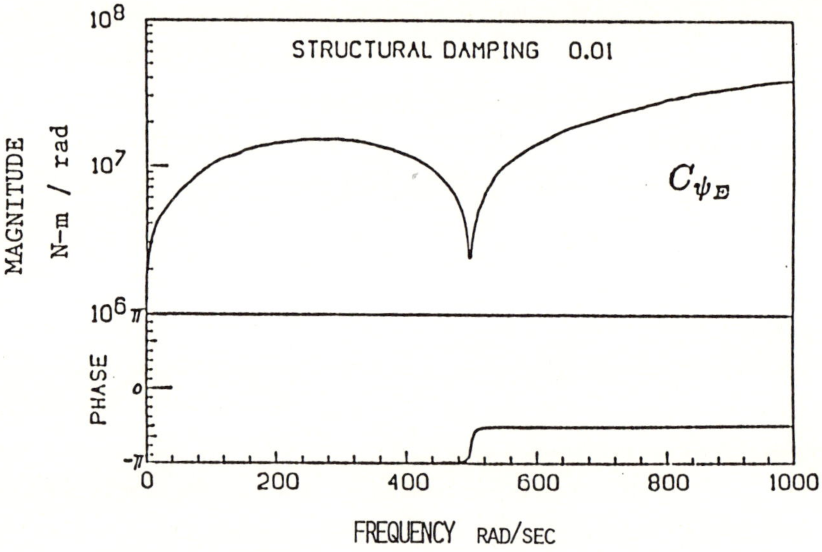

FIGURE 6 Frequency dependence of one of the eight compensators of equation (24). This compensator feeds local rotation at the junction depicted in Figure 5 back to a co-located external torque. The control task is to set $b_1 = 0$.

traveling wave mode in each direction. The governing differential equation permits another type of response; near fields which decay exponentially with axial distance and have a simple harmonic temporal behaviour. With reference to Figure 7 the deflection field can be written

$$v = a_t e^{ikx+i\omega t} + a_n e^{kx+i\omega t} + b_t e^{-ikx+i\omega t} + b_n e^{-kx+i\omega t} \tag{25}$$

where $k = \sqrt{\omega\sqrt{\rho A/EI}}$ is known as the wave number, EI is the bending stiffness, ρA is the mass per length of the beam and a_t, a_n, b_t and b_n are wave mode amplitudes at the left end of the beam, defined in Figure 7. If an external control torque, M_{EXT}, can be applied to the beam at the left end, the corresponding boundary conditions can be expressed in the form of equation (19) as

$$\begin{pmatrix} b_t \\ b_n \end{pmatrix} = \begin{pmatrix} -i & 1+i \\ 1-i & i \end{pmatrix} \begin{pmatrix} a_t \\ a_n \end{pmatrix} + \frac{1+i}{2\omega\sqrt{\rho AEI}} \begin{pmatrix} -1 \\ -i \end{pmatrix} M_{EXT} \tag{26}$$

The important entry of the scattering matrix is $\mathbf{S}(1,1)$ since this term governs the creation of outgoing traveling wave modes as a function of incoming traveling wave modes. One choice for the compensation that achieves $\mathbf{S}(1,1) = 0$ is

$$M_{EXT}(\omega) = -i\sqrt{\rho AEI}\,\omega v(x=0,\omega) \tag{27}$$

since for this choice the closed-loop reflection matrix becomes

$$\mathbf{S}_{CL} = \begin{pmatrix} 0 & i \\ -i & 0 \end{pmatrix} \tag{28}$$

The compensation of equation (27) can be seen to be velocity feedback of local deflection to applied torque. The effect of finite actuator dynamics and other sources of gain roll-off yet needs to be investigated. The question is important to this example since experimental verification is planned.

3.2.2 OPTIMAL WAVE ABSORPTION

A recent development[10] in the theory of traveling wave control is the proposal of a meaningful cost function associated with the actively controlled junction. Minimization of this cost function then leads to a control design, which since a reasonable quantity is being minimized, is termed optimal. The proposed cost function is a weighted integral over frequency of the wave power flowing out of the junction along the members and the power exerted by the control forces;

$$J = \int_{-\infty}^{\infty} ({}_{\beta}\mathbf{w}^H \mathbf{P}_{\beta}\mathbf{w} + {}_{\beta}\mathbf{f}^H \mathbf{R}_{\beta}\mathbf{f})\, d\omega \tag{29}$$

where the superscript H (Hermitian) denotes complex conjugate transpose. The first term in the above integrand is the power flow, as a function of frequency, being carried out of the junction by wave motion. The matrix $\mathbf{P}$ depends upon the characteristics of the members attached to the junction under consideration, but is always Hermitian.[10] The second term in the integrand is a quadratic penalty on control effort. The matrix $\mathbf{R}$ can be chosen by the control designer. It can be a function of frequency but must be Hermitian.

Since outgoing wave modes depend upon the control effort, this cost function can be further expanded using the substitution

$${}_{\beta}\mathbf{w} = \begin{pmatrix} {}_{\beta}\mathbf{a} \\ {}_{\beta}\mathbf{b} \end{pmatrix} = \begin{pmatrix} {}_{\beta}\mathbf{a} \\ {}_{\beta}\mathbf{S}_{\beta}\mathbf{a} + {}_{\beta}\mathbf{\Psi}_{\beta}\mathbf{f} \end{pmatrix} \tag{30}$$

The cost functional is minimized with respect to the control, ${}_{\beta}\mathbf{f}$ by

$${}_{\beta}\mathbf{f} = -[{}_{\beta}\mathbf{\Psi}^H \mathbf{P}_{bb\beta}\mathbf{\Psi} + \mathbf{R}]^{-1}{}_{\beta}\mathbf{\Psi}^H[\mathbf{P}_{ba} + \mathbf{P}_{bb\beta}\mathbf{S}]_{\beta}\mathbf{a} \tag{31}$$

where the power flow matrix has been partitioned in an obvious way into square submatrices. The optimal control, given by equation (31), differs from the control which achieves zero outgoing disturbance, $\mathbf{S}_{CL} = 0$, derivable from equation (21) as

$${}_{\beta}\mathbf{f} = -{}_{\beta}\mathbf{\Psi}^{-1}{}_{\beta}\mathbf{S}_{\beta}\mathbf{a} \tag{32}$$

In part, this difference is due to the penalty associated with control effort. If the control penalty,

R, is set to zero in equation (31), then the $\mathbf{P}_{ba}$ term still creates a difference between the two control laws of equations (31) and (32). The presence of the term $\mathbf{P}_{ba}$ in equation (31) is due to another effect. This term attempts to exploit power flow coupling between wave modes to draw the elastic energy into the junction more effectively than merely waiting for it to arrive. The off-diagonal entries of the P matrix are often zero, however, and it is not yet clear if this type of "energy vacuuming" will be useful in application. The following example is intended to clarify the situation.

Optimal Wave Cancellation at a Free Beam End

This example is a continuation of the beam example begun in the preceeding section. The junction is the termination of a single beam, so that the power flow matrix will correspond only to the four wave modes supported by the Bernoulli-Euler beam model;

$$\mathbf{P} = 2\omega^2 k\sqrt{\rho AEI}\begin{bmatrix} -1 & 0 & 0 & 0 \\ 0 & 0 & 0 & i \\ 0 & 0 & 1 & 0 \\ 0 & -i & 0 & 0 \end{bmatrix} \tag{33}$$

The wave modes are ordered $_\beta\mathbf{w} = (a_t\, a_n\, b_t\, b_n)^T$. From inspection of equation (33) one can deduce that the incoming traveling wave mode causes power to flow into the junction, the outgoing traveling wave mode cause power to flow out of the junction, while the two near fields can cause

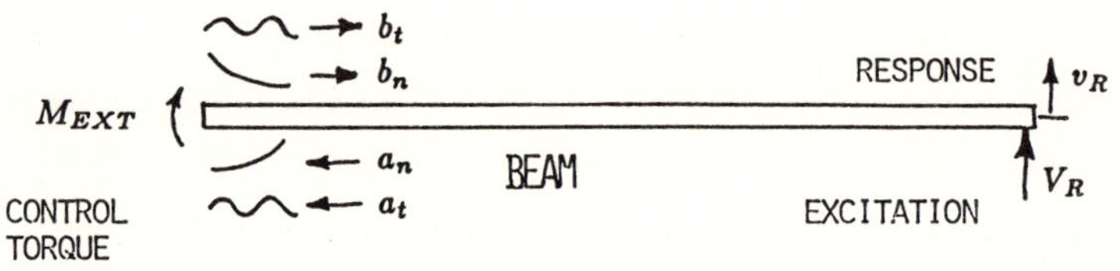

FIGURE 7 Schematic of a free-free beam in bending. The left end of beam is actively terminated by a wave absorbing controller, applying external torques, M_{EXT}, in response to local deflection or slope. The right end is excited by a lateral force.

power to flow in either direction but only through interaction with another. The in-phase portion of one near field interacts with the out-of-phase portion of the other near field to cause a net power flow. One near field is created at the junction. Equation (31) suggests that one exploit this interaction between near fields to increase the power flow into the actively controlled junction beyond that which is carried by the incoming traveling wave a_t.

When the requisite values of P, $_\beta\Psi$, and a value of $\mathbf{R} = 0$ (no control penalty) are inserted into equation (31), and the boundary conditions are used to convert to physical variables, the optimal control becomes

$$M_{EXT} = \sqrt{EI/2\sqrt{\rho AEI}}\sqrt{-i\omega}\,\frac{dv}{dx}(x = o, \omega) \tag{34}$$

which involves feedback of the beam slope to the applied torque through a temporal "half differentation".

Leaving aside, for the moment, questions of realizability, it is instructive to investigate the performance of such wave controllers in simulation. Figure 8 is a simulated transfer function of a finite length of the beam. It is clear that the effect of compensation of the form of either equation (27) or (34) would have a profound effect upon the behaviour of the beam segment. Seen from the driven (right) end, the controlled beam behaves essentially as if it extended to infinity.

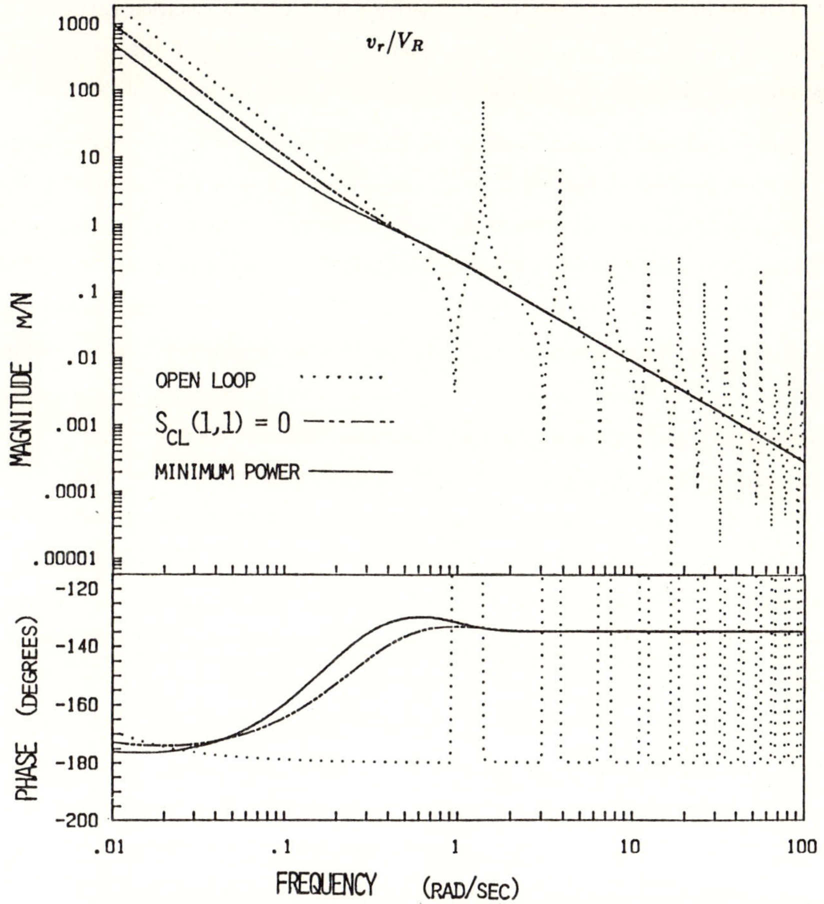

FIGURE 8 Magnitude and phase of the lateral deflection of the right end of the beam of Figure 7 to a lateral force also applied to the right end. The left end is controlled by a wave absorbing controller which either sets $S_{CL}(1,1) = 0$, or minimizes wave power departing the left end. Open-loop behaviour is shown for reference.

3.3 EXPERIMENTAL VERIFICATION OF TRAVELING WAVE CONTROL

The computational examples presented above have used very ideal assumptions in the derivations. In particular, no attempts were made to account for actuator or sensor limitations. Without exception, every compensator derived was of infinite bandwidth. The compensation is specified in the frequency domain, and is in general infinite-dimensional.

Wave absorbing compensation with finite bandwidth will always be derivable; one must simply trade performance against bandwidth. Several possibilities for doing this in a rational way come to mind. One might use a frequency-dependent control penalty matrix $\mathbf{R}(\omega)$ in the optimal control derivation with the frequency dependence chosen such that the compensation gain rolls off as desired. One might specify a closed-loop scattering matrix $\mathbf{S}_{CL}$ in equation (21) in such a way as to create the same effect. As a last resort, one might simply roll-off the compensation with low-pass filtering. The trade-offs of these techniques have yet to be thoroughly studied.

Whichever technique is used to make the bandwidth of the wave absorbing compensators finite, their infinite-dimensionality will, in general, remain. Simplified and realizable approximations of such infinite-dimensional compensation can always be derived. A rational approach for deriving such approximations has not been developed. Since the compensation is specified in the frequency domain, a frequency domain fit with a realizable transfer function seems reasonable. Such techniques, and their laboratory application to this problem, have yet to be developed.

To date, only one experiment in active control of traveling wave in structures seems to have been reported[8]. This experiment attempted the cancellation of the reflection of bending waves on a thin(1 mm) sheet of stainless steel in the frequency range from zero to 25 Hz. The tip deflection could be sensed; actuation of lateral force was possible. The ideal compensation involved feedback of the tip deflection to the lateral force through a temporal "half differentiation", as in equation (34). Several analog circuits were used to crudely approximate this ideal actuation. The compensation actually achieved was finally very similar to a previously calculated optimal velocity feedback.

The performance achieved by the approximate wave-absorbing compensation in this experiment was very comparable to that achieved by the optimized velocity feedback. Although the two concepts represent alternative approaches to the problem of active damping of structural vibrations, this experiment did not permit a solid conclusion of their relative merits.

3.4 POTENTIAL EXTENSIONS OF TRAVELING WAVE CONTROL

The active control of traveling waves in structures is a very recent development. Only relatively crude theory has been developed, and an even cruder experiment has been conducted. Many extensions to this work can be considered:

1 The sensitivity of traveling wave controllers to errors in the local acoustical models should be investigated. A first order sensitivity to first order perturbations in the model is expected.

2 Explicit account might be taken of actuator and sensor dynamics during the design of wave controllers. It is anticipated that the specified wave control compensator will then include an inversion of these dynamics.

3 The approximation of specified wave control compensators with analog circuitry or digital filters should be studied. It seems likely that realizability of these compensators will be improved by measuring the approaching wave disturbance some distance "upstream" of the point of actuation. With such precognitive sensing, the inversion of actuator and sensor dynamics may even be possible.

4 Adaptation of wave controllers is possible. A signal upon which to base the adaptation might be the level of the outgoing wave mode which the control is designed to cancel.

5 Development of controllers based upon the propagation behaviour of members rather than scattering behaviour of junctions is a possibility. Such work would have to face the problem of non-existence of distributed actuators and sensors, and should perhaps focus on periodic truss beams.

6 Cooperative behaviour among several wave controllers and/or a global controller based upon a low-order modal model might be studied.

7 All the control schemes proposed must be realised in hardware to be useful. Such experimental work is perhaps the most important missing aspect of the work performed to date in active control of structural acoustics.

4. SUMMARY

This paper makes the point that high performance active control of structural dynamics over a bandwidth including many lightly damped modes is impractical if the control design is based upon a model derived via modal analysis. The fidelity demanded of the model is simply beyond that achievable with modal analysis. The analysis technique is too sensitive to small perturbations and uncertainties in the model parameters.

Alternatives exist. An important possibility is passive damping in conjunction with active control of a small subset of the lower modes. Active damping through direct velocity feedback can also be used. If active control of a lightly damped structure over a bandwidth including many modes of vibration must be accomplished, then local, structural acoustic models of the response are suitable for the control design. This leads to local traveling wave control.

The paper presents a formalism for the synthesis of traveling wave models of a particular class of structures; networks of slender elastic members. A review of control design techniques applicable to the control of disturbance propagation in such structural networks is presented. The conclusion is drawn that the theory is very incomplete, and speculations are offered as to future developments.

ACKNOWLEDGEMENTS

Many of the ideas summarized in this paper are a direct development of a dissertation written at Stanford University in 1984 under the direction of Professor Holt Ashley, and with financial support from the Canadian Natural Sciences and Engineering Research Council and from the United States Air Force Office of Scientific Research (AFOSR). The ideas were developed further with the support of the German Aerospace Research Establishment (DFVLR), and the Alexander von Humboldt Foundation, and most recently, at the Massachusetts Institute of Technology with further support of the AFOSR.

REFERENCES

1. J. W. Strutt Lord Rayleigh, *Theory of Sound*, 1894, Dover 1945
2. M. C. Junger, D. Feit, *Sound, Structures and Their Interaction*, Massachusetts Institute of Technology Press, 1986
3. D. S. Bernstein, S. W. Greeley, "Robust Controller Synthesis Using the Maximum Entropy Design Equations," *IEEE Transactions Automatic Control*, Vol. AC-31, 1986, pp. 362-364
4. G. J. Kissel, D. R. Hegg, "Stability Enhancement for Control of Flexible Space Structures," *IEEE Control Systems Magazine*, Vol. 6, No. 3, June 1986, pp. 19-26
5. R. Gran, "Finite-Dimensional Controllers for Hyperbolic Systems," *Proceedings*, Third VPI-/SU/AIAA Symposium, Blacksburg, VA, June 15-17, 1981
6. A. H. von Flotow, "Control Motivated Dynamic Tailoring of Spacecraft Truss Structures," *Proceedings*, AIAA Guidance, Dynamics and Control Conference, Williamsburg, VA, August 1986
7. D. R. Vaughan, "Applications of Distributed Parameter Concepts to Dynamic Analysis and Control of Bending Vibrations," *J. Basic Engng*, June 1968, pp. 157-166
8. A. H. von Flotow, B. Schäfer, "Wave-Absorbing Controllers for a Flexible Beam," *J. Guidance, Control and Dynamics*, Vol. 9, No. 6, Nov-Dec 1986, pp. 673-680
9. A. H. von Flotow, "Traveling Wave Control for Large Spacecraft Structures," *J. Guidance, Control and Dynamics*, Vol. 9, No. 4, July-August 1986, pp. 462-468
10. D. W. Miller, A. H. von Flotow, S. R. Hall,"Active Modification of Wave Reflection and Transmission in Flexible Structures," *Proceedings*, American Control Conference, June 10-12, 1987, Minneapolis, Minnesota
11. P. C. Hughes, presentation comments, NASA Workshop on Applications of Distributed System Theory to the Control of Large Space Structures, JPL, Pasadena, CA, July 14-16, 1982
12. R. Courant, D. Hilbert, *Methods of Mathematical Physics*, Vol. 1, Interscience, 1953
13. R. H. Plaut, K. Huseyin, "Derivatives of Eigenvalues and Eigenvectors in Non-Self-Adjoint Systems," *AIAA Journal*, Vol. 11, No. 2, Feb 1973

14. C. H. Hodges, J. Woodhouse, "Theories of Noise and Vibration Transmission in Complex Structures," *Reports on Progress in Physics*, 1986, **49**, 107-170

15. R. H. Lyon, *Statistical Energy Analysis of Vibrating Systems*, Massachusetts Institute of Technology Press, 1975

16. B. A. Auld, *Acoustic Fields and Waves in Solids* (two volumes), John Wiley and Sons, 1973

17. L. Cremer, M. Heckl, E. E. Ungar, *Structure Borne Sound*, Springer-Verlag, 1973

18. K. F. Graff, *Wave Motion in Elastic Solids*, Ohio State University Press, 1975

19. A. H. von Flotow, "Disturbance Propagation in Structural Networks; Control of Large Space Structures," *Ph.D. Dissertation*, Department of Aeronautics and Astronautics, Stanford University, June 1984

20. A. H. von Flotow, "A Traveling Wave Approach to the Dynamic Analysis of Large Space Structures," *Proceedings* AIAA/ASME/ASCE/AHS 24th Structures, Structural Dynamics and Materials Conference, South Lake Tahoe, California, May 2-4, 1983

21. A. H. von Flotow, "Disturbance Propagation in Structural Networks," *J. Sound Vibration*, (1986), **106**(3), 433-450

22. A. K. Noor, C. M. Andersen, "Analysis of Beam-Like Lattice Trusses," *Computer Methods in Applied Mechanics and Engineering*, 20, 1979, 53-70

23. L. Brillioun, *Wave Propagation in Periodic Structures*, Dover Publications, 1946

24. S. Sgubini, F. Graziani, A. Agneni, "Elastic Waves Propagation in Bounded Periodic Structures," *Proceedings*, 35th Congress of the International Astronautical Federation, October 8-13, 1984, Lausanne, Switzerland

25. J. Signorelli, A. H. von Flotow, "Wave Propagation in a Periodic Truss Beam," *Proceedings*, AIAA Dynamics Specialists Conference, April 9-10, 1987, Monterey, CA

26. R. H. Ghose, *Microwave Circuit Theory and Analysis*, McGraw-Hill, 1963

27. M. J. Balas, "Trends in Large Space Structure Control Theory: Fondest Hopes, Wildest Dreams," *IEEE Transactions Automatic Control*, Vol. Ac-27, June 1982

Active Control for Vibration Damping

Lt P.J. Lynch
Siva S. Banda
Flight Dynamics Laboratory AFWAL/FIGC
Wright-Patterson Air Force Base OH 45433

ABSTRACT

Active control laws are developed for an LSS-type structure to damp vibrations. High frequency modelling uncertainties lead to the necessity for a robust control design. The Linear Quadratic Gaussian with Loop Transfer Recovery (LQG/LTR) control design technique is a particular robust design technique selected for use in designing a damping control system. A summary of LQG/LTR is given and numerical example using a two bay truss is presented.

INTRODUCTION

The vibration damping problem is an important aspect of the overall control problem of Large Space Structures (LSS). Vibrations will arise in the structure from external disturbances or from commanded maneuvers like slewing. The amplitude of the vibrations must be controlled to insure the LSS can continue to perform its prescribed mission and not incure structural damage. This requires the reduction of the structural vibrations to zero, or to an acceptable level, within a given time. Techniques to accomplish this damping are often separated into two broad categories; passive damping and active damping. Passive damping[1,2,3] refers to the LSS's ability to damp its own osscillations as a result of its structural design, material properties or the effect of additional devices like coatings and elastomers. Active control implies the use of a feedback control system which incorporates a sensing of structural vibrations and calculation of the control input signals for actuators to damp the vibrations present. While the best solution to the vibration damping problem most likely uses a combination of active and passive damping techniques, this paper assumes a fixed level of damping exists throughout the structure and then explores the use of an active control system. The Linear Quadratic Gaussian with Loop Transfer Recovery (LQG/LTR) technique is used to develop a robust vibration control system.

Use of an active control system provides significant design freedom. Active systems can be designed which not only damp vibrations, but simultaneously accomplish maneuvers like pointing and tracking, which may be requirements of the LSS. Additionally, control systems can be developed to reject disturbances that can be expected in an operational environment. These disturbances may include thermal effects, gravity gradient or the effect of rotating surfaces of the LSS.

The increased design freedom gained through the use of active control does not come without expense. A controller designed to accomplish specific control objectives must be developed

Springer Series in Computational Mechanics
S. N. Atluri, A. K. Amos (Eds.)
Large Space Structures: Dynamics and Control

based upon a mathematical model of the large space structure. Controllers are often designed to meet system objectives based on a low-order design model. This is a result of computational considerations and the inaccuracies associated with finite-element descriptions of high frequency structural behavior. The more closely the model represents the true system, the better the chance of obtaining a controller that accomplishes the stated control objectives. Differences between the design model and the actual system can lead to faulty performance or even instability. As a result, modelling error considerations and the influence of disturbances previously mentioned lead to the requirement for a robust vibration control system.

Robustness[4,5,6] of control systems examines the performance of the control system with respect to uncertainties present. These uncertainties may include differences between the design model and the actual system (plant uncertainties) and the effects of various disturbances which lead to system uncertainties. Robustness of actively controlled systems can be seperated into two broad categories; stability robustness and performance robustness. Closed-loop systems which remain stable in the face of uncertainties are stability robust. A closed-loop system that maintains an acceptable level of performance in the face of uncertainties is described as performance robust. Although both are desirable, stability robustness is paramount since performance can't even be considered without stability.

The purpose of this work is to demonstrate the application of the Linear Quadratic Gaussian with Loop Transfer Recovery (LQG/LTR) control design technique in arriving at a robust vibration control system. Following this introduction, Section 2 presents a general description of the structural modelling and control objectives used in this study. Section 3 discusses the basics of the LQG/LTR control design technique. This includes a portion addressing the mathematical tools necessary for computer implementation. References are provided for more in-depth coverage of the LQG/LTR method. Section 4 uses an LSS-type structure in the form of a cantilever beam to demonstrate the use of LQG/LTR. Conclusions appear in Section 5.

STRUCTURAL MODELLING AND CONTROL OBJECTIVES

Finite element methods are used to arrive at a mathematical description of the LSS[7]. The LQG/LTR design technique requires a linear state-space representation of the plant's dynamics. Most problems have a high-order truth model which describes the actual system reasonably well, with some degradation at high frequencies. Large order systems present a problem computationally to the control design engineer from both a design and implementation viewpoint. As a result, low-order models are often used for control design purposes. A model reduction is required to arrive at this low-order design model. Numerous techniques exist to accomplish this model reduction[8,9,10]. The scope of this work does not permit development of the many issues involved in model reduction. Rather this work focuses more on the requirements placed on control system design following a reduction. The differences between the design model and the truth model lead to additional plant uncertainties which must be tolerated by the robust control system. Control objectives must ultimately be achieved against the truth model and even more importantly, the actual system.

The standard second order matrix differential equation, developed through finite element methods, which governs the flexural vibrations of a structure is given in equation (1).

$$[m]\ddot{r}(t) + [c]\dot{r}(t) + [k]r(t) = F(t) \tag{1}$$

Here, the mass $[m]$, damping $[c]$, and stiffness $[k]$ matrices are n×n dimensional and describe the structure of the LSS. These are a function of the LSS's design and are assumed to be constant matrices for the purposes of this study. The vector $r(t)$ is n×1 dimensional and represents the structure's physical coordinates.

For this study, an active control system will be designed to damp an initial vibration in the structure. The study is simplified by considering only responses to initial conditions as opposed to persistent excitation or some combination of inputs and initial conditions. This is done simply to limit the scope of the work. A rich initial condition will serve to demonstrate the salient features of the damping control system. Since this is an unforced vibration, $[F(t)]$ on the right hand side of equation (1) is simply the control input distribution matrix $[b]$ multiplied by the control signal $u(t)$. The $[b]$ matrix is n×m dimensional and describes the placement of the m actuators and their effect on the structure. This results in equation (2) relating the structure's physical coordinates to the actuator's control inputs.

$$[m]\ddot{r}(t) + [c]\dot{r}(t) + [k]r(t) = [b]u(t) \tag{2}$$

A state-space representation for this system of n second-order differential equations may be written by selecting the (2n×1) state vector $\tilde{x}$ as

$$\tilde{x} = \begin{bmatrix} \dot{r}(t) \\ r(t) \end{bmatrix}_{2n\times 1} \tag{3}$$

The first n elements of the state vector $\tilde{x}$ are physical rates $\dot{r}(t)$ and the last n elements of $\tilde{x}$ are physical displacements $r(t)$. This results in the state-space realization of equation (4)

$$\dot{\tilde{x}} = \tilde{A}\tilde{x}(t) + \tilde{B}u(t) \tag{4}$$

with

$$\tilde{A} = \begin{bmatrix} [-m^{-1}c] & [-m^{-1}k] \\ [I] & [0] \end{bmatrix}_{2n\times 2n} \tag{4a}$$

$$\tilde{B} = \begin{bmatrix} [-m^{-1}b] \\ [0] \end{bmatrix}_{2n\times m} \tag{4b}$$

For this study, we assume that the system's actuators and sensors are colocated, although this is *not* a requirement of the LQG/LTR technique. It is also assumed that the sensors measure the physical displacements of the LSS. This results in the measurement equation being

$$y = \tilde{C}\tilde{x} \tag{5}$$

with

$$\tilde{C} = \begin{bmatrix} 0 & b^T \end{bmatrix}_{m\times 2n} \tag{5a}$$

Modal analysis[7] is used to transform the set of simultaneous equations given in (4) into an independent set of equations. To achieve decoupling of the equations, we must assume that the damping matrix [c] is a linear combination of the mass [m] and stiffness [k] matrices as shown in equation (6).

$$[c] = \alpha[m] + \beta[k] \tag{6}$$

A coordinate transform between the structure's physical coordinates (r) and the modal (or principal) coordinates (η) is given in (7).

$$r = T\eta \tag{7}$$

Here, T is the matrix of eigenvectors obtained from the solution of equation (8).

$$\omega^2[m]T = [k]T \tag{8}$$

These eigenvectors make up the modal matrix and determine the mode shapes. The corresponding values of ω which solve (8) are the modal frequencies or eigenvalues. Substitution of equation (7) into equation (4) results in the state-space representation in modal coordinates of equation (9).

$$\dot{x} = Ax + Bu \tag{9}$$

where

$$A = \begin{bmatrix} [-T^{-1}m^{-1}cT] & [-T^{-1}m^{-1}kT] \\ [I] & [0] \end{bmatrix} \tag{9a}$$

$$B = \begin{bmatrix} [-T^{-1}m^{-1}b] \\ [0] \end{bmatrix} \tag{9b}$$

The state vector (x) is defined as

$$x = \begin{bmatrix} \dot{\eta} \\ \eta \end{bmatrix} \tag{10}$$

Orthoganality of the modal vectors and the definition in equation (11)

$$[-T^{-1}m^{-1}cT] = -2\varsigma\omega_i \tag{11}$$

reduces the A matrix to the form of (12).

$$A = \begin{bmatrix} [-2\varsigma_i\omega_i] & [-\omega_i^2] \\ [I_n] & [0] \end{bmatrix} \tag{12}$$

Each of the four entries in the A matrix of (12) are n×n dimensional and are diagonal. The output equation (5) becomes

$$y = Cx \tag{13}$$

with

$$C = \begin{bmatrix} 0 & b^T T \end{bmatrix} \tag{13a}$$

The size of the plant's truth model must now be reconciled. For control design purposes, a reasonable size model is desirable. Individual perceptions of what is reasonable will vary with the system type, available computational power and the designer's experience. For our purposes, a model reduction will be performed based on an examination of the second-order modes[9]. These indicate the relative combined controllability and observability of the system modes for the given set of inputs and outputs. Modes are selected for retention in the design model that are most controllable and observable. The system is then reformulated as in (14) where the dynamics in the upper portion will be retained in the design model.

$$\begin{bmatrix} \dot{x}_1 \\ \dot{x}_2 \end{bmatrix} = \begin{bmatrix} A_{11} & A_{12} \\ A_{21} & A_{22} \end{bmatrix} \begin{bmatrix} x_1 \\ x_2 \end{bmatrix} + \begin{bmatrix} B_1 \\ B_2 \end{bmatrix} u \qquad y = \begin{bmatrix} C_1 & C_2 \end{bmatrix} \begin{bmatrix} x_1 \\ x_2 \end{bmatrix} \tag{14}$$

By assuming small changes occur in x_2 ($\dot{x}_2 = 0$), the states of x_2 may be solved for in terms of x_1. This residualization results in the reduced-order design of (15)

$$\dot{x}_r = A_r x_r + B_r u \qquad y = C_r x_r + D_r u \tag{15}$$

with

$$A_r = \begin{bmatrix} A_{11} - A_{12} A_{22}^{-1} A_{21} \end{bmatrix} \tag{15a}$$

$$B_r = \begin{bmatrix} B_1 - A_{12} A_{22}^{-1} B_2] \end{bmatrix} \tag{15b}$$

$$C_r = \begin{bmatrix} C_1 - C_2 A_{22}^{-1} A_{21} \end{bmatrix} \tag{15c}$$

$$D_r = \begin{bmatrix} -A_{22}^{-1} B_2 \end{bmatrix} \tag{15d}$$

Note that although the full-order model was assumed not to have a direct-feed term (D matrix), the reduced-order model may have the outputs directly effected by the inputs.

Control design takes place using the low-order design model of equation (15). The objective of the control system is to reduce the amplitude of vibrations to zero or to an acceptable level in a given time. This will most likely require integral action to achieve the desired tracking characteristics and zero steady-state error. The controller must take into account the fact that differences exist between the design model and the truth model. The robustness of designs are measured in terms of multivariable gain and phase margins[11].

The vibration control system in this paper will be evaluated based upon its ability to damp a vibration to .1 percent of its initial amplitude. Factors which will be considered in evaluation include the settling time, control power requirements relative to the initial condition and overall system robustness. A sufficiently rich initial condition will be used to insure significant excitation of the structure.

LQG/LTR DESIGN METHOD

The Linear Quadratic Gaussian with Loop Transfer Recovery (LQG/LTR) method is just one of several robust control design techniques that have been developed in recent years. The technique was formulated in an attempt to regain the desirable stability margins associated with linear quadratic regulators (LQR) that are lost when a Kalman filter is introduced in a linear quadratic gaussian (LQG) formulation[12]. Doyle and Stein developed a procedure called robustness recovery[13] that allows LQG designs to approach the guaranteed stability margins characteristic of full-state designs. The LQG/LTR method is based on robustness recovery, and was formalized by Stein and Athans[14]. Applications of the technique can be found in References 15 through 17.

The standard LQG problem is well known and has been described in detail[18]. LQG/LTR modifies the conventional LQG problem into a loop-shaping problem. Figure 1 and equations (16) through (19) present the problem set-up.

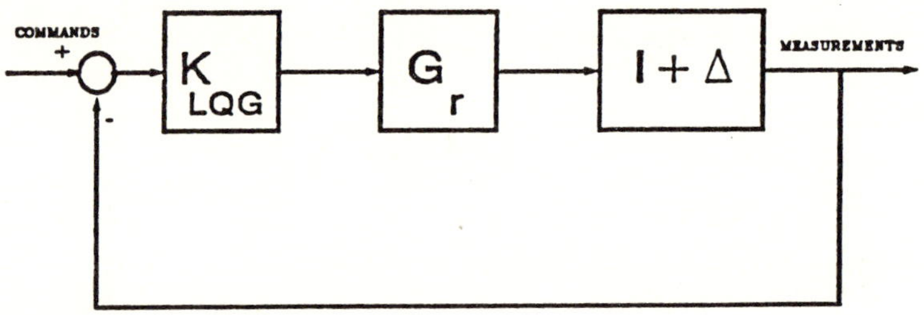

Figure 1. LQG/LTR Problem Block Diagram

$$\dot{x} = Ax + Bu + \Gamma\xi \tag{16}$$

$$y = Cx + Du + \mu In \tag{17}$$

$$z = Hx \tag{18}$$

$$J = E\left\{\lim_{T\to\infty} \frac{1}{T}\int_0^T \left(z^T z + \rho u^T u\right)\right\} \tag{19}$$

The matrix K_{LQG} in Figure 1 is the controller transfer function matrix obtained by the LQG/LTR methodology. The plant G_r is the reduced-order design model derived from the methods of Section 2.

$$G_r(s) = C(sI - A)^{-1}B + D \tag{20}$$

$$G = C\Phi B + D \tag{20a}$$

The subscipts and the Laplace variable s have been dropped for ease of notation. Similarly, Φ is used in equation (20a) in place of $(sI - A)^{-1}$ for convience. The starting point for an LQG/LTR design is the design model's quadruple [A,B,C,D] as shown in equations (16) and (17). In addition, equation (16) includes a noise input distribution matrix (Γ) and a process noise (ξ). Similarly, equation (17) includes a sensor noise distribution matrix (μI) and a sensor noise (n). LQG/LTR considers these noise distribution matrices as tuneable parameters to achieve the desired loop shapes to meet control objectives.

The performance index (J) includes the selected response variables (z) of equation (19) and another free parameter (ρ) which weights the level of control effort. In summary, the LQG/LTR procedure contains four design parameters and a recovery parameter (discussed shortly) which are tuned to achieve desired loop shapes. These include Γ, μ, H, ρ and the recovery parameter. The selection of these parameters in the design process produces the loop-shaping effect. This can be associated with the injection of noise into the system of varying strengths to achieve desired system characteristics.

Given the LSS plant description G, we desire to design a controller (K) to meet the vibration control system objectives. These objectives will be represented in the frequency domain as bounds on the loop transmission GK. The concept of representing the characteristics of a multivariable system through singular values is key to the LQG/LTR methodology[5]. By definition, the singular values of a matrix [P] are given by equation (21).

$$\sigma_i(P) = \sqrt{\lambda_i(P^* P)} \tag{21}$$

The singular values of the loop transmission σ_i(GK) are bounded in the frequency domain based on performance requirements and a representation of the system's uncertainty. System uncertainty can be represented as an output multiplicative perturbation[5] $(I + \Delta(s))$ as shown in the block diagram of Figure 1. Uncertainty is reflected at either the input or output of the plant. The LQG/LTR procedure effectively 'breaks' the loop where this uncertainty is assumed to exist and attempts to recover good stability margins at that point. Typical bounds developed from this uncertainty description and performance requirements are shown on the singular value plot of Figure 2.

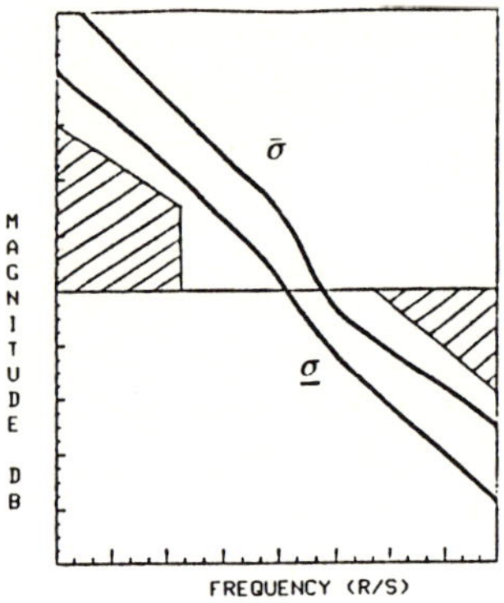

Figure 2. Desired Loop Shapes and Performance and Robustness Bounds

Only the minimum ($\underline{\sigma}$) and maximum ($\bar{\sigma}$) singular value plots of the hypothetical system are shown. These are sufficient since we are concerned with bounds and all other singular values will be bounded if the maximum and minimum singular values are. The system will actually have as many singular values as the lesser of the number of inputs and outputs. In the case of a single-input single-output system, only one singular value plot exists which is identical to the Bode magnitude plot.

The high frequency bound on the system's singular values is related to the description of system uncertainty. The major contribution to uncertainty is usually at high frequencies due to modelling errors and sensor dynamics. This is especially true in an LSS problem where high frequency modes are often neglected in a design model following model reduction. In the case of an output multiplicative System uncertainty is described by a bound $l(\omega)$ on an output multiplicative perturbation matrix ($\Delta(s)$) as shown in equation (22).

$$\bar{\sigma}(\Delta(s)) \leq l(\omega) \tag{22}$$

Figure 3 shows a typical plot of an uncertainty bound $l(\omega)$.

The uncertainty plot is low at low frequencies indicating that model confidence is high in this frequency range. As frequency increases, uncertainty grows indicating relatively less faith is placed in the system representation. In the typical LSS scenario, the uncertainty profile will increase sharply where modes have been excluded in the design model. The determination of the exact bounds of the uncertainty representation are extremely difficult. Consequently, uncertainty profiles are developed based upon general knowledge of the system's potential limitations.

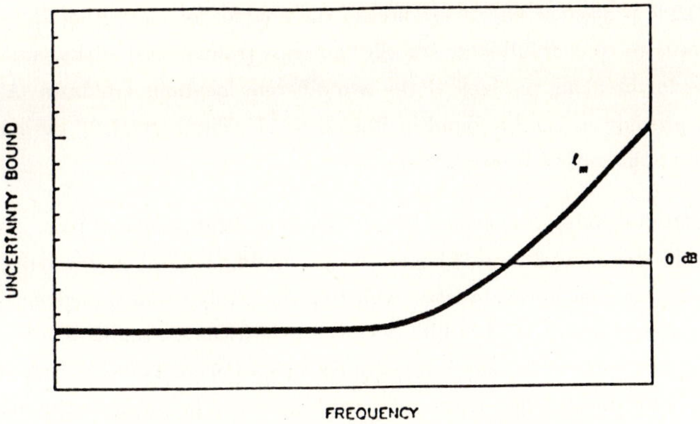

Figure 3. Typical Uncertainty Profile $l(\omega)$

The high frequency robustness barrier is taken directly from the uncertainty profile. The bound is simply $1/l(\omega)$ and essentially results in an upper bound for the bandwidth of the system. System performance and the bound on uncertainty effect the low frequency bound on the singular values of GK. Traditional control objectives like command following and disturbance rejection demand high gains at low frequencies. These objectives are incorporated into a function $p(\omega)$ resulting in a low frequency bound of $p(\omega)/1 - l(\omega)$. As in the case of the uncertainty profile, no explicit guidance is available on the selection of an appropriate performance function $p(\omega)$. Traditionally, integral control action is a desirable property to achieve zero steady-state error. This requirement and the selection of an appropriate system bandwidth often dominate the development of a low frequency bound.

In order to meet performance requirements, integral action is often required in the system. This is achieved by augmenting the plant with a bank of integrators and designing an LQG compensator (K_{LQG}) based on the new *augmented* plant. Once satisfied with the design performance, the augmented dynamics are cascaded with K_{LQG} to form the overall compensator (K) as shown in Figure 4.

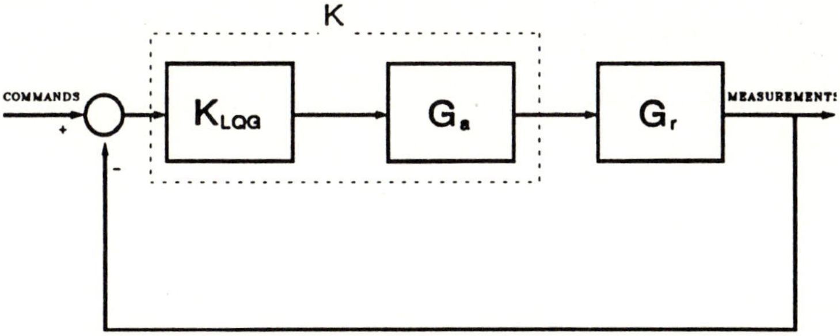

Figure 4. System with Augmented Dynamics

The LQG/LTR procedure effectively breaks the control loop at either the plant input or output and attempts to recover full-state stability margins (robustness) at the same location. The design procedures for breaking the loop at the two different locations are duals. A more detailed discussion of the procedures can be found in Reference 17. The LQG/LTR design procedure by breaking the loop at the output is summarized next.

Standard LQG formulations involve the design of a filter-regulator pair. The LQG/LTR design (with loop broken at the output) begins with a modified Kalman filter design. The design is modified in the sense that noise statistics which are normally treated as given parameters are now viewed as tunable parameters. The filter weighting matrices are tuned to achieve a filter loop shape $\sigma_i(C\Phi k_f)$ that meets the frequency domain specifications or bounds. Here, k_f is the filter gain matrix. This loop shape is deemed the desired loop shape (as it meets the performance and stability bounds) and posesses guaranteed excellent stability margins.

In order to find the desired loop shape, a Riccati equation must be solved. This can be time consuming computationally while iteratively attempting to find loop shapes that satisfy performance and robustness barriers. Consequently, the low frequency approximation, shown on the left hand side of equation (23), is used which does not involve repeated solution of the Riccati equation.

$$\sigma_i\left(\frac{1}{\sqrt{\mu}}C\Phi\Gamma\right) \approx \sigma_i\left(C\Phi k_f\right) \tag{23}$$

By selecting values for the matrix Γ and scalar μ, the approximate filter loop shapes can be adjusted to attempt to meet the design specifications. Once satisfied with the resulting approximate loop shapes, the values of Γ and μ are used in the Algebraic Riccati equation of (24).

$$0 = A\Sigma + \Sigma A^T + \Gamma\Gamma^T - \frac{1}{\mu}\Sigma C^T\Sigma \tag{24}$$

Solution of (24) yields Σ which can be used to find the filter gain k_f.

$$k_f = \frac{1}{\mu}\Sigma C^T \tag{25}$$

The singular values of the filter loop $\sigma_i(C\Phi k_f)$ are the desired loop shapes. These should now meet performance and robustness bounds. This is a full-state design, and as such has a guaranteed gain margin of at least $[\frac{1}{2}, \infty]$ and a phase margin of at least 60 degrees.

The regulator is now designed to attempt to recover to the desired loop shapes of the filter. The state weighting matrix ($[Q_c]$) must be selected as in (26)

$$Q_c = H^T H + q^2 C^T V C \tag{26}$$

and the control weighting matrix($[R_c]$) as in (27)

$$R_c = \rho[\mathrm{I_m}] \tag{27}$$

to effect recovery. The recovery parameter (q^2) in equation (26) is used to adjust how closely the singular values of GK approach the desired loop shapes. Recovery does not come without expense.

Larger control signals are required as recovery continues. The recovery process is halted short of theoretical limits based on available actuator control power. The matrix (V) in equation (26) is a weighting matrix (often selected to be identity). The control weighting (27) consists of an identity matrix of order m (the number of actuators) and a tuneable scalar parameter ρ.

Successive regulator designs with increasing values of the recovery parameter (q^2) cause the system loop shape $\sigma_i(GK)$ to approach the desired loop shape. The selection of an acceptable recovery level through (q^2) results in the controller being completely specified. The Riccati solution (P) from equation (28) can be used to obtain the regulator gain (k_c) from (29).

$$0 = PA + A^TP + Q_c - PBR_c^{-1}B^TP \tag{28}$$

$$k_c = R_c^{-1}B^TP \tag{29}$$

The controller (K_{LQG}) is now completely specified by equation (30).

$$K_{LQG} = k_c(sI - A + Bk_c + k_fC)^{-1}k_f \tag{30}$$

If dynamics (G_a) were augmented to the plant, these are now cascaded with the LQG controller (K_{LQG}) to form the overall compensator (K) as shown in Figure 4 and equation (31).

$$K = G_aK_{LQG} \tag{31}$$

The robustness properties of the resulting design can be verified by checking the multivariable stability margins.

Computer implementation of the LQG/LTR control design technique requires two basic capabilities in addition to standard matrix manipulations and simulation ability; solution of Riccati equations and graphical representaion of system singular values. Routines can be formulated that execute the design procedure interactively with the control engineer being prompted for design parameters and the necessary guidance for system development following initial data entry. Computerization of the technique allows iteration to fully exploit the robustness recovery while remaining within control power limitations.

EXAMPLE AND DISCUSSION

An example active control system for an LSS-type structure is designed using the LQG/LTR control design technique. A simple two-bay truss is used to demonstrate some of the important features of the control system design, without the handling problems associated with typical higher-order LSS systems. The design presented is by no means final. Rather, the design is simply a candidate controller which would be tuned according to specific design goals and constraints.

The cantilevered two-bay truss used for this study is shown in Figure 5. The truss was modified by AFWAL/FIB from a similar structural model[19]. The mathematical structure is 100 inches in length and 18 inches high. The problem considered is a two degree-of-freedom problem with motion allowed in the x and y directions only.

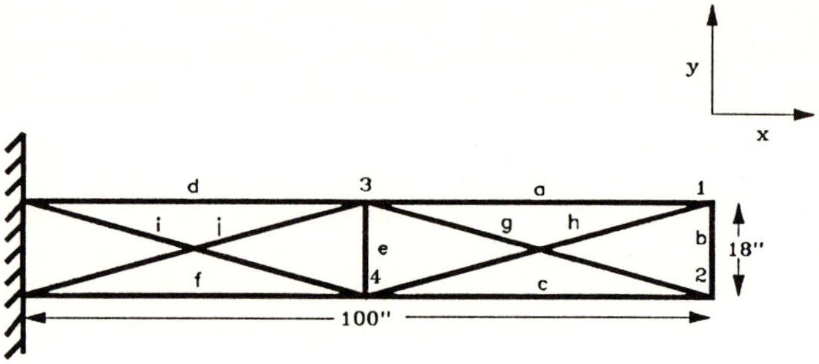

Figure 5. Two Bay Truss Model

Force actuators and position sensors are located at points 1,2,3 and 4 on the truss. Colocation of actuators and sensors is not a requirement of this particular design technique and was selected merely for model simplicity. The actuators act along the y-axis only and are assumed to be limited to $\pm 100 lb_f$. The sensors measure physical displacements in the y direction at the four locations.

The structure is assumed to be constructed from a material with a modulus $E = 10^7$ psi and a weight density of $\rho = .1 lb_f/in^3$. The cross-sectional areas of the structural members are shown in Table 1.

Table 1. Structural Member's Cross-sectional Areas

Member	Area (in^2)	Member	Area (in^2)
a	.00321	f	.01049
b	.00100	g	.00328
c	.00321	h	.00328
d	.01049	i	.00439
e	.00100	j	.00439

Non-structural masses are located at positions 1,2,3 and 4. Table 2 indicates the mass at each location. These masses can be associated with the additional mass from the actuators at the four locations. The non-structural mass is large relative to the structural mass to achieve the low frequency structural modes typical of a large space structure.

Table 2. Non-structural Mass

Location	Mass*
1	1.294
2	1.294
3	1.294
4	1.294

$* = \frac{lb_f - sec^2}{in}$

An 8-mode mathematical model of the structure was provided by the Structures Division

(AFWAL/FIB). Appendix A contains the matrices of equation (2) required to describe the structural dynamics in physical coordinates. These include the mass ([m]), stiffness([k]) and input distribution matrix ([b]). The damping matrix ([c]) is assumed to be zero temporarily. The solution of the eigenvalue problem of equation (8) results in the matrix of eigenvectors ($[T]$) also given in Appendix A. This matrix is used to transform the system to modal coordinates using (7). The resulting system is in modal form as described by (9) thru (13).

Once in modal form, some assumptions are now made regarding the level of damping in the structure. For simplicity, we assume uniform damping exists throughout the structure. This can be achieved through the selection of a particular value for damping (ς) and substition into (12). A passive damping level of $\varsigma = .005$ was chosen for this active control study. The modal frequencies and open-loop pole locations for this uniformly damped 8-mode model are shown in Table 3.

Table 3. Open-Loop Characteristics : 8 Mode Model

Mode	Freq. (Hz)	Freq. (r/s)	Real	Imag	Damping
1	.50000	3.1416	-.01571	±3.1415	.005
2	1.6529	10.3857	-.05193	±10.3856	.005
3	3.6134	22.7040	-.11352	±22.7038	.005
4	4.7020	29.5437	-.14772	±29.5435	.005
5	4.9640	31.1894	-.15595	±31.1890	.005
6	5.2315	32.8702	-.16435	±32.8698	.005
7	8.8844	55.8220	-.27911	±55.8213	.005
8	9.3551	58.7790	-.29390	±58.7790	.005

The 8-mode (16th order) truth model is now reduced for design purposes. Examination of the system's second-order modes indicate the system is most controllable and observable with respect to the first two modes. The system is reordered into the form of (14) and reduced to a 4th order system (2 modes). The resulting design model is shown in equation (32).

$$\dot{x} = Ax + Bu \qquad\qquad y = Cx + Du \tag{32}$$

with

$$A = \begin{bmatrix} -.0314 & 0 & -9.8694 & 0 \\ 0 & -.1039 & 0 & -107.86 \\ 1 & 0 & 0 & 0 \\ 0 & 1 & 0 & 0 \end{bmatrix} \tag{32a}$$

$$B = \begin{bmatrix} .3142 & .3142 & .1161 & .1161 \\ -.1040 & -.1040 & .3337 & .3337 \\ 0 & 0 & 0 & 0 \\ 0 & 0 & 0 & 0 \end{bmatrix} \tag{32b}$$

$$C = \begin{bmatrix} 0 & 0 & 1 & -.3117 \\ 0 & 0 & 1 & -.3117 \\ 0 & 0 & .3402 & 1 \\ 0 & 0 & .3402 & 1 \end{bmatrix} \tag{32c}$$

$$D = \begin{bmatrix} .0004 & -.0004 & 0 & 0 \\ -.0004 & .0004 & 0 & 0 \\ 0 & 0 & .0004 & 0 \\ 0 & 0 & -.0004 & .0004 \end{bmatrix} \qquad (32d)$$

The maximum singular values of the truth model and the reduced-order design model are shown in Figure 6.

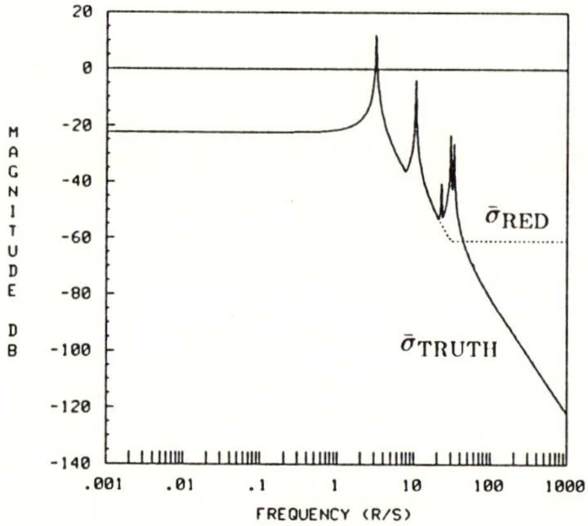

Figure 6. Singular Values of Truth and Design Models

The maximum singular values of the two models are nearly indentical at low frequency. The same is true for the minimum singular values (not shown). Open-loop initial condition responses (not shown) indicate that the models compare well.

An uncertainty profile is established based on the high frequency modelling uncertainty and the plant uncertainty between the design and truth models. This unstructured uncertainty is modelled as a bound on an output multiplicative perturbation as in equation (22). A bound $l(\omega)$ is shown in Figure 7 which approximates the variations between the design and truth models ($\Delta(s)$) as well as high frequency uncertainties. The LQG/LTR procedure uses the bound $l(\omega)$ during design.

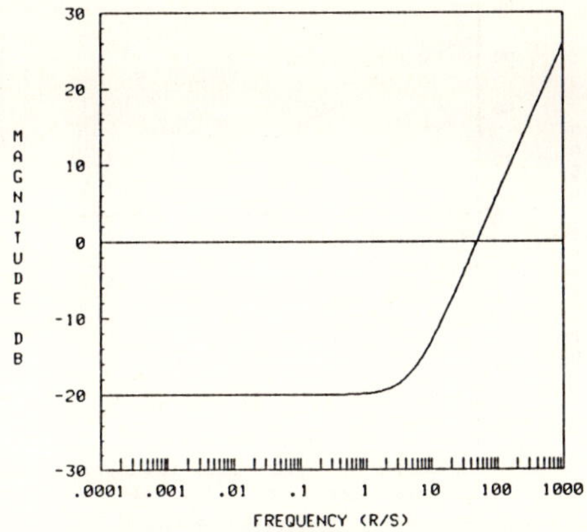

Figure 7. Uncertainty Profile

Uncertainty is low at frequencies below 10 r/s where model confidence is high. The two modes in this frequency range are included in both the truth and design models. Uncertainty increases above 10 r/s due to traditional modelling difficulties at high frequency and the absence of modes 3-8 in the design model.

Evaluation of the control system will be based upon the system response to an initial condition. The particular initial condition was selected to excite a significant number of system modes. Excitation levels were judged by the relative modal amplitudes (η). The initial condition vector of (33) given in physical coordinates excites the first, second, fifth and seventh modes of the structure. This vector corresponds to a tip displacement of approximately one inch and a mid-station displacement of nearly two inches. The initial velocity of the truss is zero.

$$x_0 = [[0]_{1\times 8} \quad .153 \quad 1.00 \quad -.153 \quad 1.00 \quad -.0785 \quad 1.97 \quad .0785 \quad 1.97\ ]^T \tag{33}$$

Examination of Figure 6 indicates that the plant does not have natural integral action. This is seen by from the flat singular values at low frequencies. To achieve zero steady-state error in the system, the design model (G_r) is augmented with a bank of integrators. A controller (K_{LQG}) will be designed upon the augmented plant (G') of equation (34).

$$G' = G_r G_a \tag{34}$$

where

$$G_a = \frac{1}{s}[\mathrm{I}]_4 \tag{34a}$$

An approximate filter loop shape is now designed by selecting Γ and μ of equation (23). The upper system bandwidth limit is approximately 10 r/s based on the uncertainty profile. Through design iteration, values of μ = .000025 and $\Gamma = B$ were found to satisfy the frequency domain bounds with the approximate filter loop shape $\sigma(\frac{1}{\sqrt{\mu}}C\Phi\Gamma)$ shown in Figure 8.

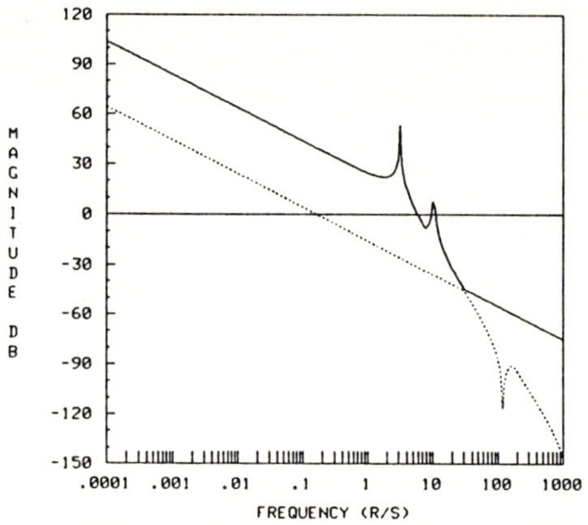

Figure 8. Approximate Filter Loop Shape ($\sigma(\frac{1}{\sqrt{\mu}}C\Phi\Gamma)$)

Because of the particular selection of $\Gamma = B$, the loop shape of Figure 8 is simply the open-loop plant augmented with integrators and scaled by $\frac{1}{\sqrt{\mu}}$. The characteristic -20 dB/decade slope from the integral action can be seen at low frequency.

Equations (24) and (25) are solved to obtain the filter gain which results in the closed-loop pole locations shown in Table 4. The seperation principle dictates that these poles are a subset of the system's closed-loop poles.

Table 4. Filter Closed-Loop Poles $\lambda_i[A - k_f C]$

Freq. (r/s)	Real	Imag	Damping
10.4716	-.6697	± 10.4502 i	.064
5.6758	-2.3628	± 5.1606 i	.416
4.7285	-4.7285	0	1
1.3428	-1.3428	0	1
.1766	-.1766	0	1
.1618	-.1618	0	1

The filter loop shape $\bar{\sigma}(C\Phi k_f)$ is plotted in Figure 9 along with its low frequency approximation. The full-state filter loop shows a characteristic -20 dB/decade rolloff at high frequency. This is the desired system loop shape which will now be recovered during the regulator design.

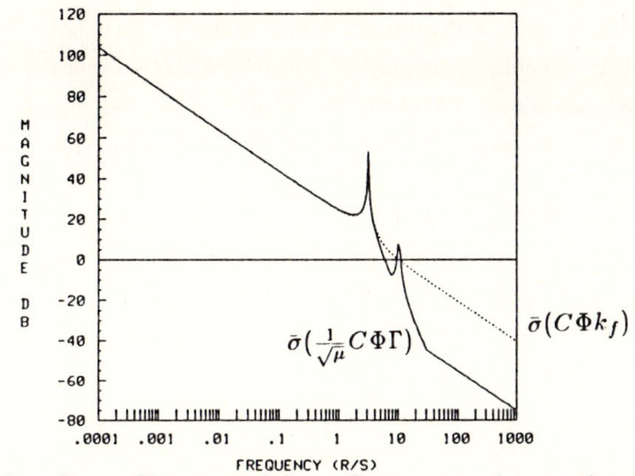

Figure 9. Filter Loop Shape and Low Frequency Approximation

The matrices $[Q_c]$ and $[R_c]$ are selected as in equations (26) and (27). The response matrix ([H]) is selected as $H = C$ and the scalar (ρ) is chosen to be unity. Designs were accomplished for increasing values of the recovery parameter (q^2). Figure 10 shows how $\bar{\sigma}(GK)$ compares with the desired loop shape $\bar{\sigma}(C\Phi k_f)$ for increasing values of q^2. Only the maximum singular value is plotted for figure clarity. Selection of the final value for the recovery parameter represents a tradeoff between competing issues.

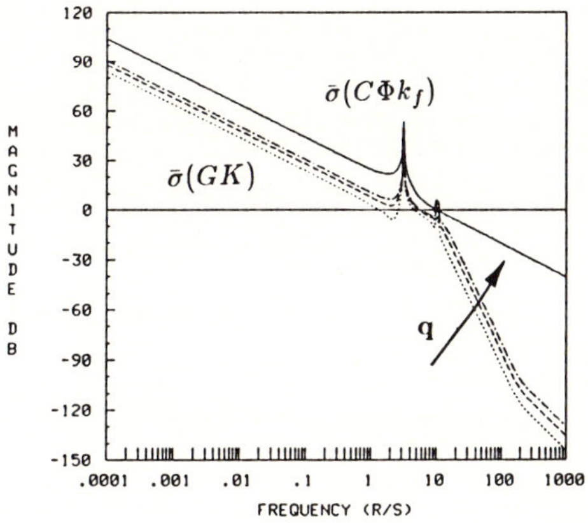

Figure 10. Recovery of Maximum Singular Value

As q^2 is increased during recovery, robustness (as measured by multivariable stability margins) increases. However, the control effort required to accomplish control tasks also increases. Consequently, robustness can be recovered only to that level allowable within the given control power limitations. This tradeoff between robustness recovery and required control power highlights the benefit of computerization of this technique. Computer implementation allows for numerous design iterations that readily allow comparison of performance, robustness and control requirements of competing designs. For this example, actuator limits of $\pm 100 lb_f$ have been imposed. This allows the recovery to continue until $q^2 = 500,000$ without exceeding control limits. Solution of the regulator algebraic Riccati equation (28) is used to find the regulator gain (28). The regulator gain matrix results in the closed-loop regulator poles for the design model shown in Table 5.

Table 5. Regulator Closed-Loop Poles $\lambda_i[A - Bk_c]$

Freq. (r/s)	Real	Imag	Damping
11.1837	-2.0686	± 10.9907 i	.185
8.3438	-3.8624	± 7.3964 i	.463
7.7350	-7.7350	0	1
4.1624	-4.1624	0	1
.6244	-.6244	0	1
.5719	-.5719	0	1

The poles of Table 4 and Table 5 make up the closed-loop poles of the system. The compensator poles are shown in Table 6. Notice the augmented integrators are included as part of the compensator.

Table 6. Compensator Poles

Compensator Poles
$-5.0720 \pm 12.0064i$
$-2.5881 \pm 11.4316i$
-14.7367
-5.7035
-.8010
-.7336
0
0
0
0

The final system loop shape $\sigma(GK)$ is shown in Figure 11. The bandwidth of the system can be seen to be approximately 10 r/s which is reasonable.

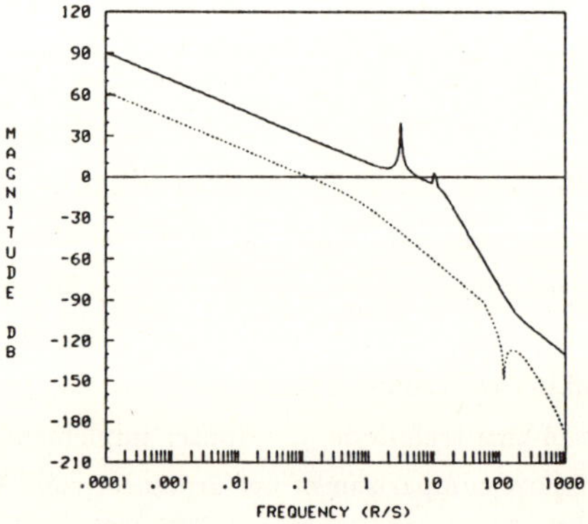

Figure 11. Final System Loop Shape ($\sigma(GK)$)

The control system was developed based upon a *design* model and a description of the differences between this model and the truth model and real world. The performance of the system must be tested against the truth model which is the closest thing to the real world used in this study. Although the controller was *not* specifically designed for the truth model, the system should remain stable and demonstrate acceptable performance. Simulation plots of the *truth model* are shown in Figures (12) through (15). Open and closed-loop plots are shown for comparison. The responses and the control signals at the upper and lower locations are nearly identical. This is a result of the initial condition significantly excited only the longitudinal modes. Breathing (lateral) modes are not directly excited. For convienience, only a single plot is shown at the tip and mid-station locations.

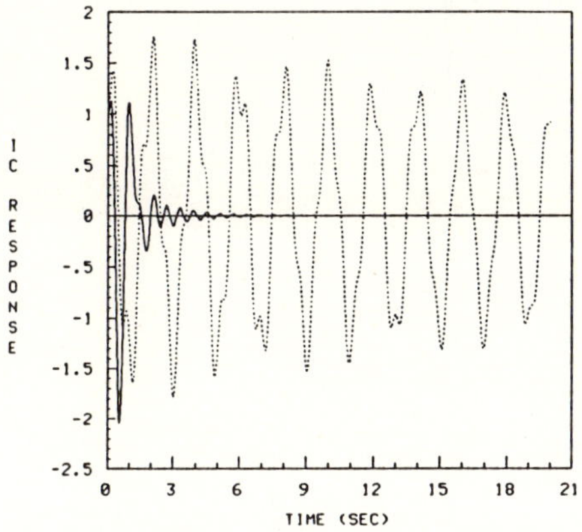

Figure 12. Open and Closed-loop Initial Condition Response (Tip)

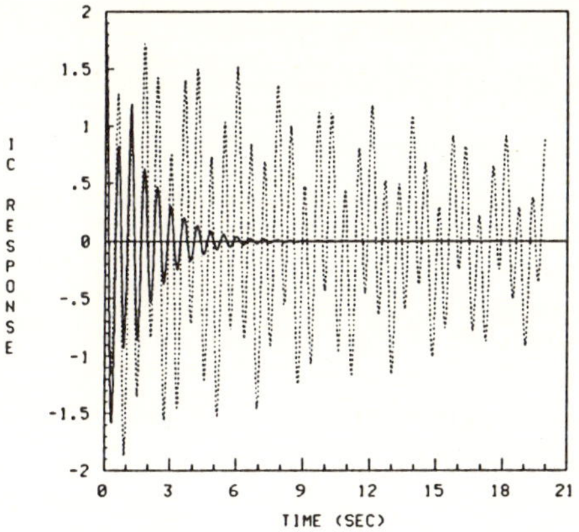

Figure 13. Open and Closed-loop Initial Condition Response (Mid)

The actuator activity required by the control system is linearly related to the magnitude of the initial conditions. The system is driven to relatively high gains by the stringent settling requirements typical of LSS problems. Control effort can be decreased by lengthening the settling time, or reducing the settling criteria and redesigning the control system. Required control power can also be reduced by decreasing the initial condition magnitude. The control signals at the tip and mid-station for this simulation are shown in Figures 14 and 15.

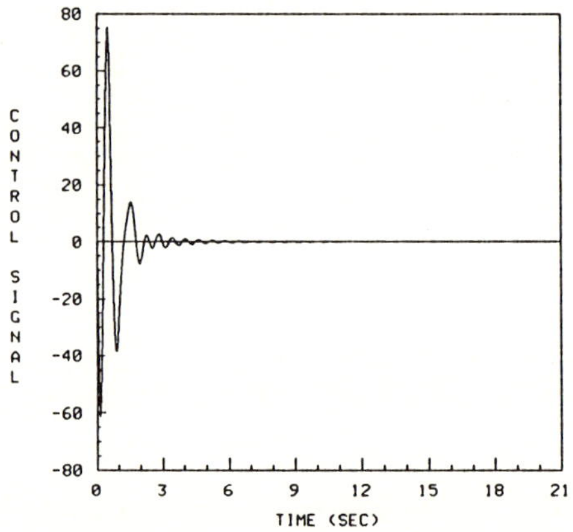

Figure 14. Control Signal for Tip Actuators

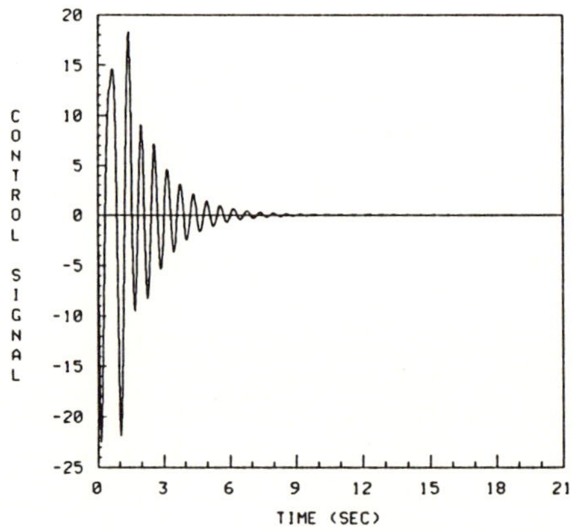

Figure 15. Control Signal for Mid-station Actuators

The initial vibrations are damped to within .1 percent of in initial amplitude in approximately 12 seconds. This is accomplished within the specified control power limits. The design can be altered (fine tuned) for different problem specifications or when design priorities are altered. The design process is always a trade-off between competing requirements. In this problem, a desired loop shape was specified that was expected to give good system performance and robustness. The recovery to that loop shape was halted (full recovery exists only in theory) based on the specified control power requirements. The resulting system robustness is measured in terms of multivariable stability margins.

The vibration control system has a gain margin of [-3.2 dB,+2.75 dB] and a 17.75 degree phase margin. These margins are acceptable for the purposes of this study. Specific application requirements will dictate what acceptable stability margins are for different problems.

CONCLUSIONS

This study has considered the use of active control for vibration damping in the framework of the LSS problem. A uniform level of passive damping was assumed throughout the structure. The authors recognize the potential of both passive and active control techniques and their ultimate synergistic application to solve the vibration damping problem. Assumptions were made concerning the level of passive damping in the structure to limit the scope of the work.

The main contribution of this paper is the demonstration of the use of an active control technique to address the vibration damping problem. The Linear Quadratic Gaussian with Loop Transfer Recovery (LQG/LTR) is an example of a robust design technique that allows the incorporation of a description of system uncertainty in the design. Unfortunately, the available descriptions of system uncertainties often fall short of the potential benefits available from these design methods. However, uncertainty is an important issue in the vibration damping and should be addressed. Continued examination of these issues may provide an impetus for obtaining improved descriptions of uncertainty.

With the use of robustness recovery, the importance of actuation control power in the vibration damping problem is highlighted. As additional robustness is desired, control requirements also increase. A design balance must be struct depending on the specific application. Again, we look for improvements in actuation to parallel the advances in modelling and control techniques to more adequately address the vibration damping problem.

Acknowledgements

The authors appreciate the assistance of Dr. V.B. Venkayya and Ms. V.A. Tischler (AFWAL/FIB) /FIB) for providing the structural model and technical discussions.

REFERENCES

1. J.M. Hedgepeth and M. Mobrem, "Investigation of Passive Damping of Large Space Truss Structures," Vibration Damping Workshop II, Las Vegas, NV, 5-7 March 1986.

2. H. Ashley and D.L. Edberg, "On the Virtues and Prospects for Passive Damping in Large Space Structures," Vibration Damping Workshop II, Las Vegas Nv, 5-7 March 1986.

3. D.R. Morgenthaler and R.N. Gehling, "Design and Analysis of the PACOSS Representative System," Vibration Damping Workshop II, Las Vegas, NV, 5-7 March 1986.

4. J.C. Doyle and G. Stein, "Multivariable Feedback Design Concepts for a Classical/Modern

Synthesis," IEEE Transactions on Automatic Control, Vol. AC-26, No. 1, Febuary 1981.

5. M. Vidyasagar and H Kimura, "Robust Controller for Uncertain Linear Multivariable Systems," Proc. IFAC World Congress, Budapest, Hungary, 1984.

6. R.K Yedavalli and S.S. Banda, "Robust Stability and Regulation in the Control of Large Space Structures," Proc. AIAA Guidance and Control Conference, Snowmass, CO. AIAA-85-1969-CP, August 1985.

7. L. Meirovich, "Elements of Vibrational Analysis," McGraw Hill, 1975.

8. R.E. Skelton, "Cost Decomposition of Linear Systems with Application to Model Reduction," International Journal of Control, Vol32, No. 6, pp.1031-1055, 1980.

9. D. Enns, "Model Reduction for Control System Design," PhD Thesis, Stanford University, June 1984.

10. K. Glover, "All Optimal Hankel-Norm Approximations of Linear Multivariable Systems and their L^∞-Error Bounds," International Journal of Control, Vol. 39, No. 6, pp 1115-1193, 1984.

11. M.G. Safonov and M. Athans, "Gain and Phase Margins for Multiloop LQG Regulators," IEEE Transactions on Automatic Control, Vol AC-22, pp 173-179, April 1977.

12. J.C. Doyle, "Guaranteed Margins for LQG Regulators," IEEE Transactions on Automatic Control, Vol. AC-23, No. 4, pp 756-757, August 1978.

13. J.C. Doyle and G. Stein, "Robustness with Observers," IEEE Transactions on Automatic Control, Vol AC-24, No. 4 pp 607-611, August 1977.

14. G. Stein and M. Athans, "The LQG/LTR Multivariable Control System Design Method," Proc. American Control Conference, San Diego, June 1984, also available as MIT LIDS-P-1384, May 1984.

15. M. Athans, P. Kappasouris, E. Kappas, and H. Spang III, "Linear-Quadratic-Gaussian with Loop-Transfer-Recovery Methodology for the F-100 Engine," Journal of Guidance, Control and Dynamics, Vol 9, No. 1, pp 45-52, Jan-Feb 1986.

16. J.A. Mette Jr., "Multivariable Control of a Submarine using the LQG/LTR Methodology," LIDS-TH-1468, M.S. Thesis, MIT, May 1985.

17. D.B. Ridgely and S.S. Banda, "Introduction to Robust Multivariable Control," AFWAL-TR-85-3102, Wright-Patterson Air Force Base, Dayton, OH.

18. H. Kwakernaak and R. Sivan, Linear Optimal Control Systems," Wiley Interscience, New York, 1972.

19. V.B Venkayya and V.A. Tischler, "Frequency Control and the Effects on the Dynamic Response

of Flexible Structures," 25th Structures, Structural Dynamics and Materials Conference, AIAA-84-1044-CP, Palm Springs, CA, May 1986.

APPENDIX A

MASS =

1.2940	0.0000	0.0000	0.0000	0.0000	0.0000	0.0000	0.0000
0.0000	1.2940	0.0000	0.0000	0.0000	0.0000	0.0000	0.0000
0.0000	0.0000	1.2940	0.0000	0.0000	0.0000	0.0000	0.0000
0.0000	0.0000	0.0000	1.2940	0.0000	0.0000	0.0000	0.0000
0.0000	0.0000	0.0000	0.0000	1.2941	0.0000	0.0000	0.0000
0.0000	0.0000	0.0000	0.0000	0.0000	1.2941	0.0000	0.0000
0.0000	0.0000	0.0000	0.0000	0.0000	0.0000	1.2941	0.0000
0.0000	0.0000	0.0000	0.0000	0.0000	0.0000	0.0000	1.2941

EVECTORS =

-0.1150	0.2463	1.0000	-0.0003	-1.0000	0.3753	-0.0318	-0.3672
1.0000	-0.3117	-0.2021	1.0000	-0.1812	0.5025	0.0477	-0.0818
0.1150	-0.2463	1.0000	-0.0003	1.0000	0.3753	0.0318	-0.3672
1.0000	-0.3117	0.2021	-1.0000	-0.1812	-0.5025	0.0477	0.0818
-0.0575	0.0047	0.3494	0.0799	-0.0266	0.1823	1.0000	1.0000
0.3402	1.0000	-0.3374	-0.5170	0.1899	1.0000	0.0181	-0.0035
0.0575	-0.0047	0.3494	0.0799	0.0266	0.1823	-1.0000	1.0000
0.3402	1.0000	0.3374	0.5170	0.1899	-1.0000	0.0181	0.0035

STIFF =

1.0D+03 *

1.1885	0.1966	0.0000	0.0000	-0.6424	0.0000	-0.5461	-0.1966
0.1966	0.6263	0.0000	-0.5556	0.0000	0.0000	-0.1966	-0.0708
0.0000	0.0000	1.1885	-0.1966	-0.5461	0.1966	-0.6424	0.0000
0.0000	-0.5556	-0.1966	0.6263	0.1966	-0.0708	0.0000	0.0000
-0.6424	0.0000	-0.5461	0.1966	4.0191	0.0669	0.0000	0.0000
0.0000	0.0000	0.1966	-0.0708	0.0669	0.7212	0.0000	-0.5556
-0.5461	-0.1966	-0.6424	0.0000	0.0000	0.0000	4.0191	-0.0669
-0.1966	-0.0708	0.0000	0.0000	0.0000	-0.5556	-0.0669	0.7212

SMALLB =

0.	0.	0.	0.
1.	0.	0.	0.
0.	0.	0.	0.
0.	1.	0.	0.
0.	0.	0.	0.
0.	0.	1.	0.
0.	0.	0.	0.
0.	0.	0.	1.

Optimal Projection for Uncertain Systems (OPUS): A Unified Theory of Reduced-Order, Robust Control Design

Dennis S. Bernstein and David C. Hyland
Harris Corporation
Government Aerospace Systems Division
Melbourne, Florida 32902

This research was supported in part by the Air Force Office of Scientific Research under contracts F49620-86-C-0002 and F49620-86-C-0038.

Abstract

OPUS (Optimal Projection for Uncertain Systems) provides new machinery for designing active controllers for suppressing vibration in flexible structures. The purpose of this paper is to review this machinery and demonstrate its practical value in addressing the structural control problem.

1. Introduction

For many years it has been widely recognized that the desire to orbit large, lightweight space structures possessing high-performance capabilities would require active feedback control techniques. More generally, the need for such techniques may arise due to the combinations of either 1) moderate performance requirements for highly flexible structures with low-frequency modes or 2) stringent performance requirements for semi-rigid structures with relatively high-frequency modes (Figure 1). Applications include pointing, slewing, and aperture shape control for optical and RF systems.

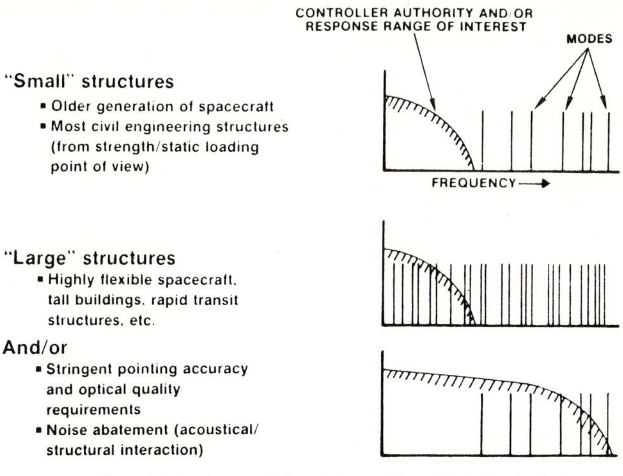
CONTROLLER AUTHORITY AND/OR RESPONSE RANGE OF INTEREST
MODES
"Small" structures
• Older generation of spacecraft
• Most civil engineering structures (from strength/static loading point of view)
FREQUENCY
"Large" structures
• Highly flexible spacecraft, tall buildings, rapid transit structures, etc.
And/or
• Stringent pointing accuracy and optical quality requirements
• Noise abatement (acoustical/structural interaction)

Figure 1. The Need for Active Structural Control Arises From Stringent Performance Requirements or Low-Frequency Modes

Springer Series in Computational Mechanics
S. N. Atluri, A. K. Amos (Eds.)
Large Space Structures: Dynamics and Control

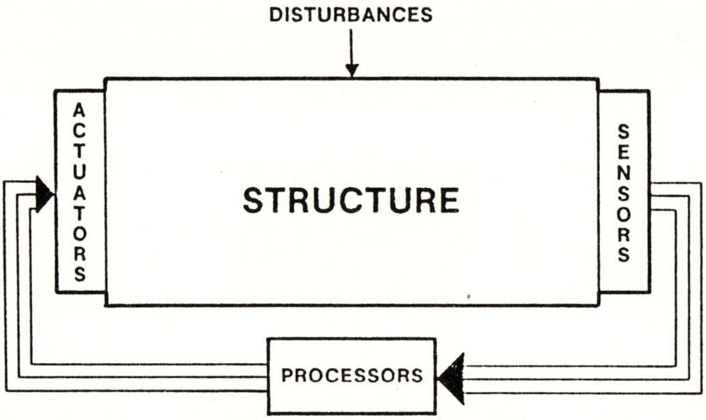

Figure 2. Vibration Control Systems Utilize Sensors, Processors and Actuators to Suppress Disturbances

The problem of active vibration suppression (Figure 2) entails the following considerations:

1. Multiple, highly coupled feedback loops. The potentially large number of sensors and actuators leads to a fully coupled multi-input, multi-output feedback control system.

2. Limited actuator power. The control authority available from on-board actuators is limited by weight, size, cost and power considerations.

3. High-dimensional models. Large structures subjected to broadband disturbances are typically represented by high-order finite element models.

4. Limited processor capacity. Reliability and cost considerations limit the processor capacity available for on-board real-time implementation of the control system.

5. Highly uncertain models with structured uncertainty. Finite element models often exhibit significant error particularly as modal frequency increases. Although modal testing and related identification methods may be used to improve modeling accuracy, residual uncertainty always remains and unpredictable on-orbit changes due to aging, thermal effects, etc., must be tolerated.

6. Stringent performance requirements. Since active space structure control is most relevant in precision applications, it can readily be expected that performance specifications will be particularly stringent.

7. Design efficiency. Because of implementation complexity due to the presence of multiple loops, high dimension, and high levels of uncertainty, the control design approach should efficiently utilize both synthesis and analysis techniques (Figure 3).

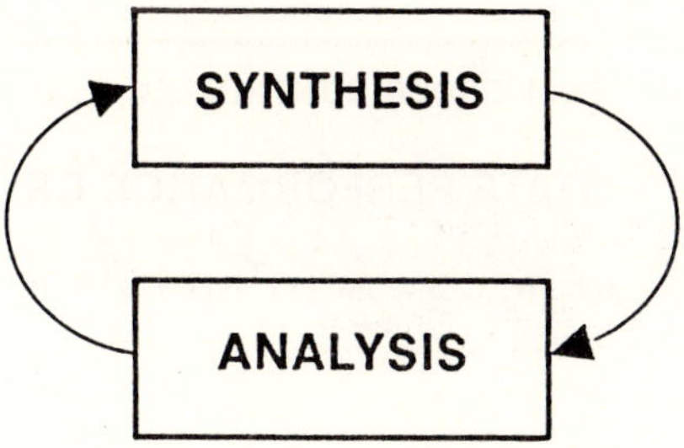

Figure 3. Control-System Design Must Efficiently Utilize Both Synthesis and Analysis Techniques

These considerations pose a considerable challenge to the state-of-the-art in control-design methodologies. For example, the presence of multiple, coupled feedback paths essentially precludes the effectiveness of single-loop design techniques. The sheer number of loops, their interaction, and the need to address a host of other issues render such methods inefficient and unwieldy.

In addition to the presence of multiple loops, the high dimensionality of dynamic models places a severe burden on control-design methodologies. For example, although LQG (linear-quadratic-Gaussian) design is applicable to multi-loop problems, such controllers are of the same order as the structural model (Figures 4 and 5). Thus LQG and similar high-order controllers can be expected to place an unacceptable computational burden on the real-time processing capability. Hence it is not surprising that a variety of techniques have been proposed to reduce the order of LQG controllers. A comparison of several such methods is given in [1].

All of the above difficulties are severely exacerbated by the fact that the dynamic (i.e., finite element) model upon which the control design is predicated may be highly inaccurate in spite of extensive modal identification. Hence, applicable control-design methodologies must account for modeling uncertainties by providing robust (i.e., insensitive) controllers. Furthermore, because of stringent

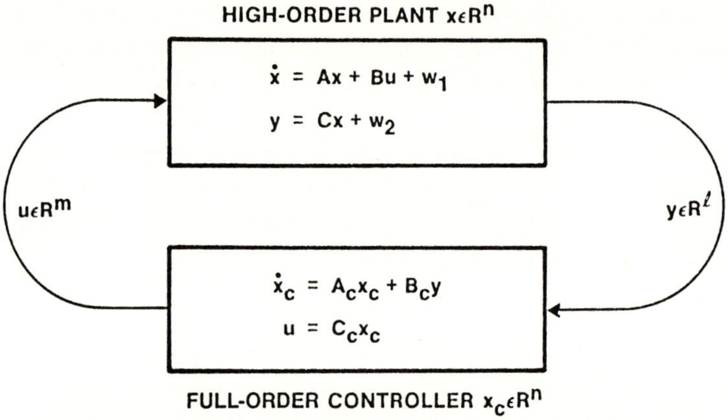

STEADY-STATE PERFORMANCE CRITERION

$$J(A_c, B_c, C_c) = \lim_{t \to \infty} E[x^T R_1 x + u^T R_2 u]$$

Figure 4. LQG Theory Addresses the Problem of Designing a Quadratically Optimal, Full-Order Dynamic Compensator

FULL-ORDER CONTROLLER GAINS

$$A_c = A - Q\bar{\Sigma} - \Sigma P$$

$$B_c = QC^T V_2^{-1}$$

$$C_c = -R_2^{-1} B^T P$$

SEPARATED RICCATI EQUATIONS

$$0 = AQ + QA^T + V_1 - Q\bar{\Sigma}Q \qquad \text{(Kalman Filter)}$$

$$0 = A^T P + PA + R_1 - P\Sigma P \qquad \text{(Regulator)}$$

$$\Sigma \equiv BR_2^{-1} B^T \qquad \bar{\Sigma} \equiv C^T V_2^{-1} C$$

Figure 5. The Optimal Full-Order (LQG) Controller Is Determined by a Pair of Separated Riccati Equations

performance requirements, robust control design must avoid conservatism with respect to modeling uncertainty which may unnecessarily degrade performance. A salient example of conservatism is illustrated in Figure 6. If uncertainty in the modal frequency is complexified in a transfer function setting, then the resulting pole location uncertainty has the form of a disk. This disk, however, intersects the right half plane in violation of energy dissipation. Hence one source of conservatism is the inability to differentiate between physically distinct parameters such as modal frequency and modal damping.

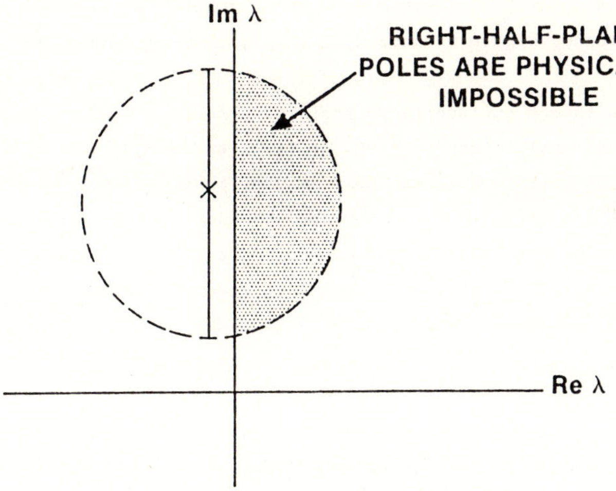

Figure 6. Complexification of Real Parameters May Lead to Robustness Conservatism

Although classical methods are inappropriate for vibration control, a wide variety of modern techniques are available. These include both multi-loop frequency-domain methods and time-domain techniques. A comprehensive review of such methods will not be attempted here. Rather, we shall merely point out aspects of several methods which motivate the philosophy of OPUS development.

As is well known, dynamic models can be transformed (at least in theory) between the frequency and time domains. Significant differences arise, however, in attempting to represent modeling errors. Specifically, model-error characterization of a particular type, which is natural and tractable in one domain, may become extremely cumbersome when transformed into the other domain. For example, consider a state space model with parameter uncertainties arising in the system matrices (A,B,C). Upon transforming to a frequency domain model $G(s) = C(sI-A)^{-1}B$ the parametric uncertainties may perturb the transfer function coefficients in a complicated manner. A more natural measure of uncertainty for transfer functions has been developed in [2] where system uncertainty in the frequency domain is modeled by means of normed neighborhoods in the H-infinity topology. There are limitations with this approach, however, in designing controllers for vibration suppression. For example, as shown in Figure 6, complexification of real-parameter uncertainties such as modal frequencies may yield unnecessary conservatism, while norm bounds often fail to preserve the physical structure of parameter variations. A case in point is the lightly damped oscillator. As shown in [A42], norm bounds predict stability over a frequency range on the order of the damping while in fact the oscillator is unconditionally stable. Furthermore, with regard to processor throughput tradeoffs, modern frequency-domain methods typically yield high-order controllers.

Although LQG addresses performance/actuator and performance/sensor tradeoffs in a multi-loop setting, it fails to incorporate modeling uncertainty. Thus it is not surprising, as shown in [3], that LQG designs fail to possess guaranteed gain margin. Since LQG designs lack such margins, attempts have been made to apply frequency-domain techniques to improve their characteristics. One such method, known as LQG/LTR ([4,5]) seeks to recover the gain margin of full-state-feedback controllers. Specifically, full-state-feedback LQR controllers are guaranteed to remain stable in the face of perturbations of the input matrix B of the form αB where $\alpha \in [1/2, \infty)$. As shown in [6,7], however, the full-state-feedback gain margin fails to provide robustness with respect to perturbations which are not of this form. For instance, the example given in [6] with $B = [0\ 1]^T$ can be destabilized for suitable performance weightings with perturbation $B(\varepsilon) = [\varepsilon\ 1]^T$ for arbitrarily small ε in spite of the 6 dB margin. Furthermore, since LQG/LTR loop shaping is based upon singular value norm bounds, treatment of physically meaningful real parameter variations may lead to unnecessary conservatism. Several approaches have been proposed for circumventing these difficulties (see, e.g., [8]).

The importance of addressing the problem of structured uncertainty in finite element models cannot be overemphasized. Structural characteristics such as modal frequencies, damping ratios, and mode shapes appear explicitly in (A,B,C) state-space models as physically meaningful parameters. Uncertainty in mode shapes, for example, which appear as columns of the B matrix, cannot in general be expected to be of a multiplicative form in accordance with traditional gain-margin specifications. This is precisely the problem illustrated by the example of [6] discussed above. Furthermore, uncertainties in modal frequencies and damping ratios must be carefully differentiated since, roughly speaking, modal frequency uncertainties affect only the imaginary part of the pole location while damping uncertainty affects the real part. Although these and related observations concerning uncertainty in the dynamic characteristics of lightly damped structures may be self evident, they have remained largely unexploited in standard control-design methods.

2. OPUS: New Machinery for Control-System Design

In view of the ability of LQG theory to synthesize dynamic controllers for multi-input, multi-output controllers, it is not surprising that LQG forms the basis for a variety of structural control methods. However, as discussed previously, LQG lacks the ability to address performance/processor and performance/robustness tradeoffs. This situation has thus motivated the development of numerous variants of LQG which entail additional procedures which attempt to remedy these defects. OPUS, however, is distinctly different. Rather than append additional procedures to LQG design, OPUS extends LQG theory itself by generalizing the basic underlying machinery.

As shown in Figure 5, the basic machinery of LQG consists of a pair of separated Riccati equations whose solutions serve to directly and explicitly synthesize the gains of an optimal dynamic compensator. The contribution of OPUS is to directly expand this machinery. The overall approach is illustrated in Figure 7 which portrays two distinct generalizations of the basic LQG machinery. As Figure 7 illustrates, these generalizations can be developed individually when either low-order or robust controllers are desired. The appealing aspect of OPUS, however, is the ability to extend LQG to address both problems simultaneously in a unified manner.

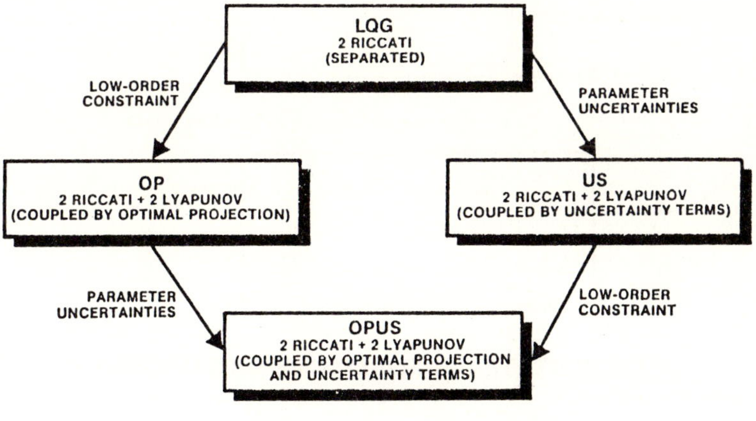

Figure 7. The Standard LQG Result Is Generalized by Both the Fixed-Order Constraint and Modeling of Parameter Uncertainties

In the following sections the generalizations depicted in Figure 7 will be reviewed following the left branch. That is, the optimal projection approach to reduced-order controller design will first be discussed in Section 3 without introducing plant uncertainties. In Section 4 the reduced-order constraint will be retained while considering, in addition, uncertainties in the system model. In each case the discussion will focus on the underlying ideas with a minimum of technical detail.

Clearly, in order for a novel design methodology to be of practical value it must be computationally tractable. Hence Section 5 will present an overview of the current state of algorithm development for solving the OPUS design equations. Finally, Section 6 will briefly summarize further OPUS generalizations of LQG theory which are relevant to structural control.

3. Extensions of LQG to Reduced-Order Dynamic Compensation

The simplest, most direct way to obtain optimal reduced-order controllers is to redevelop the standard LQG result in the presence of a constraint on controller dimension (Figure 8). The mathematical technique required to do this is remarkably straightforward. Specifically, the structure and order of the controller are fixed and the performance is optimized with respect to the controller gains. The resulting necessary conditions obtained using Lagrange multipliers thus characterize the optimal gains.

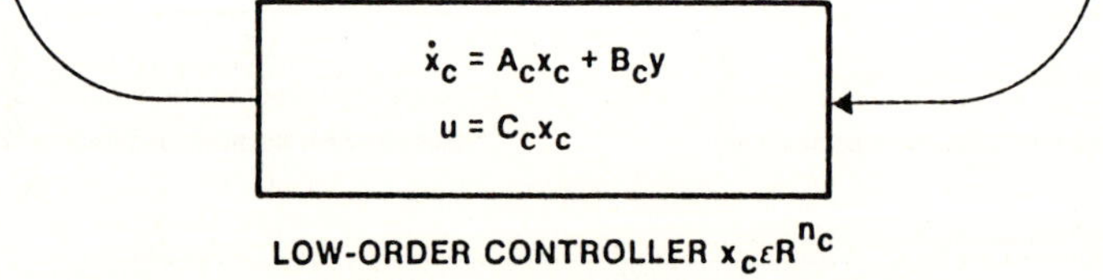

STEADY-STATE PERFORMANCE CRITERION

$$J(A_c, B_c, C_c) = \lim_{t \to \infty} E[x^T R_1 x + u^T R_2 u]$$

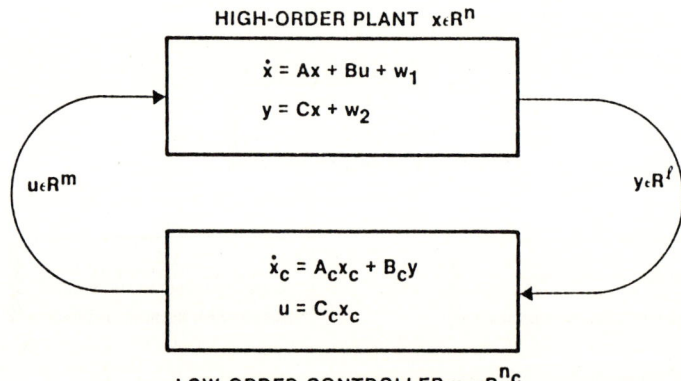

Figure 8. In Accordance With On-Board Processor Requirements, a Reduced-Order Constraint Is Imposed on the Dimension of the Dynamic Compensator

This parameter optimization approach as such is not new and was investigated extensively in the 1970's. Typically, however, the optimality conditions were found to be complex and unwieldy while offering little insight and requiring gradient search methods for numerical solution.

One curious aspect of the parameter optimization literature is that no attempt was made to actually use this direct method to rederive the LQG result itself. Such an exercise, it may be surmised, might reveal hidden structure within the optimality conditions which would shed light on the reduced-order case. Indeed, such an approach led to the realization that an oblique projection (idempotent matrix) is the key to unlocking the unwieldy optimality conditions ([A7,A17]). Although the result is mathematically straightforward, it is by no means obvious since in the full-order (LQG) case the projection is the identity and hence not readily apparent.

By exploiting the presence of the projection, the necessary conditions can be transformed into a coupled system of four algebraic matrix equations consisting of a pair of modified Riccati equations and a pair of modified Lyapunov equations (Figure 9). The coupling is via the oblique projection τ which appears in all four equations and which is determined by the solutions $\hat{Q}$ and $\hat{P}$ of the modified Lyapunov equations. A satisfying feature of the optimality conditions is that in the full-order case the projection becomes the identity, the modified Lyapunov equations drop out, and, since $\tau_\perp = 0$, the modified Riccati equations specialize to the usual separated Riccati equations of LQG theory. Since, furthermore, $G = \Gamma = n \mathrm{x} n$ identity, the standard LQG gain expressions are recovered.

Although the modified Riccati equations specialize to the standard Riccati equations in the full-order case, the modified Lyapunov equations have no counterpart in the standard theory. The role of these equations can be understood by considering the problem of optimal model reduction alone. For this problem the optimal reduced-order model is characterized by a pair of coupled modified Lyapunov equations (see [A22]). Thus the modified Lyapunov equations arising in the reduced-order dynamic-compensation problem are directly analogous to the modified Lyapunov equations arising in model reduction alone. The modified Lyapunov equations arising in the control problem, however, are intimately coupled with the modified Riccati equations. Hence it cannot be expected that reduced-order control-design techniques based upon LQG will generally yield optimal fixed-order controllers (Figure 10). It is interesting to note that several such methods discussed in [1] are based upon balancing which was shown in [A22] to be suboptimal with respect to the quadratic (least squares) optimality criterion.

REDUCED-ORDER CONTROLLER GAINS

$$A_c = \Gamma(A - Q\bar{\Sigma} - \Sigma P)G^T$$

$$B_c = \Gamma Q C^T V_2^{-1}$$

$$C_c = -R_2^{-1} B^T P G^T$$

COUPLED RICCATI/LYAPUNOV EQUATIONS

$$0 = AQ + QA^T + V_1 - Q\bar{\Sigma}Q + \tau_\perp Q\bar{\Sigma}Q\tau_\perp^T$$

$$0 = A^TP + PA + R_1 - P\Sigma P + \tau_\perp^T P\Sigma P\tau_\perp$$

$$0 = (A - \Sigma P)\hat{Q} + \hat{Q}(A - \Sigma P)^T + Q\bar{\Sigma}Q - \tau_\perp Q\bar{\Sigma}Q\tau_\perp^T$$

$$0 = (A - Q\bar{\Sigma})^T\hat{P} + \hat{P}(A - Q\bar{\Sigma}) + P\Sigma P - \tau_\perp^T P\Sigma P\tau_\perp$$

$$\text{rank}\,\hat{Q} = \text{rank}\,\hat{P} = \text{rank}\,\hat{Q}\hat{P} = n_c$$

$$\hat{Q}\hat{P} = G^T M \Gamma \qquad \Gamma G^T = I_{n_c}$$

$$\tau = G^T\Gamma = \hat{Q}\hat{P}(\hat{Q}\hat{P})^{\#} \qquad \tau_\perp = I_n - \tau$$

$$\Sigma = BR_2^{-1}B^T \qquad \bar{\Sigma} = C^T V_2^{-1} C$$

Figure 9. The Optimal Reduced-Order Compensator Is Determined by a Pair of Modified Riccati Equations and a Pair of Modified Lyapunov Equations Coupled by the Oblique Projection τ

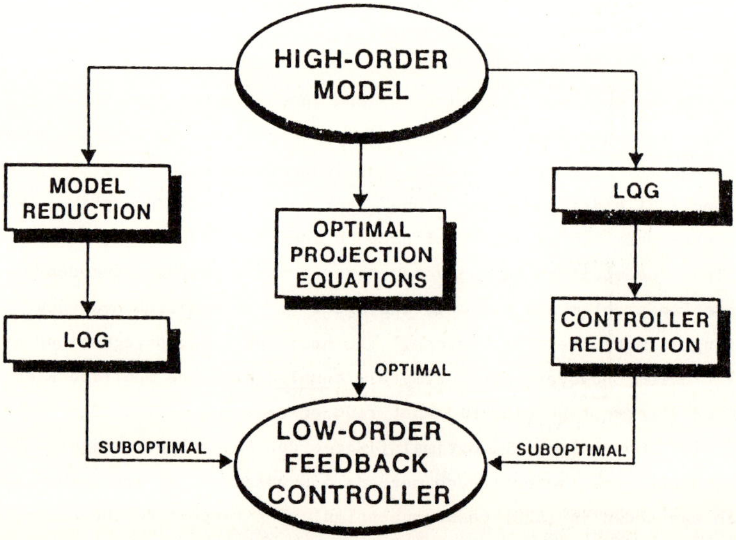

Figure 10. The Optimal Projection Equations Provide a Direct Path to Optimal Reduced-Order Dynamic Compensators

In summary, the optimal projection equations for reduced-order dynamic compensation comprise a direct extension of the basic LQG machinery to the reduced-order control problem. The design equations, which reduce to the standard LQG result in the full-order case, provide direct synthesis of optimal reduced-order controllers in accordance with implementation constraints.

4. Extensions of LQG to Uncertain Modeling

Two fundamental sources of error in modeling flexible structures are truncated modes and parameter uncertainties. Since the optimal projection approach permits the utilization of the full dynamics model, modal truncation can be largely avoided. There remains, however, a tendency to truncate poorly known modes and thus it is essential to incorporate a model of parameter uncertainty in both well-known and poorly known components of the system. Hence the problem formulation of Figure 8 is now generalized in Figure 11 to include uncertain parameters σ_i appearing in the A, B and C matrices. The parameter σ_i is assumed to lie within the interval $[-\delta_i, \delta_i]$ in accordance with identification accuracy. Clearly, when uncertainty is absent, i.e., when A_i, B_i, $C_i = 0$, the reduced-order design problem of Figure 8 is recovered.

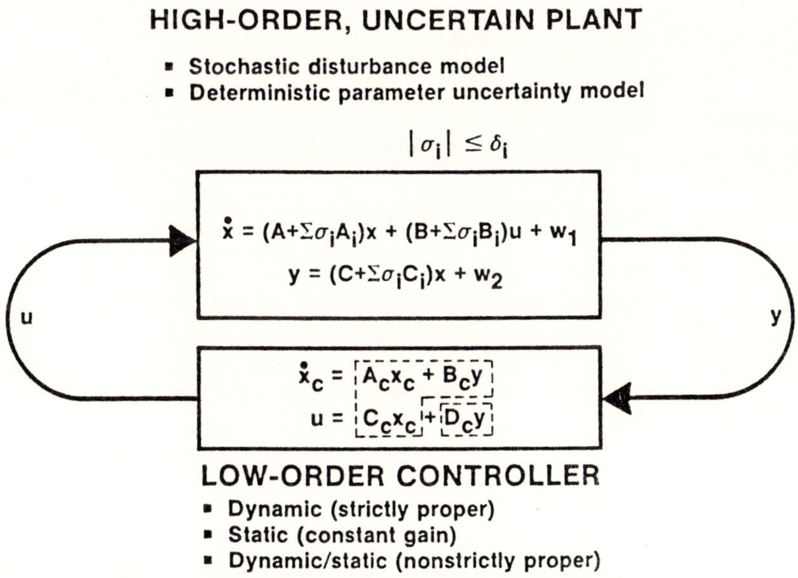

Figure 11. Robust Optimal Projection Design Is Based Upon a Hybrid Uncertainty Model Involving a Deterministic Parameter Uncertainty Model and a Stochastic Disturbance Model

A salient feature of the design model is that uncertainty is modeled in two distinctly different ways. External uncertainty appearing as additive white noise is modeled stochastically. Such a model appears appropriate for disturbances such as coolant flow for which only power spectral data are available. On the other hand, internal uncertainty appearing as parameter variations is modeled deterministically. Such a model appears appropriate for uncertainty arising from directly measurable quantities such as mass and stiffness. Thus the overall uncertainty model is hybrid in the sense that it utilizes both deterministic and stochastic characterizations of uncertainty.

A natural performance measure which accounts for both types of uncertainty characterization involves the usual LQG quadratic criterion averaged over the disturbance statistics and then maximized over the uncertain parameters (Figure 12). Hence this performance measure incorporates on the average and worst case aspects in accordance with physical considerations.

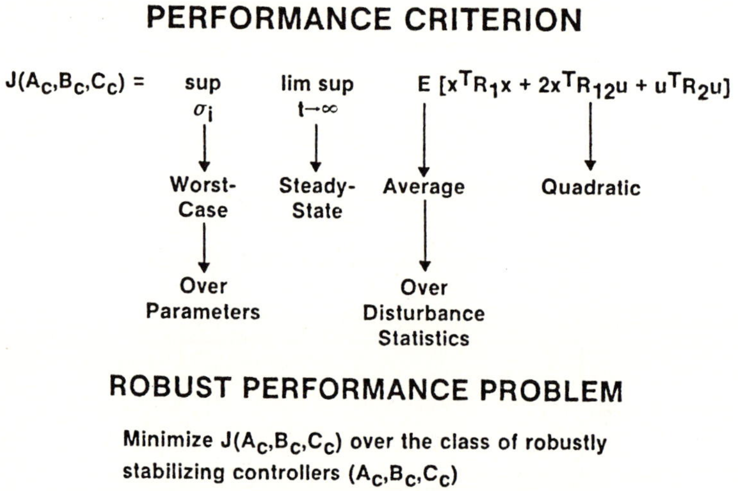
PERFORMANCE CRITERION
$J(A_c,B_c,C_c) = \sup_{\sigma_i} \limsup_{t\to\infty} E[x^TR_1x + 2x^TR_{12}u + u^TR_2u]$
Worst-Case
Steady-State
Average
Quadratic
Over Parameters
Over Disturbance Statistics
ROBUST PERFORMANCE PROBLEM
Minimize $J(A_c,B_c,C_c)$ over the class of robustly stabilizing controllers (A_c,B_c,C_c)

Figure 12. Performance Is Defined To Be Worst Case Over the Uncertain Parameters and Average Over the Disturbance Statistics

The resulting Robust Performance Problem thus involves determining the gains (A_c, B_c, C_c) to minimize the performance J. The static gain D_c can also be included but will not be discussed here. Despite the apparent complexity of the problem, remarkably simple techniques can be used. Specifically, first note that after taking the expected value the performance J has the form

$$J(A_c, B_c, C_c) = \sup_{\sigma_i} \limsup_{t\to\infty} \operatorname{tr} \tilde{Q}(t)\tilde{R}, \tag{4.1}$$

where "tr" denotes trace of a matrix, $\tilde{Q}(t)$ is the covariance of the closed-loop system, and $\tilde{R}$ is an augmented weighting matrix composed of R_1, R_{12} and R_2. The covariance $\tilde{Q}(t)$ satisfies the standard Lyapunov differential equation

$$\dot{\tilde{Q}} = (\tilde{A}+\sum_i \sigma_i \tilde{A}_i)\tilde{Q} + \tilde{Q}(\tilde{A}+\sum_i \sigma_i \tilde{A}_i)^T + \tilde{V}, \tag{4.2}$$

where $\tilde{A}$ is the closed-loop dynamics, $\tilde{A}_i$ is composed of A_i, B_i and C_i, and $\tilde{V}$ is the intensity of external disturbances for the closed-loop system including the plant and measurement noise.

Two distinct approaches to this problem will be considered. The first involves bounding the performance over the class of parameter uncertainties and then choosing the gains to minimize the bound. Since bounding precedes control design this approach is known as robust design via a priori performance bounds. The second approach involves exploiting the nondestabilizing nature of structural systems via weak subsystem interaction.

4.1 Robust Design Via A Priori Performance Bounds

The key step in bounding the performance (4.1) is to replace (4.2) by a modified Lyapunov differential equation of the form

$$\dot{\underline{\tilde{Q}}} = \tilde{A}\underline{\tilde{Q}} + \underline{\tilde{Q}}\tilde{A}^T + \Psi(\underline{\tilde{Q}}) + \tilde{V}, \tag{4.3}$$

where the bound Ψ satisfies the inequality

$$\sum_i \sigma_i (\tilde{A}_i\underline{\tilde{Q}}+\underline{\tilde{Q}}\tilde{A}_i^T) \le \Psi(\underline{\tilde{Q}}) \tag{4.4}$$

over the range of uncertain parameters σ_i and for all candidate feedback gains. Note that the inequality (4.4) is defined in the sense of nonnegative-definite matrices. Now rewrite (4.3) by appropriate addition and subtraction as

$$\dot{\underline{\tilde{Q}}} = (\tilde{A}+\sum_i \sigma_i \tilde{A}_i)\underline{\tilde{Q}} + \underline{\tilde{Q}}(\tilde{A}+\sum_i \sigma_i \tilde{A}_i)^T + \Psi(\underline{\tilde{Q}}) - \sum_i \sigma_i (\tilde{A}_i\underline{\tilde{Q}}+\underline{\tilde{Q}}\tilde{A}_i^T) + \tilde{V}. \tag{4.5}$$

Now subtract (4.2) from (4.5) to obtain

$$\dot{\underline{\tilde{Q}}} - \dot{\tilde{Q}} = (\tilde{A}+\sum_i \sigma_i \tilde{A}_i)(\underline{\tilde{Q}}-\tilde{Q}) + (\underline{\tilde{Q}}-\tilde{Q})(\tilde{A}+\sum_i \sigma_i \tilde{A}_i)^T + \Psi(\underline{\tilde{Q}}) - \sum_i \sigma_i (\tilde{A}_i\underline{\tilde{Q}}+\underline{\tilde{Q}}\tilde{A}_i^T). \tag{4.6}$$

Since by (4.4) the term

$$\Psi(\underset{\sim}{\tilde{Q}}) - \sum \sigma_i (\tilde{A}_i \underset{\sim}{Q} + \underset{\sim}{Q} \tilde{A}_i^T) \tag{4.7}$$

is nonnegative definite, it follows immediately that

$$\tilde{Q} \leq \underset{\sim}{\tilde{Q}} \tag{4.8}$$

over the class of uncertain parameters. Thus the performance (4.1) can be bounded by

$$J(A_c, B_c, C_c) \leq \underset{\sim}{J}(A_c, B_c, C_c) \triangleq \lim_{t \to \infty} \operatorname{tr} \underset{\sim}{\tilde{Q}}\tilde{R}. \tag{4.9}$$

The auxiliary cost $\underset{\sim}{J}$ is thus guaranteed to bound the actual cost J. This leads to the Auxiliary Minimization Problem: Minimize the auxiliary cost $\underset{\sim}{J}$ over the controller gains. The advantage of this approach is that necessary conditions for the Auxiliary Minimization Problem effectively serve as sufficient conditions for robust performance in the original problem. Since the bounding step precedes the optimization procedure, this approach is referred to as robust design via a priori performance bounds. This procedure is philosophically similar to guaranteed cost control ([9,10]). Note that since bounding precedes optimization, the bound (4.4) must hold for all gains since the optimal gains are yet to be determined.

To obtain sufficient conditions for robust stability, the bounding function Ψ must be specified. Since the ordering of nonnegative-definite matrices appearing in (4.4) is not a total ordering, a unique lowest bound should not be expected. Furthermore, each differentiable bound leads to a fundamental extension of the optimal projection equations and thus of the basic LQG machinery. In work thus far, two bounds have been extensively investigated. Only one bound, the right shift/multiplicative white noise bound, will be discussed here. The structured stability radius bound introduced in [11,12] is discussed in [A43].

The right shift/multiplicative white noise bound investigated in [A29,A41] is given by

$$\Psi(\underset{\sim}{\tilde{Q}}) = \sum \delta_i (\alpha_i \underset{\sim}{\tilde{Q}} + \alpha_i^{-1} \tilde{A}_i \underset{\sim}{\tilde{Q}} \tilde{A}_i^T), \tag{4.10}$$

where $\alpha_i > 0$ are arbitrary scalars. Note that this bound consists of two distinct parts which must appear in an appropriate ratio. The first term $\alpha_i \underset{\sim}{\tilde{Q}}$ arises naturally when an exponential time weighting $e^{\alpha_i t}$ is included in the performance measure. As is well known ([13]) this leads to a prescribed uniform stability margin for the closed-loop system (Figure 13). A uniform stability margin, no matter how large, however, does not guarantee robustness with respect to arbitrary parameter variations. The complementary second term $\alpha_i^{-1}\tilde{A}_i\underline{Q}\tilde{A}_i^T$ is crucial in this regard.

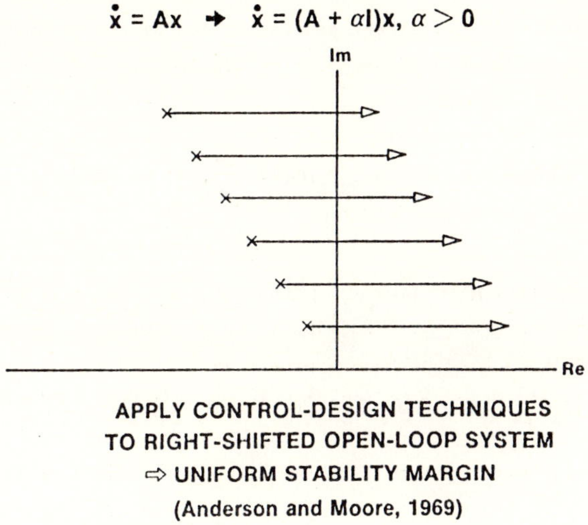

Figure 13. Open-Loop Right-Shifted Dynamics Arising From Exponential Cost Weighting Lead to a Uniform Closed-Loop Stability Margin

Although terms of the form $\tilde{A}_i\underline{Q}\tilde{A}_i^T$ are unfamiliar in robust control design, they arise naturally in stochastic differential equations with <u>multiplicative</u> white noise. That is, if the uncertain parameters σ_i are <u>replaced</u> by white noise processes entering multiplicatively rather than additively, then the covariance equation for $\tilde{Q}$ automatically includes terms of the form $\tilde{A}_iQ\tilde{A}_i^T$. The literature on systems with multiplicative white noise is quite extensive; see [A38] for references. It should be stressed, however, that for our purposes the multiplicative white noise model is <u>not</u> interpreted literally as having physical significance. Rather, multiplicative white noise can be thought of as a useful <u>design model</u> which correctly captures the impact of uncertainty on the performance functional via the state covariance. Furthermore, just as the right shift term alone does not guarantee robustness, neither does the multiplicative white noise term. Both terms must appear simultaneously. Roughly speaking, since multiplicative white noise disturbs the plant though uncertain parameters, the closed-loop system is automatically desensitized to <u>actual</u> parameter variations.

After incorporating the right shift/multiplicative white noise bound (4.10) into (4.3) to obtain a bound $\underline{J}$ for the performance, the optimal projection equations can be rederived following exactly the same parameter optimization procedure discussed in Section 3. Again, the mathematics required is but a straightforward application of Lagrange multipliers. The additional bounding terms are carried through the derivation to yield a direct generalization of the optimal projection equations shown in Figure 14 with gains given in Figure 15.

$$0 = A_s Q + Q A_s^T + \mathcal{A} Q \mathcal{A}^T + V_1 + (\mathcal{A} - \mathcal{B} R_{2s}^{-1} \mathcal{P}_s) \hat{Q} (\mathcal{A} - \mathcal{B} R_{2s}^{-1} \mathcal{P}_s)^T - \mathcal{Q}_s V_{2s}^{-1} \mathcal{Q}_s^T + \tau_\perp \mathcal{Q}_s V_{2s}^{-1} \mathcal{Q}_s^T \tau_\perp^T$$

$$0 = A_s^T P + P A_s + \mathcal{A}^T P \mathcal{A} + R_1 + (\mathcal{A} - \mathcal{Q}_s V_{2s}^{-1} \mathcal{C})^T \hat{P} (\mathcal{A} - \mathcal{Q}_s V_{2s}^{-1} \mathcal{C}) - \mathcal{P}_s^T R_{2s}^{-1} \mathcal{P}_s + \tau_\perp^T \mathcal{P}_s^T R_{2s}^{-1} \mathcal{P}_s \tau_\perp$$

$$0 = (A_s - B_s R_{2s}^{-1} \mathcal{P}_s) \hat{Q} + \hat{Q} (A_s - B_s R_{2s}^{-1} \mathcal{P}_s)^T + \mathcal{Q}_s V_{2s}^{-1} \mathcal{Q}_s^T - \tau_\perp \mathcal{Q}_s V_{2s}^{-1} \mathcal{Q}_s^T \tau_\perp^T$$

$$0 = (A_s - \mathcal{Q}_s V_{2s}^{-1} C_s)^T \hat{P} + \hat{P} (A_s - \mathcal{Q}_s V_{2s}^{-1} C_s) + \mathcal{P}_s^T R_{2s}^{-1} \mathcal{P}_s - \tau_\perp^T \mathcal{P}_s^T R_{2s}^{-1} \mathcal{P}_s \tau_\perp$$

Figure 14. The Robustified Optimal Projection Design Equations Account for Both Reduced-Order Dynamic Compensation and Parametric Uncertainty

GAINS

$$A_c = \Gamma (A_s - B_s R_{2s}^{-1} \mathcal{P}_s - \mathcal{Q}_s V_{2s}^{-1} C_s) G^T$$

$$B_c = \Gamma \mathcal{Q}_s V_{2s}^{-1}$$

$$C_c = -R_{2s}^{-1} \mathcal{P}_s G^T$$

NOTATION

$$\hat{Q}\hat{P} = G^T M \Gamma, \quad \Gamma G^T = I_{n_c} \quad (\Rightarrow \tau = G^T \Gamma = \tau^2)$$

$$\mathcal{A} Q \mathcal{A}^T = \sum_{i=1}^{p} A_i Q A_i^T, \quad \mathcal{A} Q \mathcal{B} = \sum_{i=1}^{p} A_i Q B_i, \text{ etc.}$$

$$R_{2s} = R_2 + \mathcal{B}^T (P + \hat{P}) \mathcal{B} \qquad V_{2s} = V_2 + \mathcal{C} (Q + \hat{Q}) \mathcal{C}^T$$

$$\mathcal{Q}_s = Q C_s^T + V_{12} + \mathcal{A} (Q + \hat{Q}) \mathcal{C}^T \qquad \mathcal{P}_s = B_s^T P + R_{12}^T + \mathcal{B}^T (P + \hat{P}) \mathcal{A}$$

Figure 15. The OPUS Controller Cains Are Explicitly Characterized as a Direct Generalization of the Classical LQG Gains

The robustified optimal projection equations comprise a system of four matrix equations coupled by both the optimal projection and uncertainty terms. When the uncertainty terms are absent, the optimal projection equations of Figure 9 are immediately recovered. On the other hand, if the order of the controller is set equal to the order of the plant, then all terms involving $\tau_{\perp}$ can be deleted. However, in this case the modified Lyapunov equations do not drop out since $\hat{Q}$ and $\hat{P}$ still appear in the modified Riccati equations. Hence the basic machinery of LQG is again extended to include a pair of Lyapunov equations coupled to a generalization of the standard LQG equations. It is interesting to note that a related result in the context of multiplicative noise also appeared in the Soviet literature ([14]). It should also be pointed out that although the modified Lyapunov equations arising in the reduced-order control-design problem have analogues in model reduction, the modified Lyapunov equations appearing in the full-order robustified equations represent new machinery not anticipated in robustness theories. Hence using straightforward mathematical techniques, the basic LQG machinery has again been extended in novel directions.

Solving the design equations shown in Figures 14 and 15 yields controllers with guaranteed levels of robustness. The actual robustness levels may, however, be larger than specified by a priori bounds. Thus, to achieve desired robustification levels for the uncertainty structure specified by the a priori bounds, the design procedure may be utilized within an iterative synthesis/analysis procedure (Figure 16).

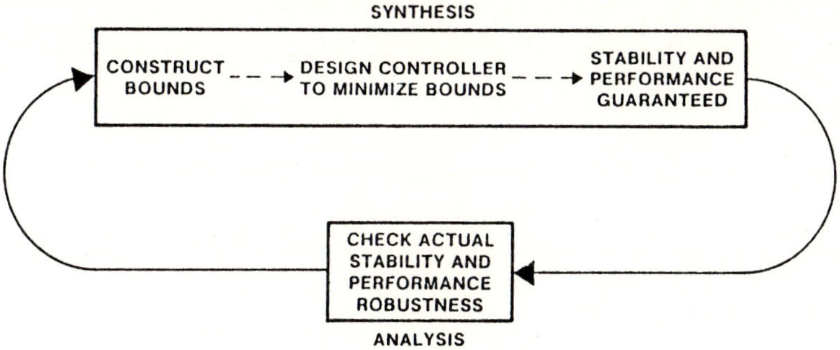

Figure 16. Optimal Projection/Guaranteed Cost Control Provides Direct Synthesis of Robust Dynamic Compensators

4.2 Robust Design Via Weak Subsystem Interaction

The mechanism by which LQG was robustified in Section 4.1 involved bounding the performance over the class of parameter uncertainties and then determining optimal controller gains for the performance bound. As discussed in Section 2, however, flexible structures possess special properties which may, in addition, be exploited to achieve robustness. Specifically, aside from rigid-body modes, energy dissipation implies that mechanical structures are open-loop stable regardless of the level of uncertainty. That is, flexible structures possess only nondestabilizing uncertainties. Hence, in the closed loop, a given controller may or may not render a particular uncertainty destabilizing. A priori bounds on controller performance must, however, be valid for all gains since bounding precedes optimization. Hence, a priori bounding may in certain cases fail to exploit nondestabilizing uncertainties.

A familiar example of a nondestabilizing uncertainty involves uncertain modal frequencies. Such an uncertainty will not, of course, destabilize an uncontrolled (open-loop) structure. If particular modal frequencies are poorly known then it is clearly advisable to avoid applying high authority control. Hence, rather than the right-shift approach of Figure 13, it appears advantageous (although, at first, counterintuitive) to utilize just the opposite, namely, a left shift (Figure 17). Furthermore, in view of the fact that uncertainty usually increases with modal frequency (Figure 18), a variable left shift appears to be more appropriate than a uniform left shift. By left-shifting high-frequency poorly known modes, the control-system design procedure applies correspondingly reduced authority to modes "perceived" as highly damped. Hence the variable left shift can be roughly thought of as a device for achieving suitable authority rolloff. As will be seen, however, the underlying robustification mechanism, namely, weak subsystem interaction, is far more subtle than the approach of classical rolloff techniques. It is also interesting to note that the weak subsystem interaction approach to robustness is entirely distinct from classical robustness approaches which utilize high loop gain to reduce sensitivity.

$$\dot{x} = Ax \rightarrow \dot{x} = (A + \frac{1}{2}\sum_{i=1}^{p} A_i^2)x$$

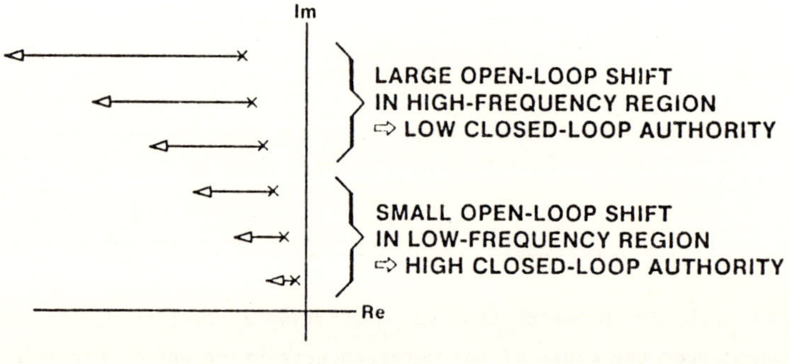

Figure 17. A Variable Left Shift Exploits Open-Loop Nondestabilizing Uncertainties

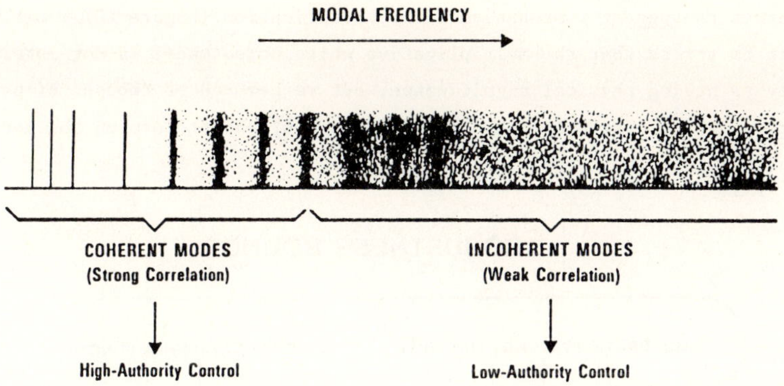

Figure 18. Modal Uncertainty Generally Increases With Frequency

A variable left shift can readily be introduced into the robustified optimal projection design equations by replacing A by

$$A_s = A + \tfrac{1}{2}\sum A_i^2, \tag{4.11}$$

where A_i denotes the structure of modal frequency uncertainty (Figure 19). Most interestingly, such a modification of the dynamics matrix arises naturally from a multiplicative white noise model defined not in the usual Ito sense but rather in the sense of Stratonovich. Thus, as in the a priori bounding approach, a stochastic

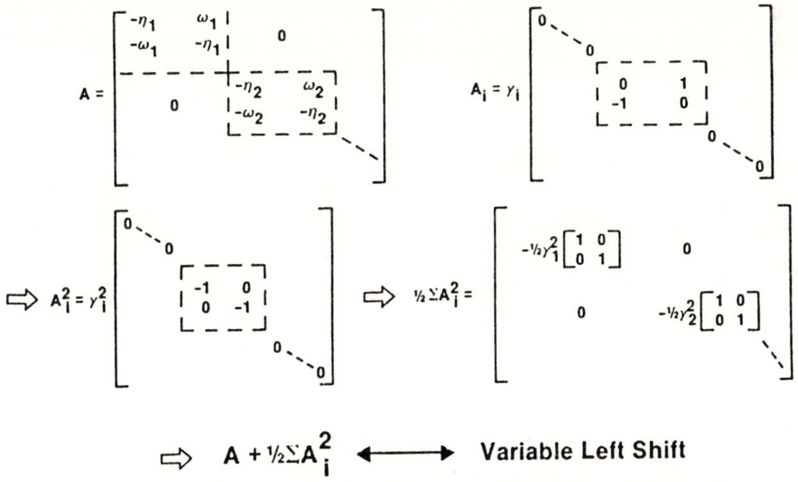

Figure 19. For Modal Systems With Frequency Uncertainty
the Stratonovich Correction Corresponds to a Variable Left Shift

model serves to suggest a mechanism for robustification (Figure 20). Again it is important to stress that the multiplicative white noise model is not interpreted literally as having physical significance, but rather can be thought of as a useful design model which correctly captures the impact of uncertainty on the performance functional via the state covariance.

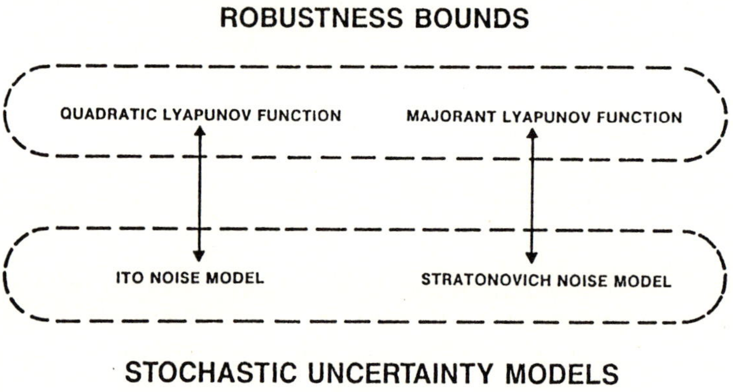

Figure 20. Stochastic Models and Robustness Bounds Are Fundamentally Related

In earlier work the Stratonovich dynamics model was justified by means of the minimum information/maximum entropy approach ([A1-A15]). A central result of the maximum entropy approach is that the high authority/low authority transition of a vibration control system from well-known low-frequency modes to poorly known high-frequency modes (Figure 18) is directly reflected in the structure of the state covariance matrix (Figure 21). A full-state feedback design applied to a simply

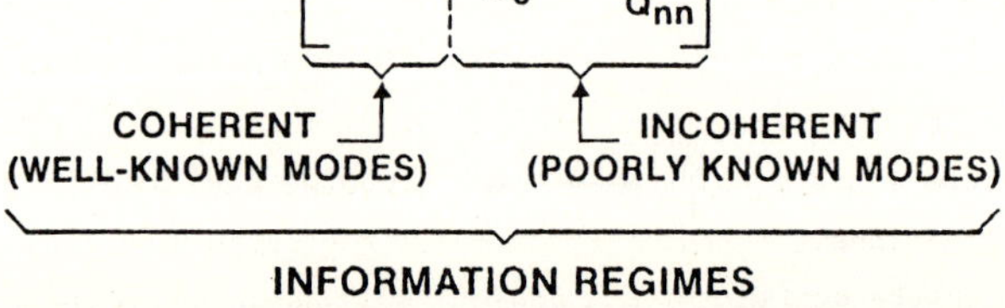

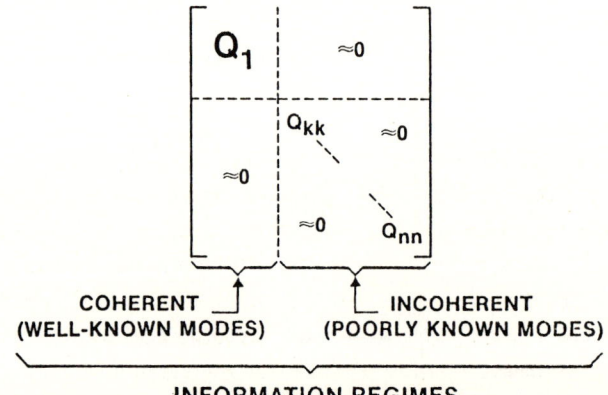

Figure 21. Frequency Uncertainties in the Stratonovich Model Lead to Suppressed Cross Correlation in the Steady-State Covariance

supported beam illustrates this point (Figure 22). By assuming that uncertainty in modal frequencies increases linearly with frequency, the structure of the covariance matrix leads directly to the control gains illustrated in Figure 23. Note that in the high-frequency region the position gains are essentially zero and thus the control law approaches positive-real energy dissipative rate feedback. This, of course, is precisely the type of structural controller expected in the presence of poor modeling information. Of course, any effective control-design theory for active vibration suppression in flexible structures should produce energy dissipative controllers when structural modeling information is highly uncertain.

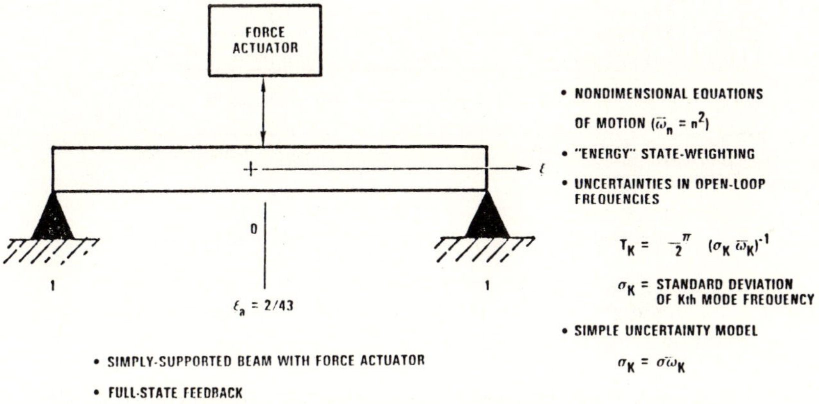

Figure 22. The Effects of Frequency Uncertainties Can Be Illustrated for a One-Dimensional Beam With Idealized Full-State Feedback

To carry out robustified optimal projection design in the presence of left-shifted open-loop dynamics, it is only necessary to utilize the left-shifted dynamics matrix (4.11) in place of the right-shifted matrix. All of the robustified optimal projection machinery, including gain expressions, can be utilized directly. It is also important to stress that the left shift must be used in conjunction with terms of the form $\tilde{A}_i\tilde{Q}\tilde{A}_i^T$.

One explanation for the mechanism by which robustification is achieved is illustrated in Figure 24. By left shifting the open-loop dynamics within the design process, the compensator poles are similarly left-shifted. Thus the compensator poles are effectively moved further into the left half plane away from the actual plant poles. Since the interaction between compensator and plant poles is weakened, the closed-loop system is correspondingly robustified with respect to uncertainties in the plant pole locations. A sensitivity analysis of this mechanism utilizing a uniform left shift in the context of LQG design is given in [15].

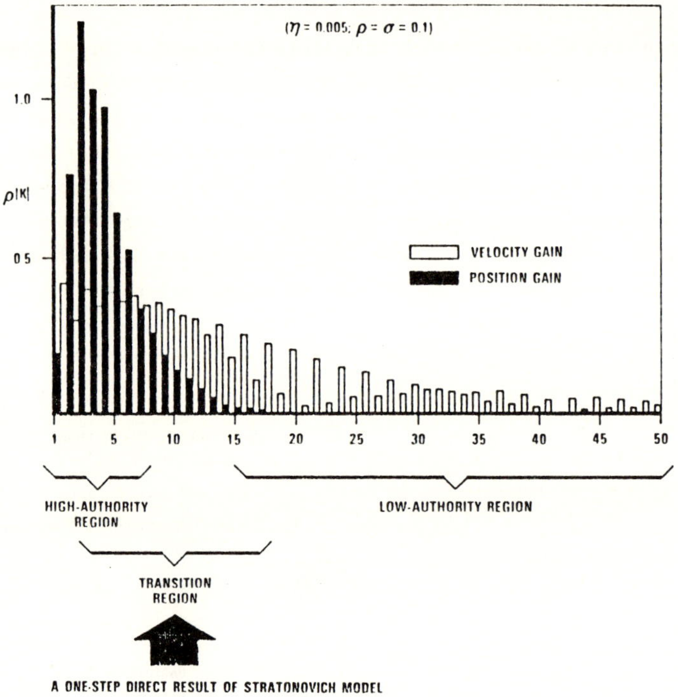

Figure 23. The Maximum Entropy Controller Approaches Rate Feedback in the Limit of Poor Modeling Information (High Uncertainty)

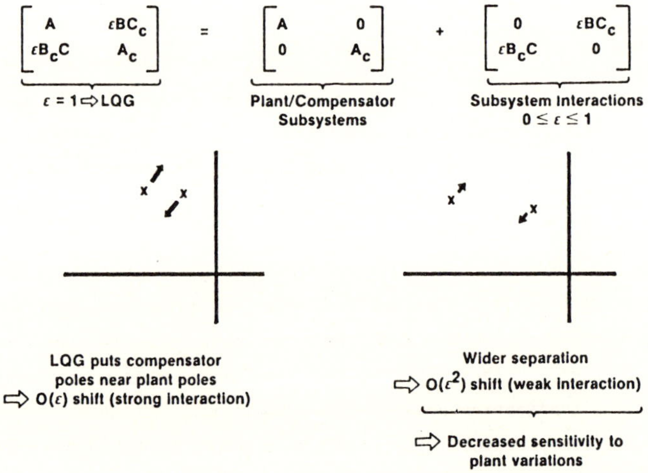

Figure 24. The Stratonovich Variable Left-Shift Model Effectively Places the Compensator Poles Further Into the Left Half Plane Where Plant/Compensator Interaction Is Weakened

As discussed above, the left-shift approach exploits open-loop nondestabilizing uncertainties and thus cannot operate through a priori bounding. Thus the actual level of robustification achieved from the robustified optimal projection equations for a given level of uncertainty modeling cannot be predicted a priori, i.e., in advance of control design. Indeed, this situation is to be expected when nondestabilizing uncertainties are exploited in a nonconservative design theory. Thus a suitable robust analysis technique is required for nonconservatively determining the robustification of the closed-loop system with respect to open-loop nondestabilizing uncertainties.

A suitable robustness analysis technique, known as majorant Lyapunov analysis, has indeed been developed ([A42]). Essentially, this technique employs a new type of Lyapunov function for assessing robustness due to weak subsystem interaction. The underlying machinery consists of the block-norm matrix which is a nonnegative matrix each of whose elements is the norm of a block of a suitably partitioned matrix (Figure 25). A matrix which bounds the block-norm matrix in the sense of nonnegative matrices, i.e., element by element, is known as a majorant. Majorants were introduced in [16] and were applied to stability analysis of integration algorithms for ODE's in [17].

(Ostrowski, 1961; Dahlquist, 1983)

$$M = \begin{bmatrix} M_1 & M_{12} & \cdots \\ M_{21} & M_2 & \\ \vdots & & \ddots \end{bmatrix}$$

$$\mathcal{M} = \begin{bmatrix} \|M_1\| & \|M_{12}\| & \cdots \\ \|M_{21}\| & \|M_2\| & \\ \vdots & & \ddots \end{bmatrix}$$

NONNEGATIVE CONE ORDERING

$$\mathcal{M} \leq\leq \hat{\mathcal{M}}$$

Figure 25. The Matrix Majorant Is a Bound for the Matrix Block Norm, i.e., the Nonnegative Matrix Each of Whose Elements Is the Norm of the Corresponding Block of a Given Matrix

To apply majorants to dynamical systems, the model is written in the form shown in Figure 26. The matrix A is block diagonal and consists of subsystem dynamics. The subsystem interactions represented by the partitioned matrix G are assumed to be uncertain. By suitable manipulation, uncertainties in the diagonal blocks of A can also be captured by G. By assuming that the spectral norm (largest singular value) of the blocks of G satisfy given bounds, the covariance block-norm inequality is obtained (Figure 27). This inequality is interpreted in the sense of nonnegative matrices, i.e., element-by-element, and * denotes the Hadamard (element-by-element) product.

$$\dot{x} = (A + G)x + w \qquad \dot{Q} = (A + G)Q + Q(A + G)^T + V$$

$$A = \begin{bmatrix} A_1 & 0 & --- \\ 0 & A_2 & \\ | & & \ddots \\ | & & \end{bmatrix} \qquad G = \begin{bmatrix} 0 & G_{12} & --- \\ G_{21} & 0 & \\ | & & \ddots \\ | & & \end{bmatrix}$$

Known Subsystem Dynamics — Uncertain Subsystem Interactions

$$V = \begin{bmatrix} V_1 & V_{12} & --- \\ V_{21} & V_2 & \\ | & & \ddots \\ | & & \end{bmatrix} \qquad Q = \begin{bmatrix} Q_1 & Q_{12} & --- \\ Q_{21} & Q_2 & \\ | & & \ddots \\ | & & \end{bmatrix}$$

Noise Intensity — State Covariance

Figure 26. The Large-Scale System Model Involves Known Local Dynamics and Uncertain Interactions

$$\dot{x} = (A + G)x + w \qquad J = E[x^T R x] = \mathrm{tr}\, QR$$

$$0 = (A + G)Q + Q(A + G)^T + V \qquad \mathcal{A} = [\underline{\sigma}(A_i \oplus A_j)]$$

$$\mathcal{V} = \begin{bmatrix} \|V_1\|_F & \|V_{12}\|_F & --- \\ \|V_{21}\|_F & \|V_2\|_F & \\ | & & \ddots \\ | & & \end{bmatrix} \qquad \mathcal{Q} = \begin{bmatrix} \|Q_1\|_F & \|Q_{12}\|_F & --- \\ \|Q_{21}\|_F & \|Q_2\|_F & \\ | & & \ddots \\ | & & \end{bmatrix}$$

$$\begin{bmatrix} 0 & \bar{\sigma}(G_{12}) & --- \\ \bar{\sigma}(G_{21}) & 0 & \\ | & & \ddots \\ | & & \end{bmatrix} \leq\leq \mathcal{G}$$

$$\Downarrow$$

$$\mathcal{A} * \mathcal{Q} \leq\leq \mathcal{G}\mathcal{Q} + \mathcal{Q}\mathcal{G}^T + \mathcal{V}$$

Figure 27. The Block-Norm Matrix of the State Covariance Satisfies a Lyapunov-Type Inequality Involving Nonnegative Matrices

To achieve robustness, the covariance block-norm inequality is replaced by the majorant Lyapunov equation (Figure 28). The solution of the majorant Lyapunov equation provides a bound (majorant) for the block norm of the covariance thereby guaranteeing both robust stability and performance.

MAJORANT LYAPUNOV EQUATION

$$\mathcal{A} * \widehat{\mathcal{Q}} = \mathcal{G}\widehat{\mathcal{Q}} + \widehat{\mathcal{Q}}\mathcal{G}^{T} + \mathcal{V}$$

$$\bar{\sigma}(G_{ij}) \leq \mathcal{G}_{ij}$$

$$\Downarrow$$

$$\mathcal{Q} \leq\leq \widehat{\mathcal{Q}}$$

$$\Downarrow$$

- **Robust Stability**
- **Robust Performance**

Figure 28. The Corresponding Nonnegative Matrix Equation Yields a Majorant for the State Covariance and Hence Robust Stability and Performance

It is interesting to note that numerical solution of the majorant Lyapunov equation requires no new techniques. Utilizing properties of M matrices, the solution can be obtained monotonically by means of a straightforward iterative technique (Figure 29).

MLE has a unique solution iff $\{\widehat{\mathcal{Q}}_K, K=0, 1, \ldots, \infty\}$ where:

$$\widehat{\mathcal{Q}}_0 = 0$$

$$\widehat{\mathcal{Q}}_{K+1} = \mathcal{A}^{III} * (\mathcal{G}\widehat{\mathcal{Q}}_K + \widehat{\mathcal{Q}}_K\mathcal{G}^{T} + \mathcal{V})$$

$$(\mathcal{A}^{III}_{mn} \triangleq \mathcal{A}^{-1}_{mn})$$

converges. If so, then:

$$\widehat{\mathcal{Q}} = \lim_{K\to\infty} \widehat{\mathcal{Q}}_K$$

$$J - J_0 \leq 2\sum_{K=1}^{r} (\mathrm{tr}\,\hat{P}_K)(\mathcal{G}\widehat{\mathcal{Q}})_{KK}$$

$$(0 = A_K^{T}\hat{P}_K + \hat{P}_K A_K + R_K)$$

Figure 29. By Exploiting the Properties of M-Matrices, the Majorant Lyapunov Equation Can Be Solved Monotonically by Means of a Simple Iterative Technique

An illustrative application of the majorant Lyapunov equation involves lightly damped subsystems (Figure 30). As shown in [A42] (and expected intuitively), robustness with respect to uncertain subsystem interaction is proportional to the frequency separation between the subsystems. The ability to capture this robustification mechanism is a unique feature of the majorant Lyapunov function not available from quadratic (i.e., scalar) Lyapunov functions or vector Lyapunov functions ([18,19]).

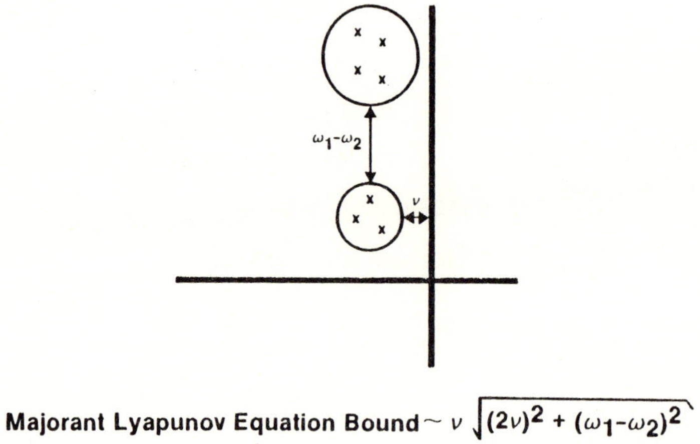

Figure 30. Robustness Bounds for Uncertain Coupling in Modal Systems Are Proportional to the Frequency Separation Between Subsystems

The next step in the majorant development involves a hierarchy of finer and finer robustness bounds which account for higher order subsystem interactions, e.g., the interaction between the ith and jth subsystems via the kth subsystem. The second member of the hierarchy (Figure 31) provides robustness guarantees with respect to frequency uncertainties. The interesting aspect of this robustness test is the fact that the performance bound is characterized precisely by a Stratonovich model. Hence the Stratonovich model can be viewed as an approximation to a robustness bound, while exploiting the Stratonovich/majorant relationship leads to a natural synthesis/analysis scheme (Figure 32) which nonconservatively exploits open-loop nondestabilizing uncertainties.

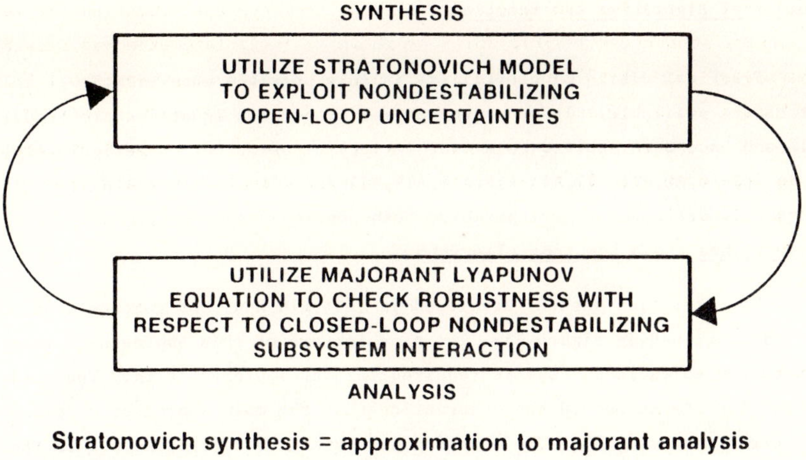
SYNTHESIS
UTILIZE STRATONOVICH MODEL TO EXPLOIT NONDESTABILIZING OPEN-LOOP UNCERTAINTIES
UTILIZE MAJORANT LYAPUNOV EQUATION TO CHECK ROBUSTNESS WITH RESPECT TO CLOSED-LOOP NONDESTABILIZING SUBSYSTEM INTERACTION
ANALYSIS
Stratonovich synthesis = approximation to majorant analysis

Figure 31. The Stratonovich Synthesis Model Provides a First Approximation to the Majorant Analysis Bounds

Second member of the hierarchy:

$$\mathcal{A} * \hat{\mathcal{Q}} + \mathcal{H}[\hat{\mathcal{Q}}] = \mathcal{G}\langle\hat{\mathcal{Q}}\rangle + \langle\hat{\mathcal{Q}}\rangle\mathcal{G}^{T} + \mathcal{V}$$

$$J - \mathrm{tr}[\hat{Q}R] \leq 2\sum_{K=1}^{r} (\mathrm{tr}\,\hat{P}_K)(\mathcal{G}\langle\hat{\mathcal{Q}}\rangle)_{KK}$$

$$0 = A\hat{Q} + \hat{Q}A^T + \mathcal{H}[\hat{Q}] + V$$
$$0 = A^T\hat{P} + \hat{P}A + \mathcal{H}^{\dagger}[\hat{P}] + R$$

where:

$\langle\hat{\mathcal{Q}}\rangle \triangleq$ off-diagonal part of $\hat{\mathcal{Q}}$

$\mathcal{H}[.]$ = Stratonovich model operator

- Tighter bound—incorporates more information on A and G
- Predicts stability when $(A + A^T)$ stable, $G = -G^T$
- "Nominal" performance, tr $[\hat{Q}R]$, given by Stratonovich model

Figure 32. The Refined Majorant Bound Incorporates a Stratonovich Covariance Model

5. Numerical Algorithms and Examples

Practical design of controllers is only possible when efficient, reliable algorithms are available. Indeed, the optimal projection equations are readily solvable and have been applied to a wide variety of examples. Numerical results appear in [A3-A6, A8, A11, A12, A14-A16, A18, A19, A21-A24, A26-A28, A30-A33, A39, A42, A44, A46]. Two distinctly different algorithms have been developed thus far, namely, an iterative method and a homotopy algorithm.

The iterative method, developed in [A14, A16, A44] and further studied in [20, 21], is outlined in Figure 33. The nice feature of this approach is that only a standard LQG software package is required for its implementation. The basic motivation for the method is the observation that the main source of coupling is via the terms involving $\tau_\perp$. The coupling is absent, of course, when τ is the identity, i.e., LQG. Note also that the terms involving $\tau_\perp$ are small when R_2 and V_2 are large, i.e., when control cost is high and the measurement noise is significant. This case, which yields low-authority controllers, is approximately characterized by decoupled control-design and controller-reduction operations. Thus it is not surprising that LQG reduction techniques are most successful when controller authority is low.

Since the $\tau_\perp$ terms occasion the greatest difficulty, it appears advantageous to bring them into play gradually. This can be accomplished by fixing τ after each iteration to yield updated values of Q, P, $\hat{Q}$ and $\hat{P}$. Furthermore, τ is introduced gradually by means of α to reduce its rank.

The crucial step of the algorithm concerns the construction of the projection τ from the pseudogramians $\hat{Q}$ and $\hat{P}$. Specifically, τ can be characterized (see [A22]) as the sum of eigenprojections of $\hat{Q}\hat{P}$, where each choice of eigenprojections may correspond to a local extremal. However, the necessary conditions do not specify which eigenprojections are to be selected for obtaining a particular local solution. Nevertheless, there do exist useful methods for constructing τ. For example, component-cost decomposition methods ([22]) when applied within the optimal projection framework often permit efficient identification of the global optimum.

Although the iterative method is convenient to use because it utilizes readily available software, it is suboptimal in the sense that it does not fully exploit the structure of the equations. Specifically, while the iterative method addresses a system of four nxn matrix equations, careful analysis reveals that because of the rank deficiency of the projection the problem can be recast as four $n_c$xn equations. Hence, when n_c is much smaller than n, which is clearly the most

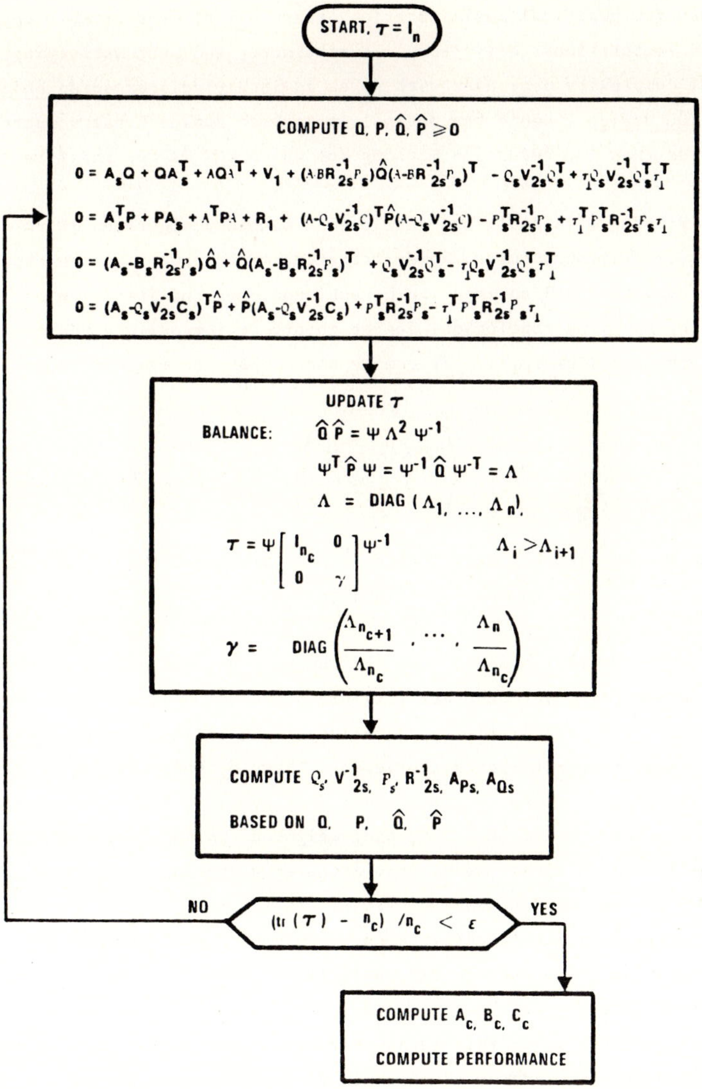

Figure 33. The Iterative Method for Solving the Robustified OPUS Design Equations Requires Only an LQG Software Package and Involves Refinement of the Optimal Projection τ

desirable case for practical implementation, there exists considerable opportunity for increased computational efficiency. Furthermore, and most satisfying, the computational complexity decreases with n_c as is intuitively expected below that required by LQG design. Hence the optimal projection approach has computational complexity less than LQG reduction methods for which LQG is but the first step.

S. Richter ([23,A46]) has developed a homotopy algorithm which fully exploits this crucial structure. Numerical experiments thus far have shown that considerable computational savings can be achieved over the iterative method. Furthermore, by applying topological degree theory to investigate the branches and character of the local extremals, it can be shown that the maximum number of possible extremals is

$$\binom{\min(n,m,\ell)}{n_c}$$

if $n_c \le \min(n,m,\ell)$ or 1 otherwise. Hence in most practical cases the equations support a relatively small number of solutions.

Both the iterative method and the homotopy algorithm have been applied to a design problem involving an 8th-order flexible structure originally due to D. Enns and considered in [1]. Specifically, a variety of LQG reduction methods are compared in [1] for a range of controller authorities. These methods include:

1. Enns: This method is a frequency-weighted, balanced realization technique applicable to either model or controller reduction.

2. Glover: This method utilizes the theory of Hankel norm optimal approximation for controller reduction.

3. Davis and Skelton: This is a modification of compensator reduction via balancing which addresses the case of unstable controllers.

4. Yousuff and Skelton: This is a further modification of balancing for handling stable or unstable controllers.

5. Liu and Anderson: In place of using a balanced approximation of the compensator transfer function directly, this method approximates the component parts of a fractional representation of the compensator.

All of the above methods proceed by first obtaining the full-order LQG compensator design for a high-order state-space model and then reducing the dimension of the resulting LQG compensator.

Figure 34 summarizes the results reported in [1] for the above LQG reduction methods along with results obtained using the iterative method for solving the optimal projection equations. Here q_2 is a scale factor for the plant disturbance noise affecting controller authority. Clearly, LQG reduction methods experience increasing difficulty as authority increases, i.e., as the $\tau_{\perp}$ terms become increasingly more important in coupling the control and reduction operations. For the low authority cases, the optimal projection calculations, which were performed on a Harris H800 minicomputer, appeared to incur roughly the same computational burden as the LQG reduction methods. Although the optimal projection computational burden increases with authority, comparison with the LQG reduction methods is not meaningful because of the difficulty experienced by these methods in achieving closed-loop stability. See [A44] for further details and for comparisons involving transient response.

The homotopy algorithm was also applied to the example considered in [1]. One of the main goals of the development effort was to extend the range of disturbance intensity or, equivalently, observer bandwidth, out beyond $q_2 = 2000$. To this end, second-order ($n_c = 2$) controllers were obtained with relatively little computation for $q_2 = 10{,}000$, 100,000 and 1,000,000. In addition, the performance of each reduced-order controller was within 25% of LQG. These cases can surely be expected to present a nontrivial challenge to both the LQG reduction methods and the iterative optimal projection method.

Numerical solution of the robustified optimal projection equations has been carried out for several examples. For illustrative purposes a 2x2 example was considered in [A26] and the results illustrated in Figure 35 indicate performance/robustness tradeoffs possible. The variable left-shift technique was applied in [A19] to the NASA SCOLE problem with frequency uncertainties. The robustness of LQG and two robustified designs is shown in Figure 36. The plots illustrate the degradation in performance due to simultaneous perturbation of all modal frequencies. Note that LQG is rendered unstable by +5% frequency perturbation while a high-authority robustified design improves this region to +8%. The low-authority design increases this region significantly while sacrificing 6% nominal performance.

Method	q_2 \ n_c	0.01	0.1	1	10	100	1000	2000
Enns	7	S	S	S	S	S	S	S
	6	S	S	S	S	S	S	S
	5	S	S	S	S	S	S	S
	4	S	S	S	S	S	S	U
	3	S	S	S	S	S	S	S
	2	S	S	S	S	U	U	U
Glover	7	S	S	S	S	S	U	S
	6	S	S	S	S	U	U	U
	5	S	S	S	S	U	U	U
	4	S	S	S	S	U	U	U
	3	S	S	U	S	U	U	U
	2	S	U	S	U	S	U	U
Davis & Skelton	7	S	U	U	S	S	S	S
	6	S	S	S	S	S	S	S
	5	S	U	S	S	S	U	U
	4	S	S	U	S	S	U	U
	3	U	U	U	U	U	U	U
	2	S	U	S	U	U	U	U
Yousuff & Skelton	7	S	S	S	S	U	U	U
	6	S	S	S	S	U	U	U
	5	S	S	S	U	U	U	U
	4	S	S	S	U	U	U	U
	3	S	U	U	U	U	U	U
	2	S	S	S	U	U	U	U
Liu & Anderson	7	S	S	S	S	S	S	U
	6	S	S	S	S	S	S	U
	5	S	S	S	S	S	S	S
	4	S	S	S	S	S	S	S
	3	S	S	S	S	S	U	U
	2	S	S	S	S	S	S	S
Optimal Projection	7	S	S	S	S	S	S	S
	6	S	S	S	S	S	S	S
	5	S	S	S	S	S	S	S
	4	S	S	S	S	S	S	S
	3	S	S	S	S	S	S	S
	2	S	S	S	S	S	S	S

S - The closed-loop system is stable
U - The closed-loop system is unstable

Figure 34. The Optimal Projection Approach Was Compared to Several LQG Reduction Techniques Over a Range of Controller Authorities for an Example of Enns

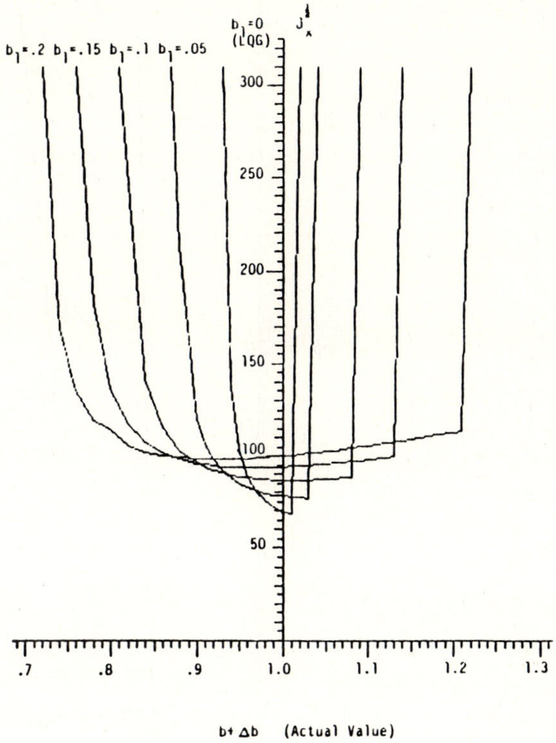

Figure 35. The Robustified Optimal Projection Equations Provide Robustness/Performance Tradeoffs for a Highly Sensitive Nominal LQG Design

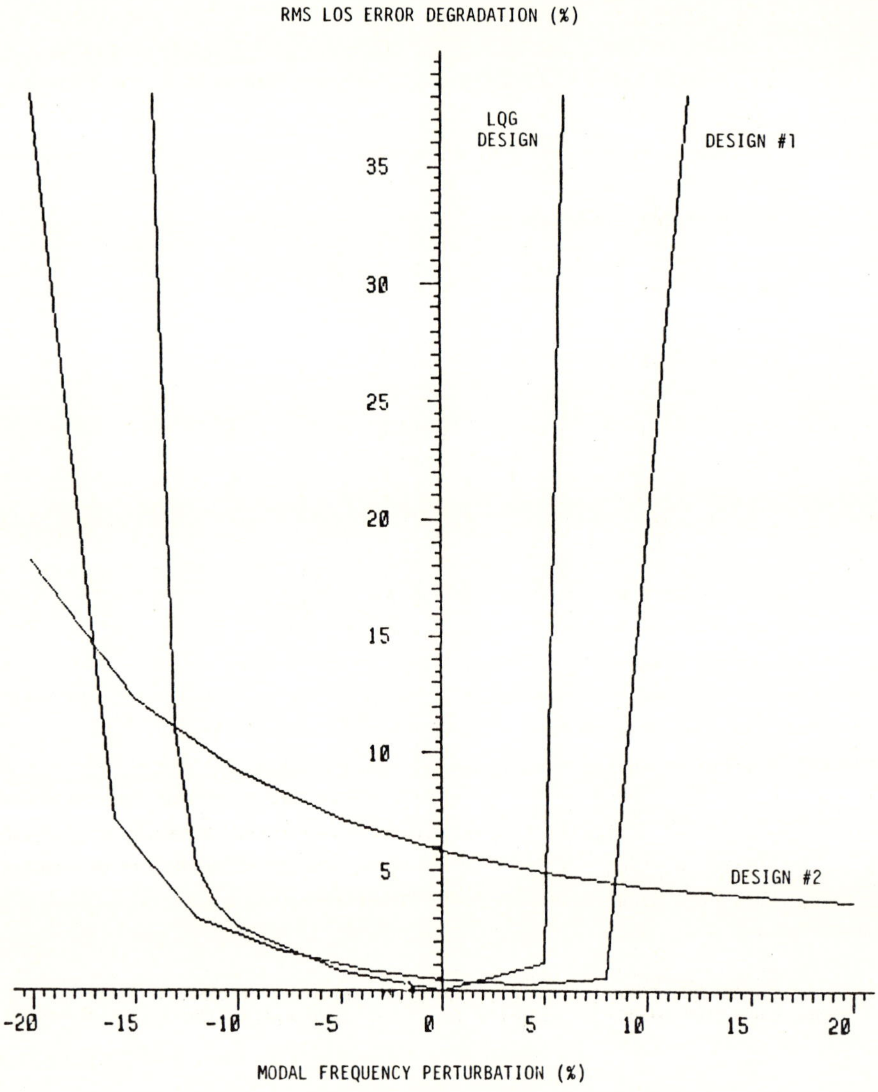

Figure 36. The Stratonovich Model Robustifies the LQG Design for the NASA SCOLE Model with Uncertain Modal Frequencies

6. Additional Extensions

The robustified optimal projection design machinery has been further extended to encompass a larger number of design cases arising in practical application. Here we merely list the extensions:

1. Discrete-time and sampled-data controllers ([A28,A30,A34,A35]).

2. Decentralized controllers ([A39]).

3. Nonstrictly proper controllers ([A37]).

4. Distributed parameter systems ([A25]).

7. Concluding Remarks

The machinery provided by OPUS for designing active controllers for flexible structures has been reviewed. The basic machinery is a system of coupled Riccati and Lyapunov equations which directly generalize the classical LQG result to include both a constraint on controller order and a model of parameter uncertainty. The overall approach thus encompasses all major design tradeoffs arising in vibration-suppression applications. Substantial numerical experience has been gained through an iterative method requiring only an LQG software package and, more recently, by means of a highly efficient homotopy algorithm developed by S. Richter. The overall approach opens the door for effective design of implementable controllers for large precision space structures.

Acknowledgment. We wish to thank Ms. Jill M. Straehla for the excellent preparation of this paper.

General References

1. Y. Liu and B. D. O. Anderson, "Controller Reduction Via Stable Factorization and Balancing," Int. J. Contr., Vol. 44, pp. 507-531, 1986.

2. G. Zames, "Feedback and Optimal Sensitivity: Model Reference Transformations, Multiplicative Seminorms, and Approximate Inverses," IEEE Trans. Autom. Contr., Vol. AC-26, pp. 301-320, 1981.

3. J. C. Doyle, "Guaranteed Margins for LQG Regulators," IEEE Trans. Autom. Contr., Vol. AC-23, pp. 756-757, 1978.

4. J. C. Doyle and G. Stein, "Multivariable Feedback Design: Concepts for a Classical/Modern Synthesis," IEEE Trans. Autom. Contr., Vol. AC-26, pp. 4-16, 1981.

5. G. Stein and M. Athans, "The LQG/LTR Procedure for Multivariable Feedback Control Design," IEEE Trans. Autom. Contr., Vol. AC-32, pp. 105-114, 1987.

6. E. Soroka and U. Shaked, "On the Robustness of LQ Regulators," IEEE Trans. Autom. Contr., Vol. AC-29, pp. 664-665, 1984.

7. U. Shaked and E. Soroka, "On the Stability Robustness of the Continuous-Time LQG Optimal Control," IEEE Trans. Autom. Contr., Vol. AC-30, 1039-1043.

8. J. C. Doyle, "Analysis of Feedback Systems with Structured Uncertainties," IEE Proc., Vol. 129, pp. 242-250, 1982.

9. S. S. L. Chang and T. K. C. Peng, "Adaptive Guaranteed Cost Control of Systems with Uncertain Parameters," IEEE Trans. Autom. Contr., Vol. AC-17, pp. 474-483, 1972.

10. A. Vinkler and L. J. Wood, "Multistep Guaranteed Cost Control of Linear Systems with Uncertain Parameters," J. Guid. Contr., Vol. 2, pp. 449-456, 1979.

11. I. R. Petersen and C. V. Hollot, "A Riccati Equation Approach to the Stabilization of Uncertain Systems," Automatica, Vol. 22, pp. 433-448, 1986.

12. D. Hinrichsen and A. J. Pritchard, "Stability Radius for Structured Perturbations and the Algebraic Riccati Equation," Sys. Contr. Lett., Vol. 8, pp. 105-113, 1986.

13. B. D. O. Anderson and J. B. Moore, Linear Optimal Control, Prentice-Hall, Englewood Cliffs, NJ, 1970.

14. G. N. Milstein, "Design of Stabilizing Controller with Incomplete State Data for Linear Stochastic System with Multiplicative Noise," Autom. Remote Contr., Vol. 43, pp. 653-659, 1982.

15. G. A. Adamian and J. S. Gibson, "Sensitivity of Closed-Loop Eigenvalues and Robustness," preprint.

16. A. M. Ostrowski, "On Some Metrical Properties of Operator Matrices and Matrices Partitioned into Blocks," J. Math. Anal. Appl., Vol. 2, pp. 161-209, 1961.

17. G. Dahlquist, "On Matrix Majorants and Minorants, with Applications to Differential Equations," Lin. Alg. Appl., Vol. 52/53, pp. 199-216, 1983.

18. D. D. Siljak, Large-Scale Dynamic Systems, Elsevier/North-Holland, 1978.

19. M. Ikeda and D. D. Siljak, "Generalized Decompositions of Dynamic Systems and Vector Lyapunov Functions," IEEE Trans. Autom. Contr., Vol. AC-26, pp. 1118-1125, 1981.

20. A. Gruzen, "Robust Reduced Order Control of Flexible Structures," C. S. Draper Laboratory Report #CSDL-T-900, April 1986.

21. A. Gruzen and W. E. Vander Velde, "Robust Reduced-Order Control of Flexible Structures Using the Optimal Projection/Maximum Entropy Design Methodology," AIAA Guid. Nav. Contr. Conf., Williamsburg, VA, August 1986.

22. A. Yousuff and R. E. Skelton, "Controller Reduction by Component Cost Analysis," IEEE Trans. Autom. Contr., Vol. AC-24, pp. 520-530, 1984.

23. S. Richter and R. DeCarlo, "Continuation Methods: Theory and Applications," IEEE Trans. Autom. Contr., Vol. 28, pp. 660-665, 1983.

OPUS References

A1. D. C. Hyland, "The Modal Coordinate/Radiative Transfer Formulation of Structural Dynamics--Implications for Vibration Suppression in Large Space Platforms," MIT Lincoln Laboratory, TR-27, 14 March 1979.

A2. D. C. Hyland, "Optimal Regulation of Structural Systems With Uncertain Parameters," MIT Lincoln Laboratory, TR-551, 2 February 1981, DDC# AD-A099111/7.

A3. D. C. Hyland, "Active Control of Large Flexible Spacecraft: A New Design Approach Based on Minimum Information Modelling of Parameter Uncertainties," Proc. Third VPI&SU/AIAA Symposium, pp. 631-646, Blacksburg, VA, June 1981.

A4. D. C. Hyland, "Optimal Regulator Design Using Minimum Information Modelling of Parameter Uncertainties: Ramifications of the New Design Approach," Proc. Third VPI&SU/AIAA Symposium, pp. 701-716, Blacksburg, VA, June 1981.

A5. D. C. Hyland and A. N. Madiwale, "Minimum Information Approach to Regulator Design: Numerical Methods and Illustrative Results," Proc. Third VPI&SU/AIAA Symposium, pp. 101-118, Blacksburg, VA, June 1981.

A6. D. C. Hyland and A. N. Madiwale, "A Stochastic Design Approach for Full-Order Compensation of Structural Systems with Uncertain Parameters," Proc. AIAA Guid. Contr. Conf., pp. 324-332, Albuquerque, NM, August 1981.

A7. D. C. Hyland, "Optimality Conditions for Fixed-Order Dynamic Compensation of Flexible Spacecraft with Uncertain Parameters," AIAA 20th Aerospace Sciences Meeting, paper 82-0312, Orlando, FL, January 1982.

A8. D. C. Hyland, "Structural Modeling and Control Design Under Incomplete Parameter Information: The Maximum Entropy Approach," AFOSR/NASA Workshop in Modeling, Analysis and Optimization Issues for Large Space Structures, Williamsburg, VA, May 1982.

A9. D. C. Hyland, "Minimum Information Stochastic Modelling of Linear Systems with a Class of Parameter Uncertainties," Proc. Amer. Contr. Conf., pp. 620-627, Arlington, VA, June 1982.

A10. D. C. Hyland, "Maximum Entropy Stochastic Approach to Control Design for Uncertain Structural Systems," Proc. Amer. Contr. Conf., pp. 680-688, Arlington, VA, June 1982.

A11. D. C. Hyland, "Minimum Information Modeling of Structural Systems with Uncertain Parameters," Proceedings of the Workshop on Applications of Distributed System Theory to the Control of Large Space Structures, G. Rodriguez, ed., pp. 71-88, JPL, Pasadena, CA, July 1982.

A12. D. C. Hyland and A. N. Madiwale, "Fixed-Order Dynamic Compensation Through Optimal Projection," Proceedings of the Workshop on Applications of Distributed System Theory to the Control of Large Space Structures, G. Rodriguez, ed., pp. 409-427, JPL, Pasadena, CA, July 1982.

A13. D. C. Hyland, "Mean-Square Optimal Fixed-Order Compensation--Beyond Spillover Suppression," paper 1403, AIAA Astrodynamics Conference, San Diego, CA, August 1982.

A14. D. C. Hyland, "The Optimal Projection Approach to Fixed-Order Compensation: Numerical Methods and Illustrative Results," AIAA 21st Aerospace Sciences Meeting, paper 83-0303, Reno, NV, January 1983.

A15. D. C. Hyland, "Mean-Square Optimal, Full-Order Compensation of Structural Systems with Uncertain Parameters," MIT Lincoln Laboratory, TR-626, 1 June 1983.

A16. D. C. Hyland, "Comparison of Various Controller-Reduction Methods: Suboptimal Versus Optimal Projection," Proc. AIAA Dynamics Specialists Conf., pp. 381-389, Palm Springs, CA, May 1984.

A17. D. C. Hyland and D. S. Bernstein, "The Optimal Projection Equations for Fixed-Order Dynamic Compensation," IEEE Trans. Autom. Contr., Vol. AC-29, pp. 1034-1037, 1984.

A18. D. C. Hyland, "Application of the Maximum Entropy/Optimal Projection Control Design Approach for Large Space Structures," Proc. Large Space Antenna Systems Technology Conference, pp. 617-654, NASA Langley, December 1984.

A19. L. D. Davis, D. C. Hyland and D. S. Bernstein, "Application of the Maximum Entropy Design Approach to the Spacecraft Control Laboratory Experiment (SCOLE)," Final Report, NASA Langley, January 1985.

A20. D. S. Bernstein and D. C. Hyland, "The Optimal Projection Equations for Reduced-Order State Estimation," IEEE Trans. Autom. Contr., Vol. AC-30, pp. 583-585, 1985.

A21. D. S. Bernstein and D. C. Hyland, "Optimal Projection/Maximum Entropy Stochastic Modelling and Reduced-Order Design Synthesis," Proc. IFAC Workshop on Model Error Concepts and Compensation, Boston, MA, June 1985, pp. 47-54, R. E. Skelton and D. H. Owens, eds., Pergamon Press, Oxford, 1986.

A22. D. C. Hyland and D. S. Bernstein, "The Optimal Projection Equations for Model Reduction and the Relationships Among the Methods of Wilson, Skelton and Moore," IEEE Trans. Autom. Contr., Vol. AC-30, pp. 1201-1211, 1985.

A23. D. S. Bernstein and D. C. Hyland, "The Optimal Projection/Maximum Entropy Approach to Designing Low-Order, Robust Controllers for Flexible Structures," Proc. 24th IEEE Conf. Dec. Contr., pp. 745-752, Fort Lauderdale, FL, December 1985.

A24. D. S. Bernstein, L. D. Davis, S. W. Greeley and D. C. Hyland, "Numerical Solution of the Optimal Projection/Maximum Entropy Design Equations for Low-Order, Robust Controller Design," Proc. 24th IEEE Conf. Dec. Contr., pp. 1795-1798, Fort Lauderdale, FL, December 1985.

A25. D. S. Bernstein and D. C. Hyland, "The Optimal Projection Equations for Finite-Dimensional Fixed-Order Dynamic Compensation of Infinite-Dimensional Systems," SIAM J. Contr. Optim., Vol. 24, pp. 122-151, 1986.

A26. D. S. Bernstein and S. W. Greeley, "Robust Controller Synthesis Using the Maximum Entropy Design Equations," IEEE Trans. Autom. Contr., Vol. AC-31, pp. 362-364, 1986.

A27. D. C. Hyland, D. S. Bernstein, L. D. Davis, S. W. Greeley and S. Richter, "MEOP: Maximum Entropy/Optimal Projection Stochastic Modelling and Reduced-Order Design Synthesis," Final Report, Air Force Office of Scientific Research, Bolling AFB, Washington, DC, April 1986.

A28. D. S. Bernstein, L. D. Davis and D. C. Hyland, "The Optimal Projection Equations for Reduced-Order, Discrete-Time Modelling, Estimation and Control," J. Guid. Contr. Dyn., Vol. 9, pp. 288-293, 1986.

A29. D. S. Bernstein and S. W. Greeley, "Robust Output-Feedback Stabilization: Deterministic and Stochastic Perspectives," Proc. Amer. Contr. Conf., pp. 1818-1826, Seattle, WA, June 1986.

A30. D. S. Bernstein, L. D. Davis and S. W. Greeley, "The Optimal Projection Equations for Fixed-Order, Sampled-Data Dynamic Compensation with Computation Delay," IEEE Trans. Autom. Contr., Vol. AC-31, pp. 859-862, 1986.

A31. D. S. Bernstein, "OPUS: Optimal Projection for Uncertain Systems," Annual Report, Air Force Office of Scientific Research, Bolling AFB, Washington, DC, October 1986.

A32. B. J. Boan and D. C. Hyland, "The Role of Metal Matrix Composites for Vibration Suppression in Large Space Structures," Proc. MMC Spacecraft Survivability Tech. Conf., MMCIAC Kaman Tempo Publ., Stanford Research Institute, Palo Alto, CA, October 1986.

A33. D. C. Hyland, "An Experimental Testbed for Validation of Control Methodologies in Large Space Optical Structures," SPIE Optoelectronics and Laser Applications Conference, Los Angeles, CA, January 1987.

A34. W. M. Haddad and D. S. Bernstein, "The Optimal Projection Equations for Discrete-Time Reduced-Order State Estimation for Linear Systems with Multiplicative White Noise," Sys. Contr. Lett., 1987.

A35. D. S. Bernstein and W. M. Haddad, "The Optimal Projection Equations for Discrete-Time Fixed-Order Dynamic Compensation of Linear Systems with Multiplicative White Noise," Int. J. Contr., 1987.

A36. W. M. Haddad and D. S. Bernstein, "The Optimal Projection Equations for Reduced-Order State Estimation: The Singular Measurement Noise Case," IEEE Trans. Autom. Contr., 1987.

A37. D. S. Bernstein, "The Optimal Projection Equations for Static and Dynamic Output Feedback: The Singular Case," IEEE Trans. Autom. Contr., 1987.

A38. D. S. Bernstein and D. C. Hyland, "The Optimal Projection Equations for Reduced-Order Modelling, Estimation and Control of Linear Systems with Multiplicative White Noise," J. Optim. Thy. Appl., 1987.

A39. D. S. Bernstein, "Sequential Design of Decentralized Dynamic Compensators Using the Optimal Projection Equations," Int. J. Contr., 1987.

A40. D. S. Bernstein and W. M. Haddad, "Optimal Output Feedback for Nonzero Set Point Regulation," Proc. Amer. Contr. Conf., Minneapolis, MN, June 1987.

A41. D. S. Bernstein, "Robust Static and Dynamic Output-Feedback Stabilization: Deterministic and Stochastic Perspectives," IEEE Trans. Autom. Contr., 1987.

A42. D. C. Hyland and D. S. Bernstein, "The Majorant Lyapunov Equation: A Nonnegative Matrix Equation for Guaranteed Robust Stability and Performance of Large Scale Systems," IEEE Trans. Autom. Contr., 1987.

A43. D. S. Bernstein and W. M. Haddad, "The Optimal Projection Equations with Petersen-Hollot Bounds: Robust Controller Synthesis with Guaranteed Structured Stability Radius," submitted.

A44. S. W. Greeley and D. C. Hyland, "Reduced-Order Compensation: LQG Reduction Versus Optimal Projection," submitted.

A45. W. M. Haddad, Robust Optimal Projection Control-System Synthesis, Ph.D. Dissertation, Department of Mechanical Engineering, Florida Institute of Technology, Melbourne, FL, March 1987.

A46. S. Richter, "A Homotopy Algorithm for Solving the Optimal Projection Equations for Fixed-Order Dynamic Compensation: Existence, Convergence and Global Optimality," Proc. Amer. Contr. Conf., Minneapolis, MN, June 1987.

Adaptive Control of Large Space Structures: Uncertainity Estimation and Robust Control Calibration

Robert L. Kosut [2]

Abstract

An approach is presented to the problem of designing a robust control using on-line measurements. The idea is to use standard methods of parametric system identification to obtain a nominal estimate of the plant transfer function. Non-parametric spectral methods are then used to obtain a frequency domian expression for model uncertainty. If the model uncertainty exceeds a specified frequency bound, which has been predetermined from the nominal model and the performance criteria, then data filters used in the system identification are modified and the procedure is repeated. An analysis is presented which establishes conditions under which the procedure will actually converge to a satisfactory robust design. An example is provided which illustrates the method and supporting analysis.

1 Introduction

Large space structure (LSS) systems impose stringent performance demands, and hence, feedback designs will of necessity be based on very accurate models. However, the on-orbit dynamics will not be sufficiently like those obtained from either ground-testing or even from sophisticated computer generated modeling techniques, e.g., finite element modeling. Therefore, it is necessary to be able to identify the LSS dynamics directly from on-orbit measurements, and simultaneously, tune or re-design the control. Hence, the control design cycle is an *adaptive* process, typically starting with a nominal low-performance design based on a coarse model, and then re-designed from on-orbit data.

This paper addresses some of the issues involved in the adaptive control of LSS systems, specifically the problem of vibration suppression.

1.1 Adaptive Control

Adaptive control, as depicted in Figure 1, essentially consists of two processes, namely: (1) a model parameter estimator which uses a finite data record of input-output measurements, and (2) a control design rule which transforms the model parameter estimates into control parameters.

The procedure can go awry mainly because the model estimate is not sufficiently accurate and the control design rule effectively assumes that the estimates are perfectly accurate. Although the procedure may work well whenever the parameter estimates are in a good region, they may never get there.

One route around this problem is to prove that it can never occur, which involves *analyzing* the complete adaptive system. For example, analysis techniques based on the method of averaging can provide an assessment of the adaptive systems robustness, e.g., Anderson et al.(1986).

[1]Research supported by AFOSR, Directorate of Aerospace Sciences, under contract F49620-85-C-0094.

[2]R.L. Kosut is a Senior Scientist with Integrated Systems, Inc., 2500 Mission College Blvd., Santa Clara, CA 95054, and a Consulting Professor with the Information Systems Lab, Stanford University, Stanford, CA 95305.

Springer Series in Computational Mechanics
S. N. Atluri, A. K. Amos (Eds.)
Large Space Structures: Dynamics and Control

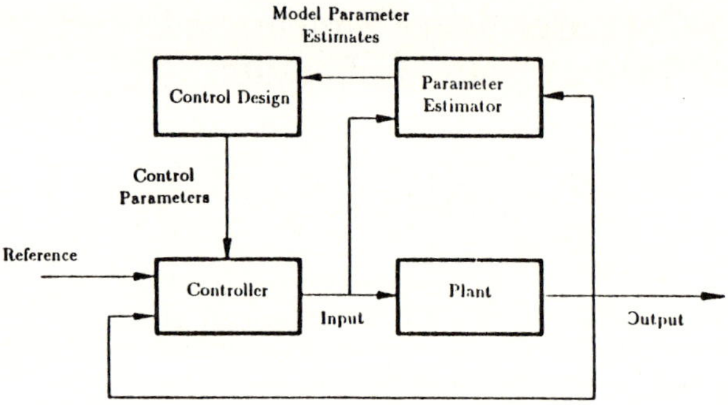

Figure 1: Adaptive Control

Another route, explored here, is to modify the process so that along with estimating model parameters, a measure of model uncertainty is also obtained. With such an estimate in hand the control design rule can be modified to account for the model uncertainty. If the resulting controller is too cautious, because the model estimate is too uncertain, then it is necessary to repeat the estimation procedure so as to obtain a better model.

In this paper we will concentrate on a particular procedure for obtaining a measure of model uncertainty for linear time invariant plants. The process is referred to as *adaptive uncertainty modeling* and the outcome is a frequency domain expression for the model uncertainty. The motivation arises from robust control theory which utilizes frequency domain expressions for characterizing a *set of uncertainty* within which lies the true plant, e.g., Francis and Zames(1984), Safonov et al.(1981), Doyle and Chu(1986), Vidyasagar(1985). An adaptive control system, modified so that it produces a set of uncertainty, is depicted in Figure 2.

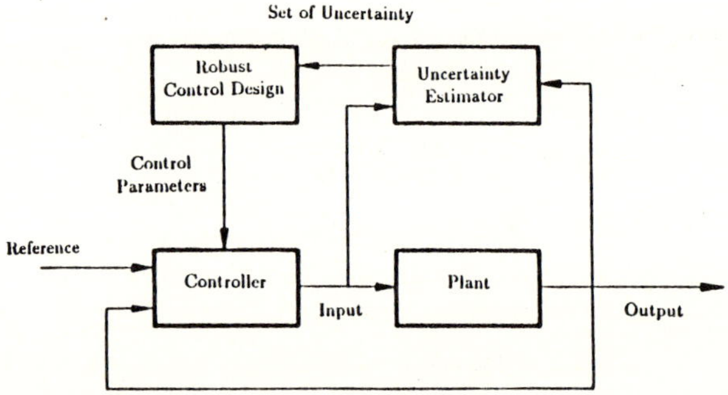

Figure 2: Adaptive Control with Uncertainty Estimator

The indicated adaptive process is referred to here as *adaptive calibration*, Kosut(1986,1987). The set of uncertainty is extracted from input-output measurements. We will describe one approach to this problem which uses a combination of parametric prediction error methods together with standard non-parametric (spectral) estimation methods. We then show how the set of uncertainty produced by the modified identifier allows for performance evaluation *before* the controller is adjusted. If the predicted performance is not satisfactory then the identification process is repeated under different conditions which will reduce model uncertainty where needed. The resulting controller is designed to be robust with respect to the estimated set of uncertainty. Thus, a large set of uncertainty requires a cautious or low authority controller, whereas a small set of uncertainty will result in a high authority controller. Observe that the order of the controller can vary with the set of uncertainty.

This paper is organized as follows: Section 2 develops the sampled-data linear transfer function model. Section 3 states some known results regarding stabilization and robustness of linear control systems. Section 4 is the main section, decribing in detail an approach to estimating the set of uncertainty and a procedure for adaptive calibration for both open-loop and closed-loop situations. Section 5 presents an example of uncertainty estimation from experimental data. Section 6 contains concluding remarks.

2 Linear Flexible Dynamics

2.1 Transfer Function Model

Under the assumptions of linear elasticity and small deflections, at any time t and position $\mathbf{r}$ on the structure, the deflections are given by

$$y(t,\mathbf{r}) = \sum_{k=0}^{\ell} \eta_k(t)\psi_k(\mathbf{r}) \tag{1}$$

where $\psi_k(\mathbf{r})$ is the kth mode shape at position $\mathbf{r}$, and $\eta_k(t)$ is the kth modal amplitude as a function of time. The upper limit ℓ in the above sum is theoretically infinite, but for all practical purposes can be considered here to be finite, but extremely large. Assuming there are point actuators located at the discrete positions $\mathbf{r}_{a1},\ldots,\mathbf{r}_{am_a}$, then the modal amplitudes each satisfy

$$\ddot{\eta}_k(t) + 2\zeta_k\Omega_k\dot{\eta}_k(t) + \Omega_k^2\eta_k(t) = \sum_{i=1}^{m_a} u_i(t)\psi_k(\mathbf{r}_{ai}),\ k = 1,2,\ldots,\ell \tag{2}$$

where Ω_k and ζ_k are the kth modal frequency and damping, respectively, and where $u_i(t)$ is the ith actuator force or torque on the structure. A position sensor located at $\mathbf{r}$ would measure

$$y_p(t,\mathbf{r}) = \sum_{k=0}^{\ell} \eta_k(t)\psi_k(\mathbf{r}) \tag{3}$$

Likewise, a velocity sensor at $\mathbf{r}$ would measure

$$y_v(t,\mathbf{r}) = \sum_{k=0}^{\ell} \dot{\eta}_k(t)\psi_k(\mathbf{r}) \tag{4}$$

Suppose that actuators are placed at $\mathbf{r}_{a1},\ldots,\mathbf{r}_{am_a}$, position sensors at $\mathbf{r}_{p1},\ldots,\mathbf{r}_{pm_p}$, and velocity sensors at $\mathbf{r}_{v,1},\ldots,\mathbf{r}_{v,m_v}$. Then, the transfer function from the ith applied force to the jth position or velocity sensor output is then, respectively

$$G_{y_{pj}u_i}(s) = \sum_{k=0}^{\ell} \psi_k(\mathbf{r}_{pj})\psi_k(\mathbf{r}_{ai})\frac{1}{s^2 + 2\zeta_k\Omega_k s + \Omega_k^2} \tag{5}$$

(6)

$$G_{y_{vj}u_i}(s) = \sum_{k=0}^{\ell} \psi_k(\mathbf{r}_{vj})\psi_k(\mathbf{r}_{ai})\frac{s}{s^2 + 2\zeta_k\Omega_k s + \Omega_k^2} \tag{7}$$

For purposes of illustration, suppose that there is only one position sensor at $\mathbf{r}_p$ and one actuator, not located at the same position as the sensor, but at $\mathbf{r}_a$. Let $y(t)$ denote the output of the position sensor. Since disturbance forces act on the structure and the sensors have some noise component, the actual sensor output will read [3]

$$y(t) = G(s)u(t) + d(t) \tag{8}$$

where $d(t)$ is the cumulative effect of disturbances and noise sources as seen at the output, and where $G(s)$ is the transfer function from actuator to sensor location. Thus,

$$G(s) = \sum_{k=1}^{\ell} \frac{\beta_k}{s^2 + 2\zeta_k\Omega_k s + \Omega_k^2} \tag{9}$$

where

$$\beta_k = \psi_k(\mathbf{r}_p)\psi_k(\mathbf{r}_a) \tag{10}$$

2.2 Sampled-Data Representation

Assume that control commands and data acqusition occur at discrete sampling instances separated by uniform sampling intervals of duration $t_{sa} = 1/\Omega_{sa}$ seconds, where Ω_{sa} is the sampling frequency in Hz. Thus, $u(t)$ is constant between samples, i.e., for any integer k,

$$u(t) = u(t_k),\ t \in [t_k, t_{k+1}) \tag{11}$$

where $t_k = kt_{sa}$. If, in addition, the sensor output is sampled at the same instances, then the discrete-time representation of (8) is

$$y(t_k) = P_0(q)u(t_k) + d(t_k) \tag{12}$$

where $P_0(q)$ is the *zero-order hold equivalent* of $G(s)$, expressed as a function of the *shift operator* q defined by $qx(t_k) = x(t_{k+1})$ and $q^{-1}x(t_k) = x(t_{k-1})$. Thus, in terms of the usual "z-transform" operator $\mathcal{Z}(\cdot)$, we have

$$P_0(q) = (1 - q^{-1})\mathcal{Z}\{P(s)/s\} \tag{13}$$

Using (9) gives

$$P_0(q) = \sum_{k=1}^{\ell}(1 - q^{-1})\mathcal{Z}\{\frac{1}{s}\frac{\beta_k}{s^2 + 2\zeta_k\Omega_k s + \Omega_k^2}\} \tag{14}$$

In order to simplify the notation in the sequel, time will be normalized with respect to the sampling instants. That is, the discrete-time representation of the LSS will be written compactly as

$$y(t) = P_0(q)u(t) + d(t) \tag{15}$$

where t takes on integer values only, i.e., $t = 1, 2, \ldots$, and the shift operator is normalized so that $qx(t) = x(t+1)$ and $q^{-1}x(t) = x(t-1)$. We will take (15) as the *true* LSS system and the measured data record as

$$z^N = \{y(t), u(t) : t = 1, \ldots, N\} \tag{16}$$

[3]The variable s is used to denote either the Laplace transform variable or the differential operator d/dt, depending on the context.

2.3 Disturbance Spectrum

Assume that $d(t)$ is a zero-mean sequence with ***spectral density*** $S_{dd}(\omega)$, where ω is normalized frequency in the interval $[-\pi, \pi]$. The spectral density is defined as the Fourier transform of the ***auto-correlation*** function $R_{dd}(\tau)$, that is

$$S_{dd}(\omega) = \sum_{\tau=-\infty}^{\infty} R_{dd}(\tau)e^{-j\omega\tau} \tag{17}$$

where, following Ljung(1987),

$$R_{dd}(\tau) = \lim_{N\to\infty} \frac{1}{N}\sum_{t=1}^{N} \mathcal{E}\{d(t)d(t+\tau)\} \tag{18}$$

where $\mathcal{E}\{\cdot\}$ is the usual expectation operator. Observe that if $d(t)$ is either a stationary stochastic process or a periodic deterministic sequence, then the auto-correlation function as defined above will exist. Such sequences are said to be ***quasi-stationary***.

A convenient representation of the disturbance is that it is the output of a stable system with transfer function $H_0(q)$ and input $w(t)$, a zero-mean sequence with constant spectrum $S_{ww}(\omega) = \sigma_0^2$, e.g., white noise. Hence, the auto-correlation is an impulse function, and furthermore

$$S_{dd}(\omega) = \sigma_0^2|H_0(e^{j\omega})|^2 \tag{19}$$

3 Linear Control Design

3.1 Stabilization and Robustness

Before designing an adaptive control we first consider the problem of designing a ***robust*** non-adaptive control where the following information is assumed to be known: (i) a nominal model of the LSS, and (ii) a description of the accuracy of the model, e.g., model error as a function of frequency. In this section we state some known results regarding the stabilization and robustness properties of feedback systems using the above information.

Consider the linear scalar feedback system depicted in Figure 3, with reference input $r(t)$, disturbance $d(t)$, control actuator input $u(t)$, and sensor output $y(t)$. The transfer functions of the plant and feedback compensator are $P(q)$ and $C(q)$, respectively.

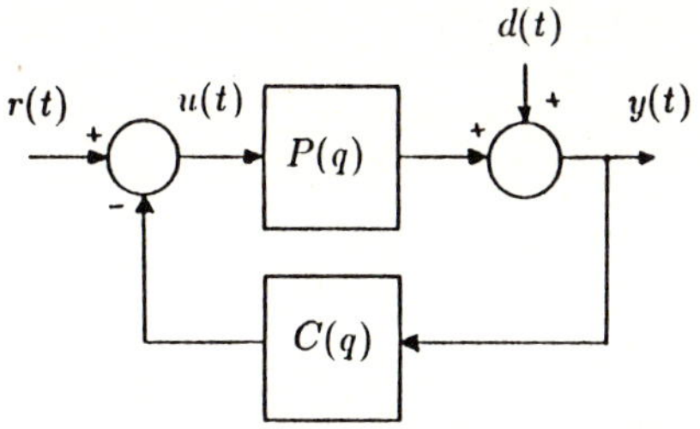

Figure 3: Linear Feedback System

Following Vidyasagar(1985), let S denote the set of transfer functions which are rational in q^{-1} (ratios of polynomials in q^{-1} with real coefficients) and stable (all poles strictly inside the unit disc). We say that C *stabilizes* P if and only if the three transfer functions $C/(1+PC)$, $1/(1+PC)$, and $P/(1+PC)$ are in $\mathbf{S}$. We now state the following known result.

THEOREM 1

(i) Stabilization

If $P \in \mathbf{S}$ then C stabilizes P if and only if $C = Q(1 - PQ)^{-1}$ for some $Q \in \mathbf{S}$.

(ii) Robustness

If

$$P = \hat{P} + \Delta \tag{20}$$

and if some compensator $\hat{C}$ stabilizes $\hat{P}$, then it also stabilizes P if and only if

$$\Delta(1 + \Delta\hat{Q})^{-1} \in \mathbf{S} \tag{21}$$

where $\hat{Q} = \hat{C}(1 + \hat{P}\hat{C})^{-1} \in \mathbf{S}$.

The first part of the theorem essentially asserts that if P is stable then the set of *all* rational compensators that stabilize P, denoted by $\mathcal{S}(P)$, is given by

$$\mathcal{S}(P) = \{Q(1 - PQ)^{-1} : Q \in \mathbf{S}\} \tag{22}$$

Thus, *any* stabilizing controller must produce a Q which is in $\mathbf{S}$.

The second part of the theorem provides conditions under which the feedback system consisting of the nominal plant $\hat{P}$ and the stabilizing control $\hat{C}$, which can be thought of as the *nominal* closed-loop system, is robustly stable to a dynamical plant perturbation Δ. Observe that in this part of the theorem, both the nominal plant model $\hat{P}$ and the perturbation Δ, may be unstable. However, when $\Delta \in \mathbf{S}$, condition (21) reduces to

$$(1 + \Delta\hat{Q})^{-1} \in \mathbf{S} \tag{23}$$

a sufficient condition for which is that

$$|\Delta(e^{j\omega})| \le \frac{1}{|\hat{Q}(e^{j\omega})|}, \ \forall\omega \in [-\pi, \pi] \tag{24}$$

Clearly then, a robust control can be designed on the basis of a nominal transfer function model $\hat{P}(q)$ and an upper bound on the model error as a function of frequency.

It is convenient to define the *set of uncertainty*

$$\mathbf{P}(\hat{P}, \delta) = \{P(q) \in \mathbf{S} : |P(e^{j\omega}) - \hat{P}(e^{j\omega})| \le \delta(\omega), \ \forall\omega \in [-\pi, \pi]\} \tag{25}$$

Thus, for any function $\delta(\omega) > |\Delta(e^{j\omega})|$, we have $P \in \mathbf{P}(\hat{P}, \delta)$. Consequently, any control $\hat{C}$ which is designed to stabilize $\hat{P}$ and simultaneously satisfy $\delta(\omega) \le 1/|\hat{Q}(e^{j\omega})|$, will also stabilize the true plant $P = \hat{P} + \Delta$.

3.2 Modal Truncation

Assume that the transfer function estimate $\hat{P}(q)$ is an estimate of the first n modes of the LSS. Typically n is not too large, say $n \approx 20$, and of course $n \ll \ell$. Hence, $\hat{P}(q)$ is an estimate of

$$P_*(q) = \sum_{k=1}^{n}(1 - q^{-1})\mathcal{Z}\{\frac{1}{s}\frac{\beta_k}{s^2 + 2\zeta_k\Omega_k s + \Omega_k^2}\} \tag{26}$$

It is convenient to express $P_*(q)$ as the rational function

$$P_*(q) = \frac{B(q,\theta_*)}{A(q,\theta_*)} \tag{27}$$

where

$$\begin{aligned} A(q,\theta_*) &= 1 + a_1^* q^{-1} + \cdots + a_{2n}^* q^{-2n} \\ B(q,\theta_*) &= b_1^* q^{-1} + \cdots + b_{2n}^* q^{-2n} \\ \theta_*{}^T &= [a_1^* \cdots a_{2n}^* \; b_1^* \cdots b_{2n}^*] \end{aligned} \tag{28}$$

The transfer function coefficients $\theta_* \in \mathbf{R}^{4n}$ are uniquely determined by the modal parameters $\{\Omega_k, \beta_k, \zeta_k : k = 1,\ldots,n\}$. The converse is not true unless the sampling frequency is large enough, i.e., if $\pi > \Omega_n/\Omega_{sa}$.

By analogy with the above desription of $P_*(q)$, the estimate $\hat{P}(q)$ can be expressed as

$$\hat{P}(q) = \frac{B(q,\hat{\theta})}{A(q,\hat{\theta})} \tag{29}$$

where $\hat{\theta} \in \mathbf{R}^{4n}$ is an estimate of θ_*. Model error is then

$$\Delta(q) = P_*(q) - \hat{P}(q) + \tilde{P}(q) \tag{30}$$

where $\tilde{P}(q)$ is the *residual mode* transfer function given by

$$\tilde{P}(q) = \sum_{k=n+1}^{\ell} (1 - q^{-1}) \mathcal{Z}\{\frac{1}{s} \frac{\beta_k}{s^2 + 2\zeta_k \Omega_k s + \Omega_k^2}\} \tag{31}$$

If $\hat{\theta} \approx \theta_*$, then the n-mode estimate $\hat{P}(q)$ is close to $P_*(q)$, and $\Delta(q) \approx \tilde{P}(q)$. Thus, a compensator $\hat{C}(q)$ which stabilizes $\hat{P}(q)$ will also stabilize the true system $P_0(q)$ provided that

$$|\hat{Q}(e^{j\omega})\tilde{P}(e^{j\omega})| < 1, \; \forall \omega \in [-\pi, \pi] \tag{32}$$

To simplify the discussion to follow suppose that the sampling frequency Ω_{sa} is sufficienly large in the sense that $\Omega_n/\Omega_{sa} \ll \pi$. Otherwise anti- aliasing filters have to be included, which although they may ultimately be required, the discussion here would be obscured. In effect, we are choosing an arbitrarily large sampling frequency.

Now, for $\Omega_n/\Omega_{sa} < \omega \leq \pi$, the control can be designed so that $|\hat{Q}(e^{j\omega})|$ is as small as required. For $\omega \ll \Omega_n/\Omega_{sa}$, the residual modes have little effect, i.e., $|\tilde{P}(e^{j\omega})|$ is small. For frequencies near Ω_n/Ω_{sa}, some care is required in the control design. However, if the flexible structure is crafted so that the frequency gap $|\Omega_{n+1} - \Omega_n|$ is large, then most any controller which has a reasonable attenuation for frequencies beyond Ω_n/Ω_{sa} is likely to be robust with respect to the unmodeled residual modes. This latter approach is preferable even though the controller can be designed independently from the structure.

If $\hat{\theta}$ is not a good estimate of θ_*, then parameter error will be a large contributor to model error magnitude $|\Delta(e^{j\omega})|$. In this case, the uncertainty set $\mathbf{P}(\hat{P}, |\Delta|)$ is too coarse of a description, because the source of uncertainty is parametric rather than unmodeled high frequency dynamics. Thus, any compensator designed to robustly stabilize this set of uncertainty will be unnecessarily cautious. However, the resulting closed loop system may still satisfy the performance goals. As will be discussed later, it is possible to design the identification experiment to preclude large parameter errors.

4 Estimating the Set of Uncertainty

The problem examined in this section is to estimate a *set of uncertainty* $\mathbf{P}(\hat{P}, \delta)$ from the data record $\{y(t), u(t) : t = 1, \ldots, N\}$. We will describe one approach and show under what conditions the true plant belongs to the estimated set of uncertainty. In this case it follows that any contol designed to robustly stabilize any plant in the estimated set of uncertainty, will also stabilize the true plant.

4.1 Parametric Model Estimation

We begin by first constructing a parametric estimate of the transfer function $P_0(q)$ using the prediction error formulation in Ljung(1985,1987). Thus, the parameter estimator is given by

$$\begin{aligned} \hat{\theta}_\alpha &= \arg\min_{\theta \in \mathcal{D}} V_N(\theta, \alpha) \\ V_N(\theta, \alpha) &= \frac{1}{N} \sum_{t=1}^{N} [L_\alpha(q)\varepsilon(t, \theta)]^2 \end{aligned} \tag{33}$$

where $\mathcal{D}$ is a subset of $\mathbf{R}^p$ and $\alpha \in \mathcal{A}$ is a set of *auxiliary parameters* which characterizes a stable filter $L_\alpha(q)$, where $\mathcal{A}$ is a subset of $\mathbf{R}^m$. The purpose of the auxiliary parameter set will be desribed in the sequel. The sequence $\varepsilon(t, \theta)$ is referred to as the *prediction error* and $L_\alpha(q)\varepsilon(t, \theta)$ as the *filtered prediction error.* Except for the auxiliary parameters α, the estimator is the usual least squares estimator.

Following Ljung(1985,1987), the prediction error is obtained from the *parametric model set*

$$y(t) = P(q, \theta)u(t) + H(q, \theta)e(t), \ \theta \in \mathcal{D} \subset \mathbf{R}^p \tag{34}$$

where $e(t)$ is an unpredictable, but bounded, function. Thus, the prediction error becomes

$$\varepsilon(t, \theta) = H^{-1}(q, \theta)[y(t) - P(q, \theta)u(t)] \tag{35}$$

The only resriction on the model set is that the *predictor*, i.e., the map $(y, u) \mapsto \varepsilon$, defined implicitly above, is stable. Thus, we require that the operators $H^{-1}(q, \theta)$ and $H^{-1}(q, \theta)P(q, \theta)$ are stable, which defines the elements of the set $\mathcal{D}$. The reason for this restriction is that despite unknown initial conditions, $\limsup_{t \to \infty} \varepsilon(t, \theta) = e(t)$, which is obviously the best the predictor can do considering the unpredictability of $e(t)$.

4.1.1 Least Squares and Linear Regression

The model set we will use here is the *equation error* model set

$$A(q, \theta)y(t) = B(q, \theta)u(t) + e(t) \tag{36}$$

where $A(q, \theta)$ and $B(q, \theta)$ are polynomials in q^{-1} whose coefficients are the elements of θ, that is, let

$$\begin{aligned} A(q, \theta) &= 1 + a_1 q^{-1} + \cdots + a_{2n} q^{-2n} \\ B(q, \theta) &= b_1 q^{-1} + \cdots + b_{2n} q^{-2n} \\ \theta^T &= [a_1 \cdots a_{2n} \; b_1 \cdots b_{2n}] \end{aligned} \tag{37}$$

This formulation coincides with (27). Hence,

$$\begin{aligned} P(q,\theta) &= \frac{B(q,\theta)}{A(q,\theta)} \\ H(q,\theta) &= \frac{1}{A(q,\theta)} \end{aligned} \tag{38}$$

and the predictor becomes

$$\varepsilon(t,\theta) = A(q,\theta)y(t) - B(q,\theta)u(t) \tag{39}$$

which is obviously stable, because $A(q,\theta)$ and $B(q,\theta)$ are stable for any $\theta \in \mathbf{R}^{4n}$, thus, $\mathcal{D} = \mathbf{R}^{4n}$. The prediction error can also be written as the *linear regression*

$$\varepsilon(t,\theta) = y(t) - \theta^T\phi(t) \tag{40}$$

where $\phi(t) \in \mathbf{R}^{4n}$ is the *regressor* given by

$$\phi^T(t) = [-y(t-1)\cdots - y(t-2n)\; u(t-1)\cdots u(t-2n)] \tag{41}$$

In this case, (33) has the well known closed form solution

$$\hat{\theta}_\alpha = R_\alpha^{-1}b_\alpha \tag{42}$$

where

$$\begin{aligned} R_\alpha &= \frac{1}{N}\sum_{t=1}^{N}[L_\alpha(q)\phi(t)][L_\alpha(q)\phi(t)]^T \\ b_\alpha &= \frac{1}{N}\sum_{t=1}^{N}[L_\alpha(q)\phi(t)][L_\alpha(q)y(t)] \end{aligned} \tag{43}$$

provided that R_α^{-1} exists. A sufficient condition is that $L_\alpha(q)\phi(t)$ is persistently exciting and the length of the data record N is sufficiently large.

DEFINITION

A sequence $f(t) \in \mathbf{R}^p$ is said to be persistently exciting if there is a positive constant β and a positive integer M such that for all $\tau \in [0,\infty)$

$$\min_{i\in[1,p]} \lambda_i \left\{ \frac{1}{M}\sum_{t=\tau+1}^{\tau+M} f(t)f^T(t) \right\} \geq \beta \tag{44}$$

Clearly if $L_\alpha(q)\phi(t)$ is persistently exciting and $N \geq M$, then R_α^{-1} exists.

4.1.2 Estimating the Nominal Transfer Function

Having found $\hat{\theta}_\alpha$, form the parametric plant transfer function estimate, denoted by $\hat{P}_\alpha(q)$, using the equation error model, i.e.,

$$\hat{P}_\alpha(q) = \frac{B(q,\hat{\theta}_\alpha)}{A(q,\hat{\theta}_\alpha)} \tag{45}$$

Thus, we obtain the *family* of parametric models

$$\{\hat{P}_\alpha(q), \alpha \in \mathcal{A}\} \tag{46}$$

Observe that the true plant $P_0(q)$ belongs to *every* member of the family of sets of uncertainty

$$\{\mathbf{P}(\hat{P}_\alpha, |\Delta_\alpha|), \alpha \in \mathcal{A}\} \tag{47}$$

where the true model error is

$$\Delta_\alpha(q) = \hat{P}_\alpha(q) - P_0(q) \tag{48}$$

Using the data record $\{y(t), u(t) : t = 1, \ldots, N\}$, we now seek to find the family of sets of uncertainty

$$\{\mathbf{P}(\hat{P}_\alpha, \delta_\alpha), \alpha \in \mathcal{A}\} \tag{49}$$

with the property that

$$\delta_\alpha(\omega) \geq |\Delta_\alpha(e^{j\omega})|, \forall\omega \in [-\pi, \pi], \forall\alpha \in \mathcal{A} \tag{50}$$

which is sufficient to guaranty that for all $\alpha \in \mathcal{A}$ we have $P_0 \in \mathbf{P}(\hat{P}_\alpha, \delta_\alpha)$.

4.1.3 Bias in Least Squares Parameter Estimate

The parameter error $\hat{\theta}_\alpha - \theta_*$, or *bias*, is a result of unmodeled dynamics and disturbances. To compute the bias, observe that (14) can be expressed as

$$P_0(q) = \frac{B(q, \theta_*)}{A(q, \theta_*)} + \tilde{P}(q) \tag{51}$$

Hence, it follows that (15) can be written as

$$y(t) = \theta_*{}^T \phi(t) + \varepsilon(t, \theta_*) \tag{52}$$

where the prediction error is

$$\varepsilon(t, \theta_*) = A(q, \theta_*)[\tilde{P}(q)u(t) + d(t)] \tag{53}$$

Substituting this expression into (42) gives the bias as

$$\hat{\theta}_\alpha - \theta_* = R_\alpha^{-1}\tilde{b}_\alpha \tag{54}$$

where

$$\tilde{b}_\alpha = \frac{1}{N}\sum_{t=1}^{N}[L(q)\phi(t)][L(q)\varepsilon(t, \theta_*)] \tag{55}$$

By carefully choosing the data filter $L(q)$ and the input spectrum $S_{uu}(\omega)$, it is clear that the bias can be made small. Hence, it is possible to design the identification experiment so that $\hat{P}_\alpha(q) \approx B(q, \theta_*)/A(q, \theta_*)$, and the inherent conservatism in the set of uncertainty $\mathbf{P}(\hat{P}_\alpha, \delta_\alpha)$ due to parameter bias is insignificant. Issues of experiment design will not be pursued here. The interested reader is referred to Ljung(1987) and the references therein.

4.2 Non-Parametric (Spectral) Estimation

The approach we propose for estimating the model error is to use standard methods of non-parametric transfer function estimation based on spectral estimation, e.g., Jenkins and Watts(1968), Ljung(1985,1987). We will assume that $u(t)$ and $d(t)$ are zero-mean sequences with spectral densities $S_{uu}(\omega)$ and $S_{dd}(\omega)$, and cross- spectral density $S_{du}(\omega)$. These are all defined analogously with (17). Assume that the input and disturbance are uncorrelated, i.e.,

$$S_{du}(\omega) = 0 \tag{56}$$

This situation arises, for example, when the LSS is operating without a stabilizing feedback, that is, in open-loop. Recall that the LSS is open- loop stable. (The proceedure described below can be modified to account for a stabilizing feedback. A brief discussion appears later in the sequel.)

Using the parametric transfer function estimate $\hat{P}_\alpha(q)$, form the corresponding ***output error***, i.e.,

$$\begin{aligned}\eta_\alpha(t) &= y(t) - \hat{P}_\alpha(q)u(t) \\ &= \Delta_\alpha(q)u(t) + d(t)\end{aligned} \tag{57}$$

where

$$\Delta_\alpha(q) = P_0(q) - \hat{P}_\alpha(q) \tag{58}$$

is the true model error, for which we now seek an estimate based on the data records

$$\{\eta_\alpha(t), u(t) : t = 1, \ldots, N\} \tag{59}$$

for each $\alpha \in \mathcal{A}$.

Using the same notation as above, the cross-spectral density between the output error and the input is given by

$$S_{\eta_\alpha u}(\omega) = \Delta_\alpha(e^{j\omega})S_{uu}(\omega) \tag{60}$$

The standard non-parametric frequency domain estimate for $\Delta_\alpha(e^{j\omega})$ is then

$$\hat{\Delta}_\alpha(\omega) = \frac{\hat{S}_{\eta_\alpha u}(\omega)}{\hat{S}_{uu}(\omega)} \tag{61}$$

where $\hat{S}_{\eta_\alpha u}(\omega)$ and $\hat{S}_{uu}(\omega)$ are spectral density estimates obtained from the finite data record (59). There are many specific ways to generate spectral density estimates. Following Ljung(1985,1987) or Jenkins and Watts(1968) we use the *smoothed* spectral estimators:

$$\begin{aligned}\hat{S}_{\eta_\alpha u}(\omega) &= \sum_{\tau=-N+1}^{N-1} \lambda_\gamma(\tau)\hat{R}_{\eta_\alpha u}(\tau)e^{-j\omega\tau} \\ \hat{S}_{uu}(\omega) &= \sum_{\tau=-N+1}^{N-1} \lambda_\gamma(\tau)\hat{R}_{uu}(\tau)e^{-j\omega\tau}\end{aligned} \tag{62}$$

where the correlation estimates are

$$\begin{aligned}\hat{R}_{\eta_\alpha u}(\tau) &= \frac{1}{N}\sum_{t=1}^{N-|\tau|} \eta_\alpha(t)u(t+|\tau|) \\ \hat{R}_{uu}(\tau) &= \frac{1}{N}\sum_{t=1}^{N-|\tau|} u(t)u(t+|\tau|)\end{aligned} \tag{63}$$

The function $\lambda_\gamma(\tau)$ is referred to as the *lag window of length* γ, where typically $\lambda_\gamma(\tau) = 0$ whenever $|\tau| > \gamma$. The above spectral estimates can be equivalently expressed as the following frequency domain convolutions:

$$\begin{aligned}\hat{S}_{\eta_\alpha u}(\omega) &= \int_{-\pi}^{\pi} W_\gamma(\omega - \sigma)I^N_{\eta_\alpha u}(\sigma)\, d\sigma \\ \hat{S}_{uu}(\omega) &= \int_{-\pi}^{\pi} W_\gamma(\omega - \sigma)I^N_{uu}(\sigma)\, d\sigma\end{aligned} \tag{64}$$

where $W_\gamma(\omega)$ is the *spectral window*, defined as the Fourier transform of the lag window, i.e.,

$$W_\gamma(\omega) = \sum_{\tau=-N+1}^{N-1} \lambda_\gamma(\tau) e^{-j\omega t} \tag{65}$$

and where $I^N_{\eta_\alpha u}(\omega)$ and $I^N_{uu}(\omega)$ are the *periodograms* given by

$$\begin{aligned} I^N_{\eta_\alpha u}(\omega) &= \frac{1}{2\pi N}\left(\sum_{t=1}^{N} \eta_\alpha(t) e^{-j\omega t}\right)\overline{\left(\sum_{t=1}^{N} u(t) e^{-j\omega t}\right)} \\ I^N_{uu}(\omega) &= \frac{1}{2\pi N}\left|\sum_{t=1}^{N} u(t) e^{-j\omega t}\right|^2 \end{aligned} \tag{66}$$

By combining the above expressions, the frequency domain estimate of model error can be compactly expressed as

$$\hat{\Delta}_\alpha(\omega) = \frac{\int_{-\pi}^{\pi} W_\gamma(\omega-\sigma) I^N_{\eta_\alpha u}(\sigma)\, d\sigma}{\int_{-\pi}^{\pi} W_\gamma(\omega-\sigma) I^N_{uu}(\sigma)\, d\sigma} \tag{67}$$

Observe that as γ increases the spectral window becomes more narrow. In addition, the spectral window is usually characterized by the following properties:

$$\begin{aligned} &\textstyle\int_{-\pi}^{\pi} W_\gamma(\omega) d\omega = 1 && \textstyle\int_{-\pi}^{\pi} \omega W_\gamma(\omega) d\omega = 0 \\ &\textstyle\int_{-\pi}^{\pi} \omega^2 W_\gamma(\omega) d\omega = M(\gamma) && \textstyle\int_{-\pi}^{\pi} W_\gamma^2(\omega) d\omega = \frac{1}{2\pi} K(\gamma) \end{aligned} \tag{68}$$

where as γ increases, $M(\gamma)$ decreases and $K(\gamma)$ increases. Under these conditions, it can be shown [see, e.g., Jenkins and Watts(1968) or Ljung(1985,1987)], that for large N, large γ, and small $K(\gamma)/N$:

$$|\hat{\Delta}_\alpha(\omega) - \Delta_\alpha(e^{j\omega})|^2 \approx M^2(\gamma)|R_\alpha(\omega)|^2 + \frac{K(\gamma)}{N}\frac{S_{dd}(\omega)}{S_{uu}(\omega)} \tag{69}$$

where

$$R_\alpha(\omega) = \frac{1}{2}\Delta''_\alpha(e^{j\omega}) + \Delta'_\alpha(e^{j\omega})\frac{S'_{uu}(\omega)}{S_{uu}(\omega)} \tag{70}$$

with $'$ and $''$ denoting differentiation with respect to ω, once and twice, respectively. Hence, for some large N, large γ, and small $K(\gamma)/N$, the error $|\hat{\Delta}_\alpha(\omega) - \Delta_\alpha(e^{j\omega})|$ can be made arbitrarily small. This fact justifies the proposed uncertainty modeling scheme for on-line robust control design. Specifically, let $T(P,C)$ denote the closed-loop system corresponding to the configuration of Figure ??, let $\mathbf{T}_*$ denote the set of all acceptable closed-loop transfer matrices, let P_0 denote the true plant, and for each $\alpha \in \mathcal{A}$, let $\mathbf{P}(\hat{P}_\alpha, \delta_\alpha)$ denote the set of uncertainty defined by

$$\mathbf{P}(\hat{P}_\alpha, \delta_\alpha) = \{P(q) \in \mathbf{S} : |P(e^{j\omega}) - \hat{P}_\alpha(e^{j\omega})| \le \delta_\alpha(\omega)|\} \tag{71}$$

We can now state:

THEOREM 2

Suppose that for all $\alpha \in \mathcal{A}$ there is a function

$$\delta_\alpha(\omega) > |\hat{\Delta}_\alpha(\omega)|, \ \forall \omega \in [-\pi, \pi] \tag{72}$$

If there is an $\alpha_* \in \mathcal{A}$ and a corresponding compensator $C_{\alpha_*}(q)$, such that

$$T(P, C_{\alpha_*}) \in \mathbf{T}_*, \ \forall P \in \mathbf{P}(\hat{P}_{\alpha_*}, \delta_{\alpha_*}) \tag{73}$$

then, for some large N, large γ, and small $K(\gamma)/N$,

$$T(P_0, C_{\alpha_*}) \in \mathbf{T}_* \tag{74}$$

Theorem 2 provides conditions under which the true plant is a member of the estimated set of uncertainty. Hence, any controller which robustly stabilizes the set of uncertainty will also stabilize the true plant. In this context robust stabilization means that the closed-loop system $T(P, C_{\alpha_*}) \in \mathbf{T}_*$ for all P in the set of uncertainty.

The difficulty in applying the theorem is to properly select N, γ, and most importantly, the function $\delta_\alpha(\omega)$. A natural choice for the latter is

$$\delta_\alpha(\omega) = [1 + k(\omega)]|\hat{\Delta}_\alpha(\omega)| \tag{75}$$

where $k(\omega) > 0$. Hence, $\delta_\alpha(\omega) \geq |\Delta_\alpha(e^{j\omega})|, \forall \alpha \in \mathcal{A}$ if

$$\sigma_\alpha(\omega) < \frac{k(\omega)}{1 + k(\omega)}, \ \forall \alpha \in \mathcal{A} \tag{76}$$

where

$$\sigma_\alpha(\omega) = \frac{1}{|\Delta_\alpha(e^{j\omega})|}\left(M^2(\gamma)|R_\alpha(\omega)|^2 + \frac{K(\gamma)}{N}\frac{S_{dd}(\omega)}{S_{uu}(\omega)}\right)^{1/2} \tag{77}$$

For large γ, typical spectral window characteristics behave like

$$M(\gamma) \approx \frac{M_0}{\gamma^2}, \quad K(\gamma) \approx K_0\gamma \tag{78}$$

Hence, for large γ,

$$\sigma_\alpha \approx \frac{1}{|\Delta_\alpha(e^{j\omega})|}\left(\frac{M_0^2}{\gamma^4}|R_\alpha(\omega)|^2 + K_0\frac{\gamma}{N}\frac{S_{dd}(\omega)}{S_{uu}(\omega)}\right)^{1/2} \tag{79}$$

Observe that if $S_{uu}(\omega)$ is more or less constant for $\omega \leq \Omega_n/\Omega_{sa}$ and attenuates rapidly thereafter, which is usually the case, then $R_\alpha(\omega) \approx (1/2)\Delta''_\alpha(\omega)$. Thus, for some large γ, $\delta_\alpha(\omega) \geq |\Delta_\alpha(e^{j\omega})|$, provided that

$$\frac{M_0}{2\gamma^2}\left|\frac{\Delta''_\alpha(e^{j\omega})}{\Delta_\alpha(e^{j\omega})}\right| < \frac{k(\omega)}{1 + k(\omega)} \ \text{and} \ \frac{\gamma}{N}\frac{S_{dd}(\omega)}{S_{uu}(\omega)} \ll 1 \tag{80}$$

4.2.1 Summary of Uncertainty Estimation Procedure

The following steps summarize the above procedure for estimating model uncertainty:

Step 1 Given a preselected filter parameter $\alpha \in \mathcal{A}$, form the filtered prediction error, and then solve for $\hat{\theta}_\alpha$ from (33).

Step 2 Using the parametric transfer function estimate $P(q, \hat{\theta}_\alpha)$, form the output error data record (59), and then calculate the frequency domain model uncertainty estimate $\hat{\Delta}_\alpha(\omega)$ from (67), using standard spectral estimation procedures.

Step 3 Design a *robust* control based on the plant being a member of the estimated set of uncertainty $\mathbf{P}(\hat{P}_\alpha, k|\hat{\Delta}_\alpha|)$, as defined in (71) for some function $k(\omega) > 0, \forall \omega \in [-\pi, \pi]$.

Step 4 If, for some $\alpha_* \in \mathcal{A}$, there is a feedback compensator $C_{\alpha_*}(q)$, such that the closed-loop systems in the set $\{T(P, C_{\alpha_*}), \forall P \in \mathbf{P}(\hat{P}_{\alpha_*}, \delta_{\alpha_*})\}$ satisfy performance objectives, then *implement* the compensator. Otherwise, go to Step 5.

Step 5 Select a new value for the filter parameters $\alpha \in \mathcal{A}$ and go to Step 1. If all filter parameter values in $\mathcal{A}$ have been exhausted, then implement that compensator in the set $\{C_\alpha, \alpha \in \mathcal{A}\}$ which produces the best closed-loop performance, i.e., the nearest to the performance objectives.

4.3 Closed-Loop Uncertainty Estimation

Suppose that the LSS system

$$y(t) = P_0(q)u(t) + d(t) \tag{81}$$

is operating in closed-loop with the stabilizing feedback

$$u(t) = r(t) - C_0(q)y(t) \tag{82}$$

The least-squares parametric procedure for obtaining $\hat{P}_\alpha$ will still work, but the spectral approach for obtaining the model error estimate $\hat{\Delta}_\alpha$ needs to be modified. The reasoning is as follows: let $r(t) = 0$ and suppose that the spectral estimates $\hat{S}_{\eta_\alpha u}(\omega)$ and $\hat{S}_{uu}(\omega)$ are very close to the true spectra $S_{\eta_\alpha u}(\omega)$ and $S_{uu}(\omega)$, respectively. We then have

$$\hat{\Delta}_\alpha(\omega) - \Delta_\alpha(e^{j\omega}) \approx \frac{S_{\eta u}(\omega)}{S_{uu}(\omega)} - \Delta_\alpha(e^{j\omega}) \tag{83}$$

$$= -\left[\frac{1}{C_0(e^{j\omega})} + P_0(e^{j\omega})\right] \tag{84}$$

which can be an arbitrarily bad estimate. To correct this difficulty, suppose that $r(t)$, which is available to the user, is selected so that it is uncorrelated with the disturbance $d(t)$ and its spectrum is much larger than the disturbance spectrum, i.e., $S_{rd}(\omega) = 0$ and $S_{rr}(\omega) \gg S_{dd}(\omega)$. If the spectral estimates are close to their true values then

$$\hat{\Delta}_\alpha(\omega) - \Delta_\alpha(e^{j\omega}) \approx \frac{S_{\eta u}(\omega)}{S_{uu}(\omega)} - \Delta_\alpha(e^{j\omega}) \tag{85}$$

$$= O\left(\frac{S_{dd}(\omega)}{S_{rr}(\omega)}\right) \tag{86}$$

Thus, by careful external input selection (experiment design), it is possible to use the uncertainty estimation procedure also in closed loop with a nominal stabilizing controller.

5 Example of Uncertainty Estimation

In this section the uncertainty estimation procedure is applied to data obtained from the laser pointing experiment described in Walker et al.(1984). A schematic drawing of the apparatus is shown in Figure 4.

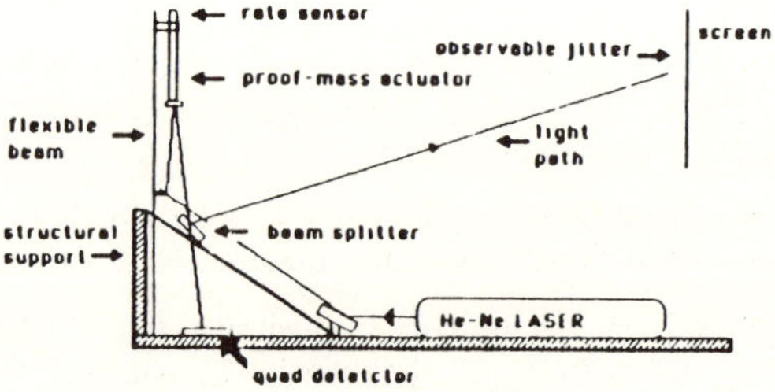

Figure 4: Laser pointing and control experiment

The objective of the experiment is to control the jitter of the laser beam. The single actuator consists of a pivoted proof-mass which exerts a reaction force on the flexible beam whenever the proof-mass is moved by an applied control current through the armature. As the flexible structure vibrates, the laser beam changes its angular direction. A quadrature detector, mounted on the structural support, registers the laser beam position, as long as it is in the field-of-view of the detector. The laser beam strikes a mirror on the flexible structure and is reflected by another mirror mounted on the proof- mass actuator. The resulting beam is split in two by a beam-splitter, with one ray going to the quadrature detector and the other to a screen where the jitter is magnified for visual inspection. The mass of the actuator is greator than the mass of the flexible structure, thereby insuring a significant interaction between the actuator and the flexible structure.

The problem is to control the jitter from 4 Hz to 20 Hz. The input $u(t)$ is chosen as a sine-sweep lasting about 16 seconds, sweeping from 4 Hz to 20 Hz and sampled at 51.2 Hz. The number of data samples is $N = 1024$.

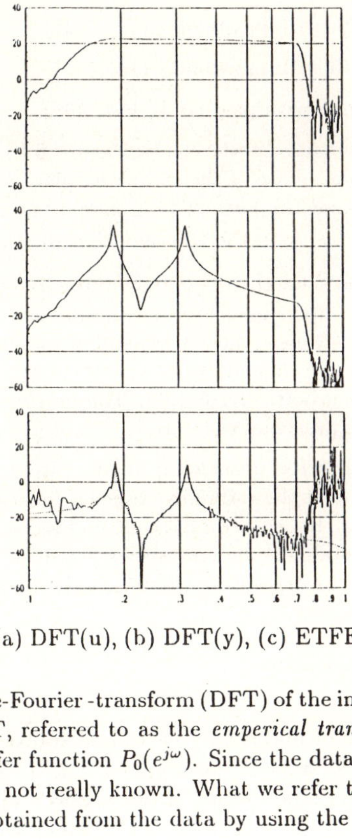

Figure 5: (a) DFT(u), (b) DFT(y), (c) ETFE and $P_0(e^{j\omega})$

Figure 5 shows the discrete-Fourier -transform (DFT) of the input and output, as well the ratio of output DFT to input DFT, referred to as the *emperical transfer function estimate* (ETFE), compared to the "true" transfer function $P_0(e^{j\omega})$. Since the data is taken from a physical device, the true system dynamics are not really known. What we refer to as the "true" system is an 8th order equation error model obtained from the data by using the least squares estimate (42) with $n = m = 8$. This gives

$$\begin{aligned} P_0(q) &= \frac{B_0(q,\theta)}{A_0(q,\theta)} \\ B(q,\theta) &= -.0044q^{-1} + .0016q^{-2} - .056q^{-3} - .015q^{-4} \\ &\quad + .013q^{-5} - .011q^{-6} - .033q^{-7} - .027q^{-8} \\ A_0(q,\theta) &= 1 - 1.11q^{-1} + .54q^{-2} + .47q^{-3} + .77q^{-4} \\ &\quad - .17q^{-5} + .08q^{-6} + .40q^{-7} + .0057q^{-8} \end{aligned}$$

The parametric model is chosen as a 4th-order equation error model as in (36), where

$$\begin{aligned} A(q,\theta) &= 1 + a_1 q^{-1} + \cdots + a_4 q^{-4} \\ B(q,\theta) &= b_1 q^{-1} + \cdots + b_4 q^{-4} \\ \theta^T &= [a_1, \ldots, a_4, b_1, \ldots, b_4] \end{aligned}$$

The auxiliary filter parameter set is the 4-tuple,

$$\mathcal{A} = \{0, 1, 2, 3\}$$

such that the corresponding filters have the following properties:

$L_0(q) = 1$, the "natural" filter.

$L_1(q) =$ 8th-order Butterworth with passband $[.11\pi, .67\pi]$.

$L_2(q) =$ 8th-order Butterworth with passband $[.11\pi, .40\pi]$.

$L_3(q) =$ 16th-order Butterworth with passband $[.15\pi, .40\pi]$.

The effect of each filter is shown, respectively, in Figures 6 to 9. Each figure shows 3 plots: the two top ones showing gain and phase of the estimated plant (the dashed lines) compared to the true plant (the solid lines), and the bottem plot comparing magnitudes of the model error estimate (the dashed line) from (67) with the true model error (the solid line).

The top and bottem plots also show the magnitude of the passband filter used to form the filtered prediction error. Observe that the model error estimates are quite accurate in the frequency range $[.15, .70]$, which is the same range where the input has a nice flat DFT. The model error estimates were obtained using a rectangular window, and no attempt was made here to adjust the lag window width to achieve a better resolution. The best model estimate is clearly for the filter $L_3(q)$, particularly over the passband of interest.

These results give confidence to the uncertainty model estimation procedure proposed here, because despite the fact the some of the estimated transfer functions are quite poor, the model uncertainty estimate is very good. Hence, a robust control designed on the basis of the estimated set of uncertainty would not de-stabilize the true system. The next step is to continue with the above example, for example, and incorporate a specific robust design control design method.

6 Concluding Remarks

A method for estimating the set of uncertainty of a plant transfer function has been proposed and analyzed. Theorem 2 provides analytic justification that the true plant can, under suitable conditions, be guaranteed to be in the estimated set of uncertainty. An example is presented which supports the theory. One of the restrictions in this paper is that the plant is stable and no feedback is present during the experiment. The results, however, can be extended to the case of an unstable plant operating in closed-loop with a stabilizing feedback compensator.

Although in this paper we restrict the auxiliary parameters to essentially index data filters, the concept can be broadened. For example, the filters can be chosen differently for the input and output data. In addition, the auxiliary parameters can index various inputs which have different spectral content and the choice made by using a local optimization on the auxiliary parameters. The index parameters are thus choices which can be made to tailor the identification experiment in accordance with the use of the model. Off- line procedures for experiment design are discussed,

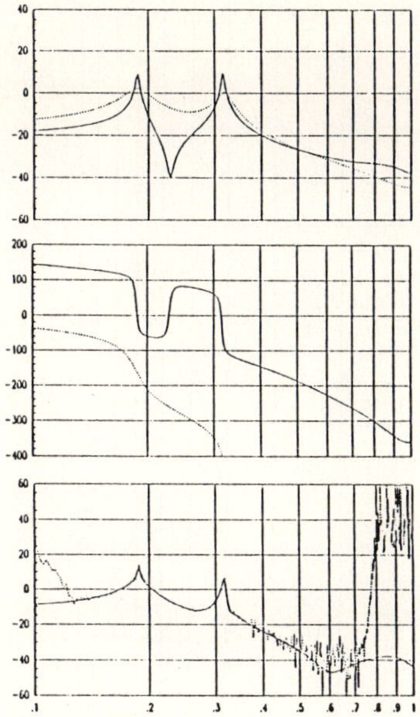

Figure 6: (a) gain, (b) phase, (c) model error magnitude

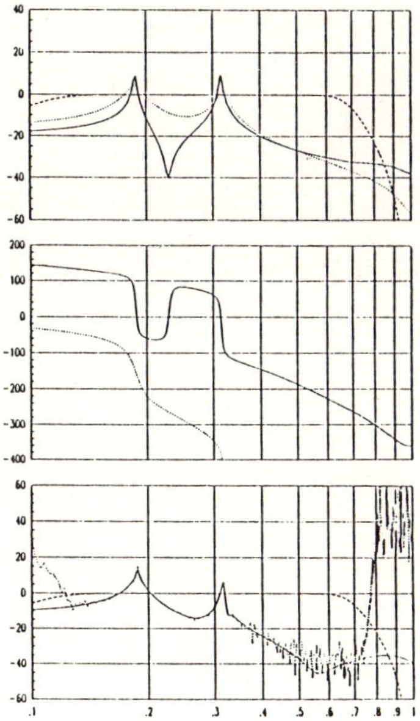

Figure 7: (a) gain, (b) phase, (c) model error magnitude

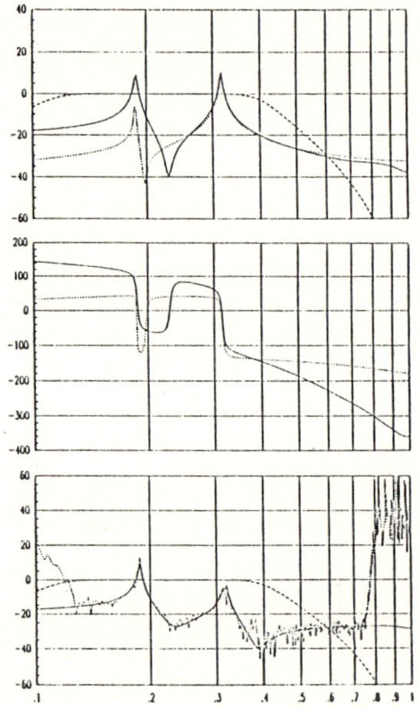

Figure 8: (a) gain, (b) phase, (c) model error magnitude

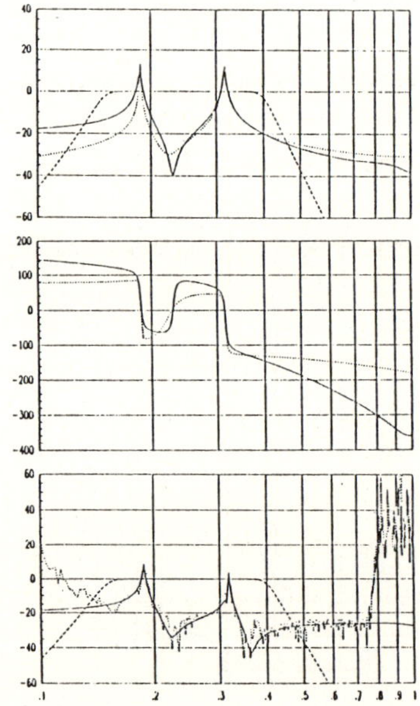

Figure 9: (a) gain, (b) phase, (c) model error magnitude

for example, in Ljung(1987) or Goodwin and Payne(1977). The issue addressed here, which might be called "on-line experiment design", remains to be studied and further developed.

The results of this paper raise some questions, like: why not use non-parametric spectral estimation to obtain the transfer function estimate in the first place? The answer is partly to be found in Ljung(1985,1987), where it is shown, as might be expected, that the parametric techniques have a smaller variance (asymptotically) as a function of data length. Here, we are using the parametric model for the critical frequency range where control is needed, whereas a cruder spectral analysis is used for the frequency range over which control is not as important. Another question: why not just increase the model order, and hence, stay with parametric methods? This approach is taken in Wahlberg(1986), where after identification a model reduction is performed. The problem is that we now have *one* plant estimate and no set of uncertainty, i.e., the estimate is assumed to be sufficiently accurate. Hopefully it is, but this can only be established by an *a priori* analysis, i.e., there is no on-line estimate of the set of uncertainty. It may be that such an estimate is unnecessary for a particular problem, provided that the experiment has been appropriately designed.

7 References

B.D.O. Anderson, R.R. Bitmead, C.R. Johnson, Jr., P.V. Kokotovic, R.L. Kosut, I.M.Y. Mareels, L. Praly, and B.D. Riedle, (1986),*Stability of Adaptive Systems: Passivity and Averaging Analysis*, MIT Press, 1986.

J.C. Doyle and C.C. Chu (1986),"Robust Control of Multivariable and Large Scale Systems", Final Tech. Report under AFOSR Contract F49620-84-C-0088, Honeywell Systems and Research Center, Minn., MN.

B.A. Francis and G. Zames (1984), "On H_∞-Optimal Sensitivity Theory for SISO Feedback Systems", *IEEE Trans. on Aut. Control*, Vol. 29, pp. 9-16.

G.C. Goodwin and R.L. Payne, (1977), *Dynamic System Identification: Experiment Design and Data Analysis*, Academic Press, New York.

G.M. Jenkins and D. G. Watts, (1968) *Spectral Analysis and its Applications*, Holden-Day, San Francisco.

R.L. Kosut, (1986), "Adaptive Calibration: An Approach to Uncertainty Modeling and On-Line Robust Control Design", *Proc. 25th IEEE CDC*, Athens, Greece, Dec. 1986.

R.L. Kosut, (1987), "Adaptive Uncertainty Modeling: On-Line Robust Control Design", *Proc. 1987 ACC*, Minneapolis, MN, June 10-12, 1987.

L. Ljung, (1985), "On the Estimation of Transfer Functions", *Automatica*, Vol. 21, No. 6, Nov.

L. Ljung, (1987), *System Identification: Theory for the User*, Prentice Hall, New Jersey.

M.G.Safonov, A.J.Laub, and G.L.Hartman(1981),"Feedback Properties of Multivariable Systems: The Role and Use of the Return Difference Matrix", *IEEE Trans. Aut. Contr.*, **Vol. 26**(1):47-65, Feb. 1981.

M. Vidyasagar (1985), *Control system Synthesis: A Factorization Approach*, MIT Press, Cambridge, MA, 1985.

B. Wahlberg (1986), "On Model Reduction in System Identification", *Proc. ACC*, Seattle, WA, June.

Unified Optimization of Structures and Controllers

J. L. Junkins and D. W. Rew
Texas A&M University

Introduction

Given a differential equation ***model*** for a dynamical system, representing for example, a flexible structure, associated sensors and actuators, and a nominally stabilizing (optimal in some sense) feedback control law, a fundamental question is the following: **will the feedback control law stabilize and near-optimally control the *actual* system**? Of course, there are many interesting and significant issues raised by this question. The modeling process is always imperfect, among the several important sources of error are the following: (i) ignorance of the actual mass, stiffness, and energy dissipation properties, as well as boundary conditions and geometric parameters, (ii) discretization and truncation errors associated with representing a continuous system in terms of a finite number of degrees of freedom, (iii) neglect of nonlinearities, (iv) ignorance of external disturbances, and (v) poorly modeled sensors and actuators.

Indeed, even if one restricts the discussion to systems adequately modeled by a finite set of linear ordinary differential equations, the effects of uncertain parameters and disturbances results in analytical and practical implementation difficulties. The additional issues raised by the possibility that the structure can be re-designed to enhance controlled performance increase the dimensionality and difficulty of dealing comprehensively with the above issues. Improved methodology for solving the high dimensioned structural control problem is of central importance for the coming generation of large flexible space structures. In the present discussion, we emphasize development of robust eigenstructure assignment methods, and summarize some optimization methods in which both the controller and selected structural parameters are re-designed to improve robustness. A multi-criterion approach is presented by which one can study the tradeoff between robustness and competing performance measures (such as average control energy, average state errors, sensitivity measures, and structural mass).

Preliminaries

Consider a finite order discrete model of a linear structure of the form

$$\boldsymbol{M}\ddot{\boldsymbol{z}} + \boldsymbol{C}\dot{\boldsymbol{z}} + \boldsymbol{K}\boldsymbol{z} = \boldsymbol{D}\boldsymbol{u} \qquad (1)$$

where the nxn symmetric mass, damping and stiffness matrices satisfy the definiteness properties $\boldsymbol{M}>0$, $\boldsymbol{C}\geq 0$, $\boldsymbol{K}\geq 0$; $\boldsymbol{z}(t)$ is an $nx1$ configuration vector, $\boldsymbol{u}(t)$ is an $mx1$ control vector; and $\boldsymbol{D}$ is an nxm control influence matrix. All quantities are considered real in Eq. (1). The structural eigenvalue problem associated with the $z = \phi e^{\lambda t}$ solutions for the undamped free vibration special case of Eq. (1)

Springer Series in Computational Mechanics
S. N. Atluri, A. K. Amos (Eds.)
Large Space Structures: Dynamics and Control

$$M\ddot{z} + Kz = 0 \tag{2}$$

is
$$det[M\lambda^2 + K] = 0, \;\Rightarrow\; eigenvalues: \{\lambda_i^2 = -\,\omega_i^2\,, i = 1,2, \ldots n\} \tag{3}$$

$$[M\lambda_i^2 + K]\phi_i = 0, \;\Rightarrow\; eigenvectors: \{\phi_i\,, \quad i = 1,2, \ldots n\} \tag{4}$$

For this case, all eigenvalues($\lambda_i = \pm\, i\omega_i$, $i^2 = -1$) are purely imaginary and eigenvectors are real. It is well known that the eigenvectors are orthogonal with respect to M and K; we normalize these orthogonality conditions in the usual way[1] as follows

$$\phi_i^T M \phi_j = \delta_{ij} \quad \text{and} \quad \phi_i^T K \phi_j = \delta_{ij}\omega_i^2\,, \; i, j = 1,2, \ldots, n. \tag{5}$$

In matrix notation, the orthonormality conditions assume the familiar forms

$$[\Phi]^T M[\Phi] = [I], \text{ and } \; [\Phi]^T K[\Phi] = [\omega^2] \tag{6}$$

where the *modal matrix* $[\Phi]$ contains the eigenvectors (normal mode shapes) as columns, $[I]$ is the identity matrix, and $[\omega^2] = diag\{\omega_1^2, \omega_2^2, \ldots, \omega_n^2\}$ is the matrix of natural frequencies squared, ordered so that $\omega_{i+1} \geq \omega_i$. It is evident from Eq. (6) that $[\Phi]^{-1} = [\Phi]^T M$.

For many purposes, it is convenient to introduce a transformation of Eq. (1) into the modal state space associated with Eq. (2) as follows:

$$z = [\Phi]\eta, \;\; \eta = [\Phi]^T M z \tag{7}$$

Use of Eqs. (6, 7) in Eq. (1) yields the modal space equation of motion

$$\ddot{\eta} + [\Phi]^T C\, [\Phi]\dot{\eta} + [\omega^2]\eta = [\Phi]^T D u \tag{8}$$

We now define the *2nx1 modal state vector* as $x = col\,\{\eta, \dot{\eta}\}$, so that the n second order system of equations, Eq. (8), can be written as a system of $2n$ first order equations in the standard form

$$\dot{x} = Ax + Bu, \;\; with\; A = \begin{bmatrix} 0 & I \\ -[\omega^2] & -[\Phi]^T C\,[\Phi] \end{bmatrix}, \quad B = \begin{bmatrix} 0 \\ [\Phi]^T D \end{bmatrix} \tag{9}$$

We note that Eq. (9) is normally preferred, for problems in structural dynamics especially, to the common alternative (configuration) state vector $s = col\,\{z, \dot{z}\}$, with the corresponding differential equation

$$\dot{s} = \hat{A}s + \hat{B}u, \;\; with\; \hat{A} = \begin{bmatrix} 0 & I \\ -M^{-1}K & -M^{-1}C \end{bmatrix}, \quad \hat{B} = \begin{bmatrix} 0 \\ M^{-1}D \end{bmatrix} \tag{10}$$

One issue should be obvious, but nonetheless is not universally appreciated, it *is material* to address which of Eqs. (1), (8), (9), or (10) is used to design feedback control laws. The algebraic equivalence of these equations is clear, but the modal space differential equations are more amenable to qualitative analysis and, as a consequence, it is easier to address such issues as order reduction, truncation, spillover, robustness etc., than is the case if one pursues control designs based upon configuration space equations. This is because the lower frequency modes are generally more accurately converged (when using, for example, a finite element discretization process to arrive at Eq. (1)) and are typically more important physically. Of course, generic developments addressed to Eqs. (1, 10) include Eqs. (8, 9) for appropriate specialization of the coefficient matrices.

In this chapter, we are primarily concerned with design of the constant *mxr* gain matrix G for a linear feedback control law of the form $u = Gy$. We assume that sensors are available to measure the *rx1* output vector y which is linearly related to the modal state vector x according to

$$y = Sx = S_1\eta + S_2\dot{\eta} \tag{11}$$

We address the general case in which there are fewer actuators than the number of modes we wish to control ($m<n$). It is important to note that introduction of real actuator hardware generally changes the mass and stiffness characteristics of the system, and also increases the order of the dynamical system(since the controller seldom directly commands force or moment, but, for example, voltage, in the case of a control moment gyro electric motor). Therefore, we note that having the number of actuators equal to the order of the controlled system can be assumed for the sake of idealized analysis, but cannot usually be realized, since the actuators themselves are dynamical systems. Obviously, the path to determine Eqs. (1) or an augmentation thereof, should include accounting for finite contributions of the apriori modeled actuators and sensors to M and K, and, at least in the "end game" of a controller optimization study, should account for the dynamics of the sensors and actuators. Furthermore, we feel that we must include the case of a relatively small number of actuators to control at least a moderately high-dimensioned system, because this pattern is present in many problems of current practical interest. The $m=n$ case will remain, of course, a special case of the more general developments.

Substituting the feedback law $u = Gy$ into Eq. (9), and using Eq. (11), the *closed loop system* becomes

$$\dot{x} = [A + BGS]x \tag{12}$$

The closed loop eigenvalue problem (associated with $x = \psi e^{\lambda t}$ solutions of Eq. (12)) is

$$det\big[[A{+}BGS] - \lambda I\big] = 0 \Rightarrow eigenvalues\text{: } \{\lambda_1, \lambda_2, \ldots \lambda_{2n}\} \tag{13}$$

$$[[A{+}BGS] - \lambda_i I]\Psi_i = 0 \Rightarrow right\ eigenvectors\text{: } \{\Psi_1, \Psi_2, \ldots \Psi_{2n}\} \tag{14}$$

where advantage can be taken of the fact that eigensolutions (λ, ψ) occur in complex conjugate pairs, for the A, B matrices of Eq. (9), and that the *2nx1* eigenvectors Ψ_i have the following structure:

$$\Psi_i = col\ \{\psi_i, \lambda_i\psi_i\}, \quad where\ \psi_i\ is\ an\ nx1\ complex\ vector. \tag{15}$$

Consider the situation wherein a subset of the structural parameters are available for re-design. Let the structural design parameter vector be denoted p, the system matrices typically have an algebraic functional dependence upon p : $M = M(p), C = C(p), K = K(p), D = D(p), S = S(p),$ and as a consequence, the eigenvalues and eigenvectors depend upon p. Using methods of [2], it is straight-forward to determine the partial derivatives of the open loop, undamped eigensolution pairs (ω_i , ϕ_i) with respect to p, through differentiation of Eqs. (4, 5). These sensitivities can be propagated, by the chain rule, through differentiation of Eqs. (13, 14) to the analogous partial derivatives of the closed loop eigensolution pairs (λ_i , ψ_i). Except for isolated events of repeated eigenvalues, the eigenvalues and eigenvectors usually behave continuously, albeit nonlinearly, as a function of $\boldsymbol{p}$. The vector $\boldsymbol{p}$ can, in principle, contain any combination of the system's mass, stiffness, damping, geometric parameters, actuator locations, etc., but we consider $\boldsymbol{p}$ to be the subset of those parameters actually available for re-design. Also, depending upon the design methodology being used, it is possible [3,23,46] to include the elements of the gain matrix $\boldsymbol{G}$ in the design vector $\boldsymbol{p}$. Several nonlinear optimization methods have been developed [3,23,46] and used successfully to optimize controlled closed loop performance by simultaneously or sequentially re-designing selected structural, sensor, and actuator parameters.

For generality of the developments below, we will only selectively specialize our results to reflect the structure of $\boldsymbol{A}, \boldsymbol{B}$, e. g., Eqs. (9), to accommodate, for example additional equations to model actuator dynamics. Therefore, we assume that the dimension of our state space is $N \geq 2n$. Due to the complexity and nonlinearity of the *simultaneous* structure/controller design process, we first address the design of a linear state feedback controller, via minimization of quadratic criteria and via a recently developed class of *robust eigenstructure assignment* methods, for the case of a *specified* structure. Since the stability and response characteristics of a given structure are fully dictated by the closed loop eigenvalues and eigenvectors, we can cast the control design problem as a quest for some optimal gain matrix *(G)* which generates a desirable set of eigenvalues and eigenvectors. In the case that we seek to impose a set of equality constraints upon the eigenvalues and eigenvectors (i., e., *eigenstructure assignment*), there is a recently developed theoretical foundation which underlies existence, uniqueness and computation of $\boldsymbol{G}$. The principle existence issues are captured in a theorem due to Srinathkumar[4,47]:

Eigenstructure Assignment Theorem. *Given a controllable and observable dynamical system described by* $\dot{\boldsymbol{x}} = \boldsymbol{A}\boldsymbol{x} + \boldsymbol{B}\boldsymbol{u}, \boldsymbol{y} = \boldsymbol{S}\boldsymbol{x}, \boldsymbol{u} = \boldsymbol{G}\boldsymbol{y}$, *{having N states, r sensors, and m actuators, with matrices **B** and **S** of full rank}, then through choosing appropriate values for the elements of **G**, max(m,r) closed-loop eigenvalues (of **A** + **BGS**) can be assigned, max(m,r) eigenvectors can be partially assigned, and min(m,r) entries in each eigenvector can be arbitrarily assigned.*

Typically, for a controllable system, a given set of eigenvalues can be assigned by an infinity of $\boldsymbol{G}$ matrices, the remaining freedom in selecting $\boldsymbol{G}$ can be used to optimize the eigenvectors to achieve robustness or some other objectives. During the past decade, a number of eigenstructure assignment (or "pole placement") formulations and algorithms have been developed and demonstrated [5-12]. The algorithms available, up until very recently, have been restricted to application to very low order systems, due to various computational difficulties. The most recent literature emphasizes gain determination which imposes eigenvalue placement constraints and result in control robustness (low sensitivity of stability and/or performance measures with respect to modeling errors and disturbances), it is fortuitous that well-conceived concepts which seek out robust controls often leads to algorithms for gain computation which are numerically robust, as well. The reason for this good fortune lies in the fact that good conditioning of the modal matrix of closed loop eigenvectors has been found to be the crucial necessary condition for robustness of the numerical processes to determine the control gains as well as low sensitivity of the resulting closed loop performance and stability indices. In the present chapter, we present several attractive concepts and associated algorithms for efficient determination of judicious gain matrices (*G*), which are "good" from one of the three viewpoints:

(i) guarantees stability of the nominal unperturbed plant, and maximizes a measure of controlled system robustness (equivalently, minimizes an appropriately defined sensitivity measure).

(ii) satisfies property (i), and also satisfies prescribed constraints on placement of the closed loop eigenvalues and eigenvectors.

(iii) satisfies conditions (ii), subject to constraints on average control energy and average state error energy.

For perspective and contrast, we first review briefly, in the following section, more familiar optimal control methodology (based upon minimizing a quadratic performance measure in the time domain). We also present a generalization of the classical developments which are more attractive computationally. Several of the results developed are subsequently used in the *multi-criterion optimization* discussions. From one viewpoint, the commonly applied steady state(i., e., constant gain) optimal linear-quadratic regulator is just one (of many) ways to parameterize a family of stable feedback controllers. We believe that the indirect connection(i.e. through the Riccati equation *and* solution of the closed loop eigenvalue problem) of the weight matrices in a performance index, to the robustness properties and eigenvalue placement are significant practical disadvantages of the conventional quadratic regulator in comparison to the *robust eigenstructure assignment* methods we discuss below. We develop eigenstructure assignment methods which, in essence, exchange the solution of the algebraic nonlinear Riccati equation for solution of strictly linear algebraic equations. Judicious use of numerically stable singular value decomposition methods lead to reliable algorithms. The implications for control of high order systems are most significant.

Generalized Quadratic Regulators

The cornerstone of *modern* control theory, judging from the frequency of application, are linear, constant gain feedback controls determined by minimizing the index

$$J = \int_0^{\infty} (x^T Q x + u^T R u)dt, \text{ with weight matrices } Q = Q^T \geq 0 \text{ , } R = R^T > 0 \qquad (16)$$

subject to satisfying $\dot{x} = Ax + Bu$. This is widely known as the linear-quadratic regulator (LQR) problem. The solution is determined from Pontryagin's Principle[13]; it requires computing the positive-definite P matrix which satisfies the algebraic Riccati equation

$$PA + A^T P - PBR^{-1}B^T P + Q = 0 \qquad (17)$$

The associated optimal control gain matrix is then

$$G = -R^{-1}B^T P \qquad (18)$$

The existence of the solution for $P = P^T > 0$ is theoretically guaranteed for controllable systems, but existing algorithms [14,15] and computer implementations limit practical applications to systems of moderate order (<~100), and of course, numerical difficulties may be encountered for even low dimensioned systems, depending upon the conditioning of the solution(which measures *how controllable* the system is, in a numerical sense). We now consider two recently developed generalizations of the quadratic regulator.

Generalized LQR - Form 1:
Recently[16], the following generalized quadratic index was introduced

$$J = \int_0^\infty \begin{Bmatrix} x \\ u \end{Bmatrix}^T W \begin{Bmatrix} x \\ u \end{Bmatrix} dt, \text{ with } W = W^T = \begin{bmatrix} Q + N^T R N & N^T R \\ RN & R \end{bmatrix} > 0,\ Q = Q^T \geq 0\ ,\ R = R^T > 0 \quad (19)$$

The Pontryagin necessary conditions(minimizing Eq. (19), subject to $\dot{x} = Ax + Bu$) lead to the generalized Riccati equation

$$P\tilde{A} + \tilde{A}^T P - PBR^{-1}B^T P + Q = 0, \text{ where } \tilde{A} = A - BN \quad (20)$$

and the optimal gain matrix

$$G = -R^{-1}B^T P - N \quad (21)$$

Reference [16] shows that an arbitrary real $m x N$ cross coupling weight matrix (N) can be chosen without de-stabilizing the closed loop system. Notice that N is essentially a pre-feedback gain, so it is evident that some redundancy must exist. In particular, [16] investigated $N = [UD^T\Phi \quad VD^T\Phi]$, with U, V positive semi-definite mxm matrices; this choice has been found attractive. We present results which demonstrate significant advantages can be realized in comparison to using the classical LQR developments.

Generalized LQR - Form 2:
The second modification of the LQR problem, introduced in [16], is obtained by minimizing the quadratic index

$$J = \int_0^\infty \begin{Bmatrix} x \\ u \end{Bmatrix}^T W \begin{Bmatrix} x \\ u \end{Bmatrix} dt, \text{ with } W = W^T = \begin{bmatrix} Q + N^T R N + PBR^{-1}B^T P & N^T R \\ RN & R \end{bmatrix} > 0, \quad (22)$$

subject to $\dot{x} = Ax + Bu$. The motivation for the final term ($PBR^{-1}B^T P$) in the *1,1* partition of W is that it results in an important simplification of the necessary conditions. In fact the the P matrix satisfies the (much simpler) linear algebraic Lyapunov equation

$$P\tilde{A} + \tilde{A}^T P + Q = 0, \text{ where, as before } \tilde{A} = A - BN, \quad (23)$$

instead of the nonlinear algebraic Riccati Eq. (20). The optimal gain matrix is still given by Eq. (21). Notice that we choose the weights $Q, R, N,$ and solve for P from Eq. (23), only then can we fully specify W in Eq. (22), since it depends upon P.

The weight matrices Q, R, and N should be parameterized in such a way as to guarantee their definiteness properties and the controllability of $\tilde{A}$, this insures the existence of a solution for a positive-definite P satisfying Eq. (23). This problem is solved[16] by the Choleski factorizations

$$Q = L_1 L_1^T, \quad R^{-1} = L_2 L_2^T, \quad U = L_3 L_3^T, \quad V = L_4 L_4^T, \quad N = [UD^T\Phi \quad VD^T\Phi] \tag{24}$$

with

$$L_1 = \begin{bmatrix} q_{11}^2 & 0 & .. & 0 \\ q_{21} & q_{22}^2 & .. & 0 \\ : & : & : & : \\ q_{NN} & q_{NN} & .. & q_{NN}^2 \end{bmatrix}, \quad L_2 = \begin{bmatrix} r_{11}^2 & 0 & .. & 0 \\ r_{21} & r_{22}^2 & .. & 0 \\ : & : & : & : \\ r_{m1} & r_{m2} & .. & r_{mm}^2 \end{bmatrix},$$

$$L_3 = \begin{bmatrix} u_{11}^2 & 0 & .. & 0 \\ u_{21} & u_{22}^2 & .. & 0 \\ : & : & : & : \\ u_{m1} & u_{m2} & .. & u_{mm}^2 \end{bmatrix}, \quad L_4 = \begin{bmatrix} v_{11}^2 & 0 & .. & 0 \\ v_{21} & v_{22}^2 & .. & 0 \\ : & : & : & : \\ v_{m1} & v_{m2} & .. & v_{mm}^2 \end{bmatrix}. \tag{25}$$

The high dimensionality of the parameter space consisting of the above, or any complete, weight matrix parameterization has proven a significant obstacle to numerical implementations with fully populated weight matrices. In [16], we introduced a *minimum modification homotopy* strategy for tuning the above matrices and demonstrated successful implementations. This method is employed where appropriate in the numerical studies reported below. While the high dimensionality of the weight matrices creates difficult problems, the assignment of these matrices is essentially the same hurdle which faces users of the classical LQR developments. However, the overwhelming majority of the historical numerical implementations have been accomplished with diagonal weight matrices (often, simply using a scalar multiple of an identity matrix), due the expense of ad hoc searches in the weight matrix space. However, faced with actuator saturation constraints, eigenvalue placement constraints, and other practical issues, constrained searches for "judicious weights" are often necessary and have been widely used. While any positive-definite choices for the weight matrices satisfy(for controllable systems) the most important necessary condition of placing all closed-loop eigenvalues in the stable left-half plane, finding the *optimal* optimal regulator (vis-a-vis robustness, stability margins, etc.) generally forces one to deal with the perhaps uncomfortable truth that the LQR weight matrices are simply one elegant, and not especially convenient, parameterization of stable controllers. On the other hand, numerical experience suggests that there are typically broad regions in the weight space which give good control laws of moderate sensitivity.

While the above modification (Form 2) of the LQR does not eliminate the necessity to find judicious weight matrices, it evident that eliminating the Riccati equation in favor of the simpler Lyapunov equation is extremely useful, especially for high dimensioned applications, since the Riccati equation solution is by far the most computationally challenging part of the LQR design process. Thus the LQR approach can be pursued without having to solve the Riccati equation, while at the same time retaining a more general family of weight matrices than the classical formulation. Numerical studies reported below and in [19] indicate that optimized, fully populated weight matrices lead to more flexible eigenvalue placement and robustness optimization than the classical LQR formulation. In [3,23], analytical methods are presented for computation of partial derivatives of the closed loop eigenvalues and eigenvectors with respect to structural parameters and weight matrix parameters, consistent with the above developments.

For controller design problems more naturally posed in the space of eigenvalues and eigenvectors of the closed loop system, the LQR approach is cumbersome, since the choice of weight matrices dictates "everything", and substantial computation lies between specification of these matrices and evaluation of the corresponding eigenvalues, eigenvectors, and robustness properties. Thus attempts to impose eigenspace or robustness constraints involves expensive nonlinear iterations in a high-dimensioned parameter space. While successful algorithms have been demonstrated, all known algorithms suffer from local versus global convergence and related difficulties. In view of these truths, it is natural to ask the question: "Is there a more direct way to parameterize a family of stable controllers, defining both the gain matrix and closed loop robustness characteristics of the system as explicit functions of the closed-loop eigenvalues and eigenvectors?" The affirmative answer to this question leads to eigenstructure assignment concepts [5-12], and in particular the methodology in the following section.

Robust Eigenstructure Assignment

In this section, an approach is developed for defining judicious sets of objective closed-loop eigenvectors, consistent with the system dynamics and prescribed placement of the corresponding closed loop eigenvalues. These objective eigenvectors are usually not exactly achievable, so we make use of orthogonal projection and least square methods to find admissible eigenvectors near the desired eigenvectors. The most general developments apply to systems having arbitrary structure in the first order system matrices$(\boldsymbol{A}, \boldsymbol{B})$, however, we develop specialized versions of the methods to take advantage of the structure introduced when we begin with a system of second order differential equations of the form of Eqs. (1) or (8).

Sylvester's Equation

For the standard first order state space system ($\dot{\boldsymbol{x}} = \boldsymbol{Ax} + \boldsymbol{Bu}$, with $\boldsymbol{u} = \boldsymbol{Gx}$), the right eigenvalue problem, for the special case that $\boldsymbol{S} = \boldsymbol{I}$ can be rearranged from Eq. (14) as

$$[\boldsymbol{A} - \lambda_i \boldsymbol{I}]\psi_i = \boldsymbol{BG}\psi_i \tag{26}$$

or, defining the $mx1$ vector $\boldsymbol{h}_i$ as

$$\boldsymbol{h}_i = \boldsymbol{G}\psi_i \tag{27}$$

then Eq. (26) is put in the form of Sylvester's equation

$$[\boldsymbol{A} - \lambda_i \boldsymbol{I}]\psi_i = \boldsymbol{Bh}_i \tag{28}$$

The matrix equivalents of Eqs. (27) and (28) are

$$\boldsymbol{H} = \boldsymbol{G}\Psi \tag{29}$$

and

$$A\Psi - \Psi\Lambda = BH, \qquad \text{with } \Lambda = diag\,\{\lambda_1 \ldots \lambda_N\}. \tag{30}$$

We can assign eigenvalues and eigenvectors using Eqs. (29) and (30) by first choosing the desired closed-loop eigenvalue matrix Λ, then selecting the parameter matrix $\boldsymbol{H}$. Eq. (30) can then be solved for the corresponding eigenvector matrix Ψ. Finally the corresponding gain matrix $\boldsymbol{G}$ is obtained through solution of Eq. (29). Unfortunately, an arbitrary choice for the $\boldsymbol{H}$ matrix does not usually generate an attractive set of closed loop eigenvectors; occasionally the eigenvectors are so poorly conditioned that computing an accurate $\boldsymbol{G}$ from Eq. (29) is not possible.

__Projection Concepts__

Since an arbitrary selection of $\boldsymbol{H}$ is not appropriate, we consider here choices which have a high probability of generating attractive gain matrices. It is well-known[17] that the condition number of the closed loop modal matrix $k(\Psi) = |\Psi||\Psi^{-1}|$ is a measure of robustness(sensitivity) of the placement of the closed loop eigenvalues. Recent research [41,46], proves that the stable system $\dot{x}=Ax$ remains stable in the presence of an arbitrary model error $E(t)$ in $\dot{x} = (A+E)x$, if E satisfies the inequality

$$|E| \leq max\,[-Real(\lambda(A))]\;/\;k(\Psi) \equiv \mu \tag{31}$$

Thus for fixed eigenvalue positions, minimizing $k(\Psi)$ is equivalent to maximizing the system robustness. More generally, we use μ in Eq. (31) as the robustness measure. This motivates our search for an $\boldsymbol{H}$ matrix to give as-small-as possible condition number $k(\Psi)$. An attractive algorithm results if we seek the $\boldsymbol{H}$ matrix which makes the Ψ matrix of closed loop eigenvectors lie as close as possible to a prescribed, well-conditioned matrix. Notice, if we select some target set of well-conditioned closed loop eigenvectors $\hat{\Psi} = [\hat{\psi}_1 \ldots \hat{\psi}_N]$, then we can use Eq. (28), or equivalently Eq. (30) to solve for the $\hat{\boldsymbol{H}}$ which most nearly (in the least squares sense, for example) produces the desired eigenvectors $\hat{\Psi}$. Upon substituting this solution for the $\boldsymbol{H}$ matrix and re-solving Eq. (30) for Ψ, we will find $\Psi \neq \hat{\Psi}$, exactly, with the degree of approximation being problem dependent. However, the Ψ matrix lies as near $\hat{\Psi}$ as possible(in the least square sense), and is typically well-conditioned. The gain matrix $\boldsymbol{G}$ calculated from solution of Eq. (29) with Ψ and $\hat{\boldsymbol{H}}$ will, however, place the eigenvalues exactly, to within arithmetic errors. We now address the crucial issue of how to select the target vectors $\hat{\Psi}$.

From Eq. (28), it is evident that the admissible eigenvectors are mapped from the m dimensioned column space of $(A-\lambda_i I)^{-1}B$. This space can be spanned by a unitary matrix U_i, so through appropriate choice of h_i, we can generate all admissible eigenvectors from the projection

$$\psi_i = U_i h_i \tag{32}$$

Assuming the target eigenvectors are denoted $\hat{\psi}_i$, then the h which minimizes the residual $|\hat{\psi}_i - Uh|$ is given by

$$\hat{h} = U_i^H \hat{\psi}_i \tag{33}$$

and so the feasible eigenvectors nearest $\hat{\psi}_i$ are obtained by substituting Eq. (33) into Eq. (32), which gives

$$\psi_i = U_i U_i^H \hat{\psi}_i \tag{34}$$

with the corresponding residual vector

$$\Delta_i = \hat{\psi}_i - \psi_i = (I - U_i U_i^H)\hat{\psi}_i \tag{35}$$

There are several computationally stable algorithms available for calculation of the unitary basis matrices U_i, three being:

(i) construct the singular value decomposition (SVD[18]) of the matrix

$$(A-\lambda_i I)^{-1}B = \begin{bmatrix} U_{1i} & U_{2i} \end{bmatrix} \begin{bmatrix} \Sigma_i & 0 \\ 0 & 0 \end{bmatrix} \begin{bmatrix} V_{1i}^H \\ V_{2i}^H \end{bmatrix} \text{ and define } U_i = U_{1i},$$

(ii) construct the QR decomposition [19] of

$$(A-\lambda_i I)^{-1}B = \begin{bmatrix} Q_{1i} & Q_{2i} \end{bmatrix} \begin{bmatrix} R \\ 0 \end{bmatrix} \text{ and define } U_i = Q_{1i}, \text{ or}$$

(iii) construct the SVD of the matrix

$$B^*(A-\lambda_i I) = \begin{bmatrix} U_{1i} & U_{2i} \end{bmatrix} \begin{bmatrix} \Sigma_i & 0 \\ 0 & 0 \end{bmatrix} \begin{bmatrix} V_{1i}^H \\ V_{2i}^H \end{bmatrix} \text{ and define } U_i = V_{2i},$$

where B^* is the orthogonal complement of B(i.e., $B^*B=0$).

For the calculations of the present discussion, we follow [16] and adopt the third approach, since it does not require inversion of $(A-\lambda_i I)$.

The above developments still do not tell us how to choose the $\hat{\psi}_i$ judiciously. Following [12,19], we state the following optimization problem:

$$\text{Find } \{\hat{\psi}_i,\ i = 1,2,\ldots \text{N}\}, \text{ to minimize } J = \sum_{j=1}^{\text{N}} \Delta_j^H \Delta_j \tag{36}$$

subject to

$$\hat{\psi}_i^H \hat{\psi}_j = \delta_{ij},\ i,j = 1,2\ldots,\text{N} \tag{37}$$

Unfortunately, solutions of this problem requires iterative numerical optimization; it can be attacked by nonlinear programming. To avoid this iteration (or to initiate it with an excellent starting solution), we seek to directly specify near-feasible and well-conditioned target eigenvectors $\hat{\psi}_i$. We begin by arranging the N sets of unitary basis matrices into a global matrix S as

$$\text{S} = [U_1, U_2, \ldots, U_{\text{N}}] \tag{38}$$

and take the singular value decomposition of S to obtain the SVD factorization

$$\text{S} = \hat{U}\,\Sigma V^H \tag{39}$$

where $\hat{U}$, V are the left and right matrices of singular vectors and Σ is the diagonal matrix of singular values. We hypothesize that the left unitary vectors of S provides an attractive set of target eigenvectors, it remains to determine which of the column vectors of $\hat{U}$ should be assigned as $\hat{\psi}_i$, to each of the N least squares problems of Eqs. (33) - (35). We summarize the algorithm as follows:

Step 1. For $i = 1, 2, \ldots, N$, compute a unitary basis matrices U_i which span the column space of $(A - \lambda_i I)^{-1}B$.

Step 2. Find the left singular vectors $\hat{U} = [\hat{u}_1, \ldots, \hat{u}_N]$ of S, and define these as the set of target vectors: $\{\hat{\psi}_i = \hat{u}_i$, for $i = 1, 2, \ldots, N\}$.

Step 3. For $i = 1, 2, \ldots, N$, determine the index k for the i *th* target eigenvector $\hat{\psi}_k$ to minimize $|\Delta_i| = |(I - U_i U_i^H)\hat{\psi}_k|$, store this index in an array *index(i)=k*, and remove $\hat{\psi}_k$ from the set of target vectors.

Step 4. Calculate the admissible eigenvectors as the least square projection

$$\psi_i = U_i U_i^H \hat{\psi}_k , \quad k = index(i) \tag{40}$$

Step 5. Determine the H matrix and the gain matrix G by solving Eqs. (30, 29).

Specializations for Second Order Equations

One conceptually attractive modification of the above for the case of mechanical systems described by Eq. (1), with $N = 2n$, is to select the open-loop eigenvectors as the target eigenvector set $\hat{\psi}_i = col\{\phi_i, \lambda_i\phi_i\}$. For a general mass matrix and for general prescription of the closed loop eigenvalues, these target eigenvectors do not constitute a unitary set and are not guaranteed to be well-conditioned. However, numerical experience reveals that the associated gain matrices usually have a relatively small (but not generally a minimum) norm, and also the condition number of the closed loop modal matrix is relatively small (but again, not generally a minimum). The algorithm for this case is obtained by replacing Steps 1 and 2 by calculation of the open loop eigenvectors ϕ_i, and the target vectors $\hat{\psi}_i = col\{\phi_i, \lambda_i\phi_i\}$.

More generally, we wish to take full advantage of the structure of the eigenspace for the case of full state feedback and a mechanical vibrating system of the form of Eq. (1), it is not necessary to limit attention to selection of the open loop eigenvectors as the target vectors; there are obviously an infinity of other possibilities. We can specialize the above developments into a $N = 2n$ dimensional state space, but can also work exclusively with *nxn* modal matrices for the original second derivative system. Let the feedback matrix be partitioned into position and velocity gains as $G = [G_1 \;\; G_2]$, so the control is given by

$$u = G_1 z + G_2 \dot{z} \tag{41}$$

Equation (1) becomes

$$M\ddot{z} + C\dot{z} + Kz = D[G_1 \;\; G_2]\begin{Bmatrix} z \\ \dot{z} \end{Bmatrix} \tag{42}$$

and the associated eigenvalue problem has the form

$$[\lambda_i^2 M + \lambda_i C + K]\psi_i = D[G_1 \;\; G_2]\begin{Bmatrix} \psi_i \\ \lambda_i \psi_i \end{Bmatrix} \tag{43}$$

Analogous to the development of Sylvester's equation for first order systems, we define

$$h_i = [\, G_1 \;\; G_2\,]\begin{Bmatrix} \psi_i \\ \lambda_i \psi_i \end{Bmatrix} \tag{44}$$

For eigenvalues assigned to positions other than their open loop locations, the inverse of Eq. (43) exists and we can determine the admissible eigenvectors generated by the infinity of possible assigned eigenvalues and h_i parameter vectors as

$$\psi_i = [\lambda_i^2 M + \lambda_i C + K]^{-1} D h_i \tag{45}$$

Notice that enforcing the structure of the modal eigenvectors results in only an $nx1$ vector ψ_i being parameterized by h_i instead of a $2nx1$ vector Ψ_i for the equivalent first order development. Upon calculating well-conditioned admissible vectors $[\psi] = [\psi_1 \; . \; . \; . \; \psi_n]$, i. e., those nearest a target unitary set $[\hat{\psi}]$ (we have used as targets the open loop eigenvectors and also, the left singular vectors from singular value decomposition of S from Eq. (38) with the new definition $U_i = [\lambda_i^2 M + \lambda_i C + K]^{-1} D$), we can determine the least square solutions for h_i and the closest admissible $[\psi]$. We can then form the $2n \; x \; 2n$ modal matrix

$$\Psi = \begin{bmatrix} [\psi] & \overline{[\psi]} \\ [\psi]\Lambda & \overline{[\psi]\Lambda} \end{bmatrix}, \; \Lambda = diag(\lambda_1 \; . \; . \; . \; \lambda_n), \tag{46}$$

where $(\bar{\;})$ denotes the complex conjugate of ( ); the control gain matrix G is then obtained by solving the linear system $H = G\Psi$, where the $m \; x \; 2n$ H matrix is constructed from the h parameter vectors as

$$H = \begin{bmatrix} h_1 & h_2 & \dots & h_n & \bar{h}_1 & \bar{h}_2 & \dots & \bar{h}_n \end{bmatrix} \tag{47}$$

For the idealized situation that the number of actuators are assumed equal to the number of controlled modes $(m=n)$, and for the special case that the target eigenvectors are taken to be the open loop undamped eigenvectors, we have established[19] analytical and numerical equivalence of the above to eigenvalue assignment by independent modal space control (IMSC[20]). We demonstrate in subsequent numerical results of the present chapter that selecting the open loop eigenvectors as the target set, while atttractive in many cases, does not generally lead to the smallest gain norm or the optimally conditioned closed loop eigenvalue problem, even for the $m=n$ case.

The above developments can be generalized and extended, as in [12], to apply in principle to the case of output feedback. While output feedback algorithms via direct eigenstructure assignment have been developed and successfully tested, the existence of solutions is not generally guaranteed. The design of gain matrices for the general output feedback problem remains an area of ongoing research, iterations are usually required to converge upon practical designs, and convergence is not guaranteed, even if the formal theoretical conditions for assignability are satisfied. Convergence of virtually all non-linear programming algorithms can be greatly enhanced by use of homotopy algorithms [21]; these approaches are also attractive, because they have been successfully demonstrated for simultaneous structure and controller design optimization. In optimization of regulator designs (for a specified structure), and especially in simultaneous structure /controller optimization problems, we must face the truth that there are typically several competing definitions or measures of optimality, and that a systematic methodology is needed to study the design tradeoffs between competing measures of optimality. This family of problems and recently developed approaches are reviewed in the following section.

Multiple Criterion Optimization

As developed above, the closed loop eigenvalues and eigenvectors can be assigned to achieve direct control over the controlled systems time constants and robustness properties. It is evident that the $\boldsymbol{H}$ matrix is usually redundant, vis-a-vis imposing eigenvalue constraints, and therefore we were free to seek small condition numbers of the closed loop modal matrix to enhance robustness properties.

More generally, there are often many competing performance issues which play a significant role in practical applications. We seek a graphical method for displaying *tradeoff surfaces* to allow the control designer visual evaluation of the nature of multiple performance compromises implicit in a given controller design problem. We develop below a method for constructing such surfaces; each point on the surface corresponds to a numerical setting on the feedback gains. It evident that implicitly we are seeking to determine and display *families* of optimal solutions (instead of optimizing a single *point* design), so that particular implementations can be selected with as much visibility of the macroscopic and microscopic performance tradeoffs as is possible. Following[45,16], the tradeoff: *robustness* versus *average control energy* versus *average state error energy* is judged to be of fundamental significance, therefore we concentrate upon developing methods to determine and display this multiple criterion tradeoff surface. In this development, *average* is taken to mean the *expected* value of a quantity, for a distribution of initial conditions having a specified covariance.

Objective Functions and Their Derivatives

Consider a closed loop system described by Eq. (12) with $\boldsymbol{S}=\boldsymbol{I}$. Assume that the initial conditions are distributed with prescribed covariance $E[\boldsymbol{x}_0\boldsymbol{x}_0^T]=X_0$. We define time *expected error energy* (J_s) and *the expected error energy* (J_u) as

$$J_s = E\left\{\int_0^\infty x^T Q_s\, x\, dt\right\},\ \text{and}\ J_u = E\left\{\int_0^\infty u^T Q_u\, u\, dt\right\},\ \text{with}\ Q_s=Q_s^T\geq 0\ ,\ Q_u=Q_u^T\geq 0 \quad (48)$$

The $\boldsymbol{Q}$ matrices should be selected to make the integrands correspond to physical energy measures. These two integrals can be evaluated as shown below by solution of Lyapunov equations. Letting Ψ denoting the matrix of closed loop eigenvectors, then the corresponding modal state vector ξ is related to $\boldsymbol{x}$ and $\boldsymbol{u}$ by $x = \Psi\xi$ and $\boldsymbol{u} = \boldsymbol{H}\xi$. The modal space closed loop system is

$$\dot{\xi} = \Lambda\xi, \text{ with } \xi_o = \Psi^{-1}x_o, \; E[\xi_o\xi_o^H] = \Psi^{-1}X_o\Psi^{-H} \tag{49}$$

Equations (48) become

$$J_s = E\left\{ \int_o^{\infty} \xi^H\Psi^H Q_s \Psi\xi \, dt \right\} \quad \text{and} \quad J_u = E\left\{ \int_o^{\infty} \xi^H\Psi^H Q_u \Psi\xi \, dt \right\} \tag{50}$$

As is developed in [19,45], these integrals are evaluated as follows

$$J_s = trace[(\Psi^{-H}P_s\Psi^{-1})X_o] \quad \text{and} \quad J_u = trace[(\Psi^{-H}P_u\Psi^{-1})X_o] \tag{51}$$

where P_s , P_u are solutions of the Lyapunov equations

$$P_s\Lambda + \Lambda^H P_s + \Psi^H Q_s \Psi = 0 \quad \text{and} \quad P_u\Lambda + \Lambda^H P_u + H^H Q_u H = 0 \tag{52}$$

The solutions are written in index form as

$$P_s(i,j) = -(\lambda_i^H + \lambda_j)^{-1}\psi_i^H Q_s \psi_j \quad \text{and} \quad P_u(i,j) = -(\lambda_i^H + \lambda_j)^{-1} h_i^H Q_u h_j \tag{53}$$

The partial derivatives of the expected state and control energy indices with respect to the parameter vectors h_i are obtained by direct differentiation[19]. Denoting the derivative with respect to a parameter p_k by Δ_k, the eigenvector derivatives are

$$\Delta_k\psi_i = U_i\Delta_k h_i \tag{54}$$

Clearly if we take p_k an an element of h_i, then $\Delta_k h_i$ is a column of zeros and ones. The derivative of the inverse of the modal matrix, where needed, is obtained as

$$\Delta_k\Psi^{-1} = -\Psi^{-1}[\Delta_k\Psi]\Psi^{-1} \tag{55}$$

As a third performance measure (robustness), we adopt the condition number J_r of the closed loop eigenvectors. The condition number is easily computed and has been found to be a good measure of closed loop robustness. We use methods of [22] to compute the partial derivatives of the condition number with respect to h_i. Additional indices, such as structural mass, stiffness, and loading measures, can be introduced to accomplish structure/controller optimization tradeoffs [23]. In the numerical studies below, we report some of these tradeoffs for an example problem.

Multiple Objective Optimization

As developed above, we have two or more performance indices (*e.g.*, J_s^*, J_u^*, J_r^*) all of which we would like to minimize. Recognizing from the onset that these minimizations are almost certainly in competition, we approach the problem from a different perspective [19,45]. Suppose that we are following the eigenstructure assignment approach as developed above; we seek to optimize, in some sense, over all H matrices for a given set of closed loop eigenvalue locations. In lieu of optimization, we first direct attention to the problem of finding H to impose the equality constraints

$$J_s^* - J_s(H) = 0, \quad s \rightarrow u,r \tag{56}$$

The values (J_s^*, J_u^*, J_r^*) are interpreted as "the best one could possibly hope for", or as design goals to be achieved as nearly as possible. Since an exact feasible solution of the problem likely won't exist, especially when aggressive, demanding goals are assigned, we begin by defining a *homotopy family* of neighboring problems with "portable" goals.

Specifically, we replace the (J_s^*, J_u^*, J_r^*) goals by a one parameter (γ) family of goals as

$$\{\gamma J_s^* + (1-\gamma)J_s(H_{start})\} - J_s(H) = 0, \quad s \to u,r,, \quad 0 \leq \gamma \leq 1 \tag{57}$$

Sweeping the homotopy parameter (γ) from zero to unity gradually drives the portable goals from their values on the starting iteration toward the specified goals. Notice H_{start} is some arbitrary starting matrix, but most usually it is determined from an eigenstructure assignment procedure similar to that described above. Since there are typically more elements in H than the number of portable goal constraints, the corrections to H are underdetermined. For most of our implementations to date [3,6,9,23], we have used the minimum correction norm algorithm. This algorithm is a generalized Newton method which finds the minimum sum square modification of H on each iteration to satisfy the currently specified goals $\{\gamma J_s^* + (1-\gamma)J_s(H_{start})\}$. This process has the advantage that each iteration, for fixed γ, seeks a minimum change of the current design and thereby increase the likelihood of achieving convergence to solutions for the sequence of neighboring problems. Notice that the current portable goals can be assigned arbitrarily near a neighboring converged design by increasing γ adaptively based upon convergence progress. As γ is increased toward unity, it is evident that successful convergences mean that the design goals are being approached. In the event that γ cannot be advanced some small assign tolerance with a convergence failure due to competition between the goal constraints, the process is terminated. Of course in the event that γ can be increased to unity without a convergence failure, the design goals have been achieved; this unusual event usually motivates one to be more aggressive in assigning at least one of the goals (J_s^*, J_u^*, J_r^*) to a smaller value. The homotopy can be applied to sweep all goals simultaneously, or to one goal at a time (while treating the remaining goals as fixed equality constraints).

Utilizing these ideas, we have developed a nonlinear programming algorithm [19] to generate the family of solutions which can be interpolated to form multiple objective performance tradeoff surfaces. We seek to establish a surface in (J_s^*, J_u^*, J_r^*, H) space, each point of which corresponds to the H (or, implicitly, the G) which optimizes robustness (e. g., minimizes the condition number J_r^*) for fixed values of the expected state error energy and control energy (J_s^*, J_u^*). The overall process is outlined as follows:

Step 1. Select the robustness index as the primary objective function, optimize it over a prescribed region of (J_s^*, J_u^*) space with (J_s^*, J_u^*) constrained only by upper and lower bounds.

Step 2. Initiating with the minimum J_r^* solution, define a neighboring grid of (J_s^*, J_u^*) values; the grid values are treated as a sequence of equality constraints, while J_r^* is minimized for each pair of fixed (J_s^*, J_u^*) points on the grid, specifically

2.1. Set values for the goals (J_s^*, J_u^*) to a point on the selected grid.

2.2. Determine minimum norm corrections to H and sweep γ to drive $J_r^*(\gamma)$ as small as possible, subject to the equality constraints
$J_s^* - J_s(H) = 0$, $J_u^* - J_u(H) = 0$
Save the resulting family of converged solutions (J_s^*, J_u^*, J_r^*, H), if the (J_s^*, J_u^*) grid is complete, then go to step 3, otherwise, move to the next point on the grid and return to step 2.1.

Step 3. Interpolate the converged (J_s^*, J_u^*, J_r^*, H) data to establish the multiple criterion tradeoff surface.

The convergence achieved in Step 1 is typically, but not always, on the boundary of the admissible (J_s^*, J_u^*) region. Also, convergence to local extrema sometimes occurs. Since the first optimization is used to establish the departure point for the construction of the performance tradeoff, it is obvious that this solution is crucial. Therefore, it is important to vary the starting H and take all obvious steps to increase confidence that the global optimum has been achieved. The family of local constrained optimizations at Step 2.2 should be done with small convergence tolerances to make the surface numerically uniquely defined. The density of points in the (J_s^*, J_u^*, J_r^*, H) space required for accurate interpolation of the performance surfaces is problem-dependent and is obviously controlled directly by the user-selected (J_s^*, J_u^*) grid. We suggest cubic spline interpolation between the converged points to approximate the surface. Prudent verifications should be done in which interpolated designs for H or G (e. g., midway between the points on the grid) are computed and used to verify that the closed loop system's actual stability and performance properties are interpolated with sufficient precision. We illustrate these ideas in the numerical studies discussed below.

Numerical Examples of Robust Controller Design and Multiple Objective Optimization

Definition of System Models for Example Sets I and II

In order to illustrate several issues associated with controller design via the LQR based formulations and the robust eigenstructure assignment algorithm, we present numerical results for two sets of models: i) Set I (various parameter values for a 6th order mass-spring system, Table 1) and ii) Set II (various reduced order models for a flexible space structure). The mass and stiffness matrices for the first set of examples are chosen as

$$M = diag[m_{11}, m_{22}, m_{33}], \quad K = k\begin{bmatrix} 2 & -1 & 0 \\ -1 & 3 & -2 \\ 0 & -2 & 2 \end{bmatrix}$$

and the open-loop eigenvalues for the second example are given in Table 2.

Eigenstructure Assignment

To study the performance of the robust eigenstructure assignment algorithm, several models are selected from Set I, in particular, we choose four variations of the system's mass matrix. For these Set I cases, the number of control inputs are chosen equal to the number of modes (i.e., $D=I$). From the second set of examples, several reduced order (modal space) models are obtained by truncating at different mode numbers of increasing frequency. The specifications of all eight models and the desired eigenvalues for each model are given in Tables 1 and 2.

For these structural systems, we tested the projection method (See associated discussion Eqs. (32)-(47)) using as target vectors a unitary basis (Algorithm I) and using as target vectors the open-loop eigenvectors (Algorithm II), and finally, using the IMSC pole placement technique (Algorithm III [20]). We summarize the condition number of the eigenvector matrix and Frobenuis norm of the gain matrix obtained by using each model in Tables 3 and 4. For the flexible structure, the result of the IMSC method is not reported since we elect to hold the number of controllers fixed (m=6) and thus IMSC can not be applied to control > 6 modes without introducing additional actuators.

As shown in Table 3, the identical results were obtained by Algorithms II and III for Models 1-4. For systems with same number of controllers as number of modes, Algorithm II is rigorously equivalent to the IMSC method, since the open-loop modal matrix (Model 1), the condition numbers obtained by all the three algorithms were exactly the same. For this case, Algorithm I, however, generated a different gain matrix and since the unitary basis vectors employed as target vectors in the projection step did not coincide with the unitary open-loop modal vectors. For Model 4, we summarize the two gain matrices as follows:

Algorithm I Gain Matrix (using unitary set of target closed loop eigenvectors):

$$G = \begin{pmatrix} -6.801 & 10.000 & 0.000 & 363.299 & 0.000 & 0.000 \\ 10.000 & -16.664 & 20.000 & 0.000 & 36.517 & 0.000 \\ 0.000 & 20.000 & 6.934 & 0.000 & 0.000 & 5.189 \end{pmatrix}$$

Algorithms II and III Gain Matrix (using open loop eigenvectors as target set):

$$G = \begin{pmatrix} 6.666 & -3.333 & 0.000 & 396.842 & -32.917 & -0.305 \\ -3.333 & 10.000 & -6.666 & -32.901 & -41.009 & -4.803 \\ 0.000 & -6.666 & 6.666 & -0.305 & -4.803 & 5.142 \end{pmatrix}$$

For all models of the first example set, Algorithm I performed superior or equal to the others in the sense that the resulting feedback system is less sensitive. The reason is that the conditioning of the open-loop modal matrices depend on the mass matrices. From these observations, we conclude that the open-loop modal matrix is not generally the optimal choice for target eigenvectors, vis-a-vis robustness, for systems with an arbitrary mass matrix.

Test results for Models 5-8 are interesting since these are consistent with many practical systems which have a smaller number of actuators than the number of modes we wish to control. As shown in Table 4, Algorithm I performs better than Algorithm II as regards conditioning (robustness) of closed-loop eigenvectors.

Table 1. Test Example Set I (Spring Mass Systems)

MODEL* NO.	MASS MATRIX	DESIRED EIGENVALUES Freq (ω)	Damping (ζ)
1	M = diag[1,1,1]	ω_1= 1.44 ω_2= 4.47 ω_3= 6.92	$\zeta_1=\zeta_2=\zeta_3=0.5$
2	M = diag[100,1,1]	ω_1= 0.89 ω_2= 2.33 ω_3= 6.76	$\zeta_1=\zeta_2=\zeta_3=0.5$
3	M = diag[1000,100,1]	ω_1= 0.09 ω_2= 0.95 ω_3= 4.70	$\zeta_1=\zeta_2=\zeta_3=0.5$
4	M = diag[10000,100,1]	ω_1= 0.03 ω_2= 0.31 ω_3= 4.49	$\zeta_1=\zeta_2=\zeta_3=0.5$

* All the models have the same stiffness matrix

Table 2 Test Example Set II (Flexible Space Structure)

MODE NO.	OPEN-LOOP EIGENVALUE		DESIRED CLOSED-LOOP EIGENVALUE	
	Freq (ω)	Damping (ζ)	Freq (ω)	Damping (ζ)
1	4.90×10^{-4}	.001	6.78	.70
2	4.46×10^{-4}	.001	8.17	.70
3	4.27×10^{-4}	.001	8.98	.70
4	2.57×10^{-4}	.001	8.79	.70
5	1.95×10^{-4}	.001	11.97	.70
6	1.07×10^{-4}	.001	15.26	.70
7	22.36	.001	16.46	.70
8	20.49	.001	18.40	.70
9	20.86	.001	20.10	.70
10	42.28	.001	21.86	.70
11	42.28	.001	23.87	.70
12	61.95	.001	67.23	.70

Table 3. Performance of Eigenstructure Assignment Algorithms on Set I Examples

MODEL NO.		Algorithm I*		Algorithm II*		Algorithm III*	
	$k(\phi^{o})$ **	$k(\phi^{c})$ **	$\|G\|_F$	$k(\phi^{c})$ **	$\|G\|_F$	$k(\phi^{c})$ **	$\|G\|_F$
1	6.92	9.26	50.91	9.26	19.83	9.26	19.83
2	7.10	9.04	68.39	10.39	45.08	10.39	45.08
3	20.79	10.08	121.79	22.31	128.96	22.31	128.96
4	72.94	31.79	367.04	73.46	402.12	73.46	402.12

* Algorithm I -Projection method using unitary basis vectors as target closed-loop eigenvectors.

Algorithm II -Projection method using open-loop eigenvectors as target closed-loop eigenvectors.

Algorithm III -IMSC method

** The superscript o and c denote open-loop and closed-loop, respectively, k( ) denotes the condition number.

Table 4. Performance of Eigenstructure Assignment Algorithms on Set II Examples

MODEL NO.	No. of States and No. of Controls n	m	$k(\phi^{o})$ **	Algorithm I* $k(\phi^{c})$ **	$\|G\|_F$	Algorithm II* $k(\phi^{c})$ **	$\|G\|_F$
5	12	6	9600	31.	234,000	32.	208,000
6	16	6	9600	347.	786,749	461.	906,380
7	20	6	9600	8,342.	11,904,002	14,753.	7,913,951
8	24	6	9600	11,406.	5,920,422	25,495.	10,730,075

* Algorithm I - Projection method based on unitary vectors.
Algorithm II - Projection method based on open-loop eigenvectors.

** The superscripts o and c denote open-loop and closed-loop respectively.

Table 5. Robustness Index Optimization

ALGORITHM	STATE ENERGY J_s	CONTROL ENERGY J_u	ROBUSTNESS INDEX J_r*
I. LQR (N=0)	3.0	0.368	3.270
I.' LQR (N≠0)	3.0	0.385	3.150
II. Modified LQR	3.0	0.842	3.050
III. Sylvester's Method	3.0	1.000	3.040

* J_r is minimized subject to the inequalities $0 \leq J_s \leq 3,\ 0 \leq J_u \leq 2.$

Multi-Criterion Eigenstructure Optimization

Multiple objective optimization and trade-off surface generation requires many nonlinear iterations and are computationally expensive processes. It is presently reasonable to carry out this conceptually attractive idea only for low to moderate order systems. To illustrate the ideas, we restrict our attention to the sixth order spring-mass system with the mass matrix, $\boldsymbol{M} = diag[1, 1, 2]$, and the stiffness matrix scaling factor k=1. This sixth order model has two controllers, i.e., the control input matrix is given by

$$\boldsymbol{D}^T = \begin{bmatrix} 1 & 0 & 0 \\ 0 & 0 & 1 \end{bmatrix}$$

For this system, we consider three feedback design algorithms: Algorithm A (generalized LQR in Form 1, Eqs. (19)), Algorithm B (LQR in Form 2, Eq. 22) and Algorithm C (Sylvester's method, Eqs. (29), (30)). For Sylvester's method, the closed loop eigenvalues are also considered design parameters and tuned in the optimization process, so that with this additional feature [16], the comparison of its performance with the two LQR algorithms will be consistent. For Algorithm A, we considered two cases: one with $N=0$ (this reduces to the classical LQR), another with $N \neq 0$. The performance of each feedback design algorithm is compared for multi-criterion optimizations of three design objectives: expected state error energy, control energy and robustness measure (condition number of the eigenvector matrix).

Arbitrary initial design parameters are selected for Algorithm B and corresponding initial parameters for Algorithms A and C are generated so that the three algorithms are initiated with the same control gains and closed loop system. The parameter values at the beginning of the iterations are summarized below:

Design Parameters for Algorithm B:

$$\boldsymbol{L}_1 = diag[.5^2, .5^2, .5^2, .5^2, .5^2, .5^2], \boldsymbol{L}_2 = \boldsymbol{L}_3 = \boldsymbol{L}_4 = diag[1,1]$$

Weight Matrices and Initial State Covariance Matrix:

$$\boldsymbol{Q}_u = \boldsymbol{Q}_s = \boldsymbol{X}_o = \boldsymbol{I}$$

Homotopy Step Size and Convergence Limit:

$$\alpha_{i+1} = \alpha_i + 0.1, |\Delta p| < 10^{-3} \text{ convergence tolerance for each } \alpha_i .$$

Using the three step numerical algorithm described above, the optimization of the primary objective, J_r (condition number of eigenvectors), was first performed subject to satisfying the specified admissible region constraining equations for the state error energy and control energy: $0 \le J_s \le 3, 0 \le J_u \le 2$. The results of this step are given in Table 5, where it can be seen that the optimum J_r's obtained by each algorithm occurred on the boundary of the feasible region at the same state error energy level J_s = 3.0, however the (J_u, J_r) values were different. Starting each algorithm at the point obtained by minimizing J_r, grids of neighboring points in the (J_s, J_u) space were defined. At each grid point,

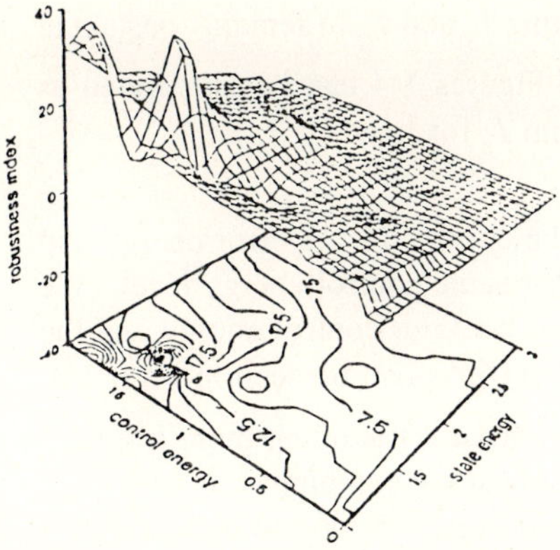

Figure 1. Multi-Criterion Tradeoff Surface: Classical LQR Control($N=0$)

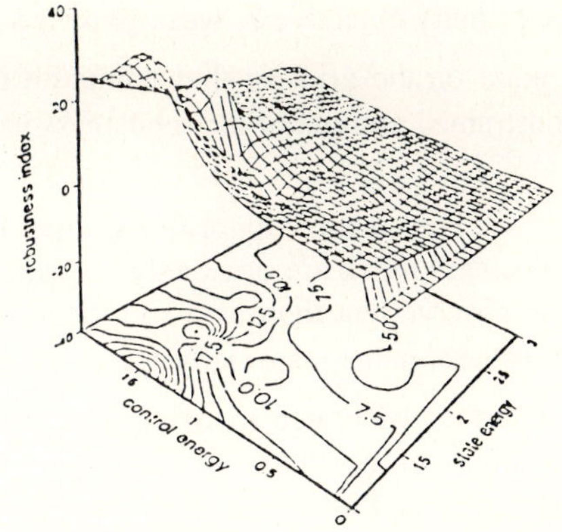

Figure 2. Multi-Criterion Tradeoff Surface: Generalized LQR Control Form 1 ($N\neq0$)

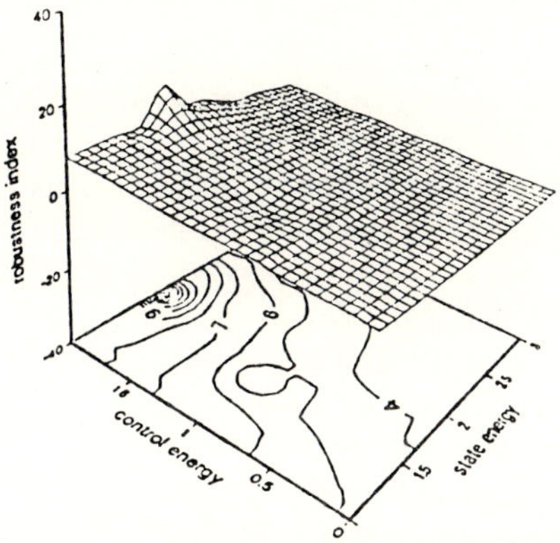

Figure 3. Multi-Criterion Tradeoff Surface: Generalized LQR Regulator Form 2

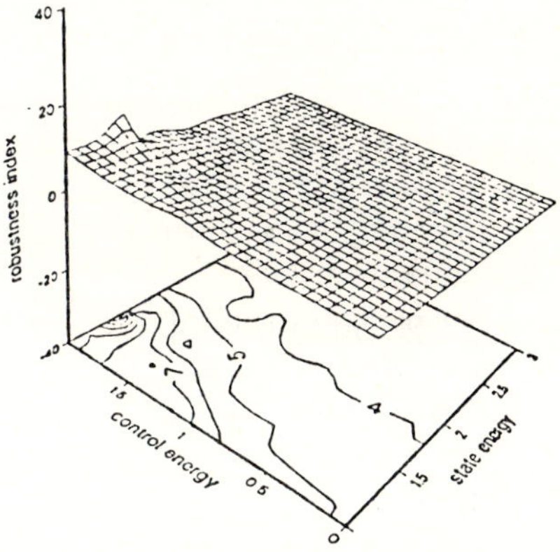

Figure 4. Multi-Criterion Tradeoff Surface: Robust Eigentructure Assignment via Sylvester's Equation

the primary objective J_r was optimized while constraining J_s and J_u to remain constant at a point on the grid, so that points on the surfaces of Figures 1-4 can be interpreted as constrained minima of the robustness index J_r (minimum J_r for fixed J_s, J_u).

For all three algorithms, Figs. 1-4 show that the expected state error energy and robustness index are obviously competing each other for same control energy level. We also observe that there is a significant trade-off relation between control energy and the robustness index, especially for small state energy (J_s<1). As can be seen in Figs. 1-4, the surface generated by using the classical LQR (Algorithm A) is much more irregular than those of the others, which may be due to more nonlinear constraining equations, e.g., the nonlinear Riccati equation instead of linear Lyapunov equation. When only diagonal elements of the ***QR*** matrices were iterated, it was not possible to generate a complete trade-off surface since the accessible region in (J_s, J_u) space was found to be severely limited. Comparing Figs. 3 and 4, we conclude that the multi-criterion surfaces of Algorithms B and C are approximately the same. In fact, the Lyapunov equation can be viewed as a special type of Sylvester equation.

Several observations made from these numerical results are summarized as follows:

1) For all the cases, the expected state error energy and robustness index were found to be strongly competing objectives.
2) The LQR design with "cross-coupling" weights (N matrix) in the criterion produces more regular tradeoff surfaces and offers more choices of the feedback gains than that standard LQR design. However, the N matrix increases the redundancy in the sense that the number of weight parameters are increased compared to the number of elements in $\boldsymbol{G}$.
3) For the cases studied, the modified LQR formulation and Sylvester's method performed approximately the same in the sense that both methods generated similar near-planar performance tradeoff surfaces, which are judged more attractive than the conventional LQR results. These two algorithms also are more easily implemented. An attractive feature of the Sylvester approach is the direct fashion in which the eigenvalues may be assigned to arbitrary points in the complex plane.

Simultaneous Structure-Controller Design Optimization

The above developments all share the underlying assumption that the nominal structural design is being held fixed, i.e., the above developments are concerned with design of robust controllers for a *fixed* structural design. Recently [3,23,26,27,29,32,33,36-38,46] there had been considerable effort to unify structural design and controller design into a unified optimization problem. This research has met with some success, but it is fair to say that the unified methodology remains in an early stage of development. It is easy to extrapolate that there is a widespread need for this methodology since a trend is evident that extensive "paper studies" and optimization of preliminary structure and controller designs will precede most missions for large space structures.

This simultaneous structure-controller optimization is complicated by the following issues, relative to separate structure or controller optimization: (i) one must simultaneously consider several competing performance measures, (ii) the dimensionality of the independent and dependent parameter and/or function spaces is higher, there are typically many inequality constraints, and (iii) the simultaneous consideration of structural mass, geometry, and stiffness parameters along with sensor/actuator locations, and control law parameters inherently leads to highly nonlinear problems. Recently [23,32,33,38,46] progress has been made in developing an attractive nonlinear programming algorithm which has proven successful on a variety of problems. These algorithms combine the homotopy idea with Sequential Linear Programming (SLP). The motivation for using linear programming (LP) lies in the fact the (i) numerous inequality constraints are handled with ease, (ii) standard, reliable algorithms are available which are applicable to LP problems with several hundred unknowns [39,40,42,43], and (iii) it has been shown [32,39,40] to be competitive or superior to more popular penalty function approaches which use quasi-Newton or conjugate gradient correction schemes. For these reasons, we will overview the SLP approach and some recently obtained results.

Unified Optimization Problem

We consider the closed loop, output-feedback controlled system of Eqs. (12)-(15). We address the most general situation wherein the $d \times 1$ global structure/controller design vector $\boldsymbol{p}$ is given by

$$\boldsymbol{p} = col\,[m_1,...,m_{n_1};\, s_1,...,s_{n_2};\, a_1,...,a_{n_3};\, g_1,...,g_{n_4}] \tag{58}$$

with $\boldsymbol{m} = col\,\{m_1,...,m_{n_1}\}$, structural model parameters, $\boldsymbol{M=M(m), K=K(m), C=C(m)}$

$\boldsymbol{s} = col\,\{s_1,...,s_{n_2}\}$, sensor location and model parameters, $\boldsymbol{S = S(s)}$

$\boldsymbol{a} = col\,\{a_1,...,a_{n_3}\}$, actuator location and model parameters, $\boldsymbol{D = D(a)}$

$\boldsymbol{g} = col\,\{g_1,...,g_{n_4}\}, = \mathrm{vec}\{\boldsymbol{G}\}$, control gains, $\boldsymbol{G = G(g)}$

consider the situation where all independent and dependent constraint functions (e.g. eigenvalue constraints) are expressed as $f(p) = f^*, f \le f^*, f \ge f^*$, all three possibilities are captured by the notation

$$f(p)\ \{\le, =, \ge\}\ f^* \tag{59}$$

The objective or desired values f^* are considered specified. In addition to performance bounds, eigenvalue constraints, and other dependent geometry constraints, Eq. (59) includes all structural constraints (i.e., negative thickness is not allowed), and requirements that sensors and actuators be located on the structure! Thus the dimension of Eq. (59) is often much higher than the $\boldsymbol{p}$ vector itself, even though only a subset of the inequalities are likely to be active at the final converged solution. In addition to the constraints of Eq. (59), we consider another set of dependent functions $\{J_1(p), J_2(p),...\}$ to be performance measures which we are interested in extremizing subject to conditions of Eq. (59). Since we do not expect to be able to extremize minimize all functions simultaneously, we anticipate from the onset the necessity of tradeoff studies. In particular, we consider three performance objectives:

$$\text{I} \quad \textit{minimize eigenvalue sensitivity} = J_1 = J_1(p) = \sum_{i=1}^{n} \sum_{j=1}^{d} \left| \frac{\partial \lambda_i}{\partial p_j} \right|^2 w_{ij}, w_{ij} > 0 \tag{60}$$

$$\text{II} \quad \textit{maximize Patel/Toda robustness} = J_2 = J_2(p) = \max\, [Re(-\lambda)] \,/k(\psi) \tag{61}$$

$$\text{III} \quad \textit{minimize total mass}: J_3 = J_3(p) = \textit{total mass} \tag{62}$$

To simplify the SLP discussion below, we will use J to refer to any one of the J_i unless we need to be specific.

Sequential Linear Programming (SLP)

We begin by defining a homotopy to replace constraint objectives by portable constraints, Eq. (59) is replaced by

$$f(p)\ \{\leq, =, \geq\}\ (1-\gamma) f(p_{start}) + \gamma f^* \tag{63}$$

Obviously as γ s swept from zero to unity, we define a family of neighboring objectives which initiate at $f(p_{start})$, associated with our starting design vector (p_{start}), and terminating with the prescribed objective f^*. Clearly we have $p=p(\gamma)$, if we seek to minimize $J=J(p)$ subject to Eq. (63), we suppress the dependence of p on γ, for notational convenience. Without loss of generality, we seek to maximize $J(p)$.

Equation (63) and $J(p)$ can be linearized about the current p vector to obtain the following local linear optimization problem

$$\textit{Maximize} \quad \sum_{j=1}^{d} \frac{\partial J}{\partial p_j} \Delta p_j = \{\frac{\partial J}{\partial p}\}^T \Delta p \tag{64}$$

$$\textit{Subject to} \quad [\frac{\partial f}{\partial p}]\ \Delta p \{\leq, =, \geq\}\ (1-\gamma) f(p_{start}) + \gamma f^* - f(p) \tag{65}$$

$$-\varepsilon \leq \Delta p \leq \varepsilon \tag{66}$$

Equation (66) is a step size constraint (notice ε has positive elements and is an assigned vector, Eq. (66) holds element-by-element). In [23,46], the above problem is arranged in standard form for using the Simplex Linear programming algorithm, as implemented in [43]. By slowly sweeping γ from zero to unity and iteratively solving the above linear program to convergence for each γ, we have a nonlinear programming method which applies to the family of problems under consideration. In [32], the above approach is compared to more popular nonlinear programming (penalty function methods using the CONMIN quasi-Newton algorithm); it is shown to be superior vis-a-vis reliable convergence. We now consider an illustrative example due to and Lim[46] and Junkins [23].

Numerical Example of Structure/Controller Design via SLP

Figure 5 shows a two-body problem consisting of a rigid body with a flexible beam attached by a torsional spring. The nominal configuration is specified in Tables 6 and 7.

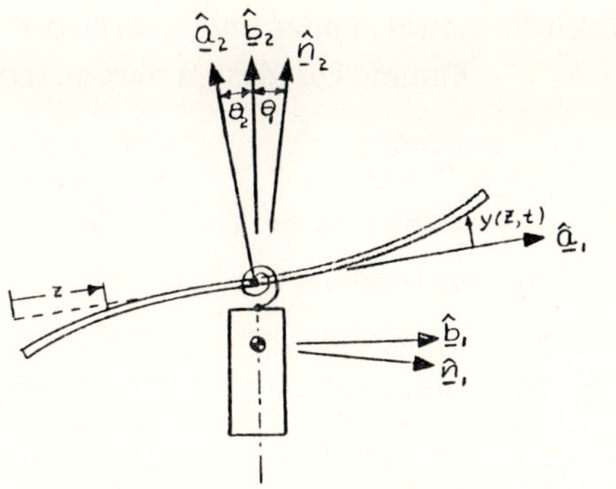

Figure 5 Two-Body Structure

Table 6. Nominal Design Variables

DESIGN VARIABLE	SYMBOL	VALUE
actuator 2 location	a_1	5m
actuator 3 location	a_2	10m
actuator 4 location	a_3	15m
stiffness of torsional spring	k	500 n-m/rad
thickness of flexible beam	t_F	.1m
Young's modulus of beam	E	$.1482 \times 10^9$ N/m^2
mass density of rigid body	ρ_R	300 kg/m^3
output gain elements:		
G(1,1), G(2,1), G(3,1), G(4,1)		-1
all other G(i,j) elements		0

Table 7. Fixed structural parameters

PARAMETER	SYMBOL	VALUE
width of rigid body	w_R	1m
thickness of rigid body	t_R	3m
depth of rigid body	d_R	2m
width of flexible beam	w_F	20m
depth of flexible beam	d_F	1m
mass density of flexible beam	ρ_F	1799 kg/m^3
sensor 1 location	s_1	3m
sensor 2 location	s_2	7m
sensor 3 location	s_3	13m
sensor 4 location	s_4	17m

Table 8. Open loop and desired closed loop damped frequencies and damping factors

MODE	OPEN LOOP		DESIRED CLOSED LOOP	
#	ω_d(rad/s)	ζ	ω_d^*(rad/s)	ζ^*
1	.0056	.4819E-10	.1	.7
2	.2803	.1402E-5	.3	.1
3	.3443	.1718E-5	.45	.1
4	1.241	.6204E-5	1.0	.05
5	1.768	.8839E-5	1.5	.05
6	3.981	.1990E-4	4.0	.05
7	5.004	.2502E-4	$>\omega_{d_6}^* + .1$	.02
8	8.295	.4147E-4	unconstrained	.02
9	9.902	.4951E-4	unconstrained	.02
10	14.34	.7171E-4	unconstrained	.02

Table 9. Lower, upper and local step size bounds on design parameters

PARAMETER	SYMBOL	VALUE
lower bounds on actuator location	a_1^l, a_2^l, a_3^l	0m
upper bounds on actuator location	a_1^u, a_2^u, a_3^u	20m
lower bounds on spring stiffness	k^l	5 N-m/rad
lower bound on beam thickness	t_F^l	.01m
upper bound on beam thickness	t_F^u	2m
lower bound on beam stiffness	E^l	$.1480\times10^9$ N/m^2
upper bound on beam stiffness	E^u	$.1496\times10^9$ N/m^2
lower bound on rigid body density	ρ_R^l	50 kg/m^3
upper bound on rigid body density	ρ_R^u	1000 kg/m^3
local step size bounds:		
actuator location	$\Delta a_1, \Delta a_2, \Delta a_3$	.1m
spring stiffness	Δk	30 N-m/rad
beam thickness	Δt_F	.01m
beam stiffness	ΔE	$.1\times10^5$ N/m^2
rigid body density	$\Delta \rho_R$	40 kg/m^3
gain elements $i=1,..,4; j=1,..,12$	$\Delta G(i,j)$	10
frequency separation between mode 7 and 6	$\Delta\omega_{76}$	.1 rad/s

Table 10. Computer execution times for various subproblems

SUBPROBLEM	CP secs. (CYBER 170)
real, nonsymmetric left and right eigenvalue problems (20x20) using IMSL routine EIGRF	2.67
Solution of linear program (55 variables, 83 constraints) using IMSL routine ZX4LP	22.73
Compute eigenvalue derivatives $\dfrac{\partial \lambda_i}{\partial p_j}$; $i=1,...,6$; $j=1,...,55$	5.05
Compute second eigenvalue derivatives $\dfrac{\partial^2 \lambda_i}{\partial p_k \partial p_j}$; $i=1,...,6$; $j=1,...,55$; $k=1,...,55$	159.06

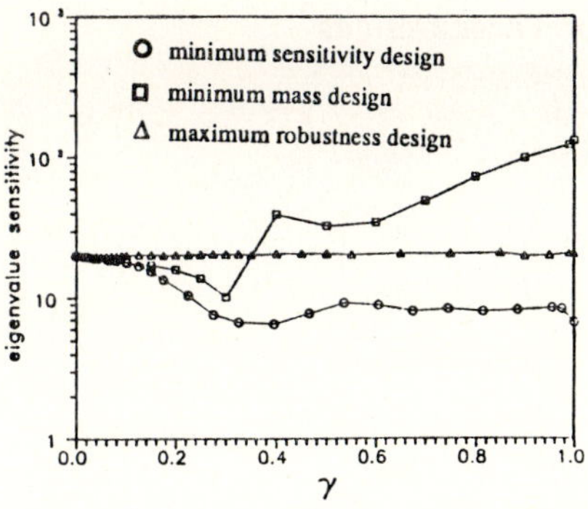

Figure 6 Convergence of Eigenvalue Sensitivity

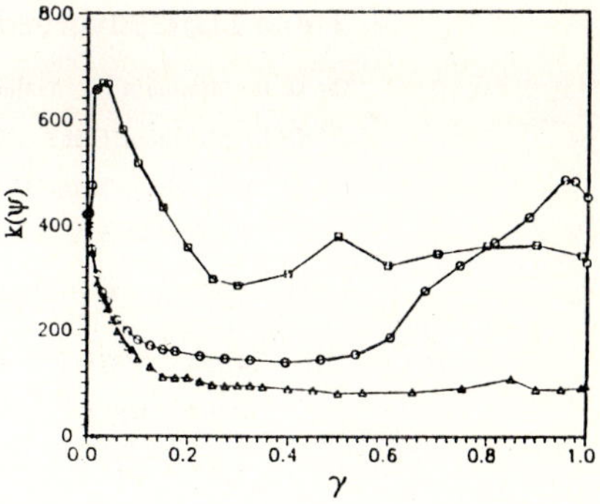

Figure 7 Convergence of Condition Number

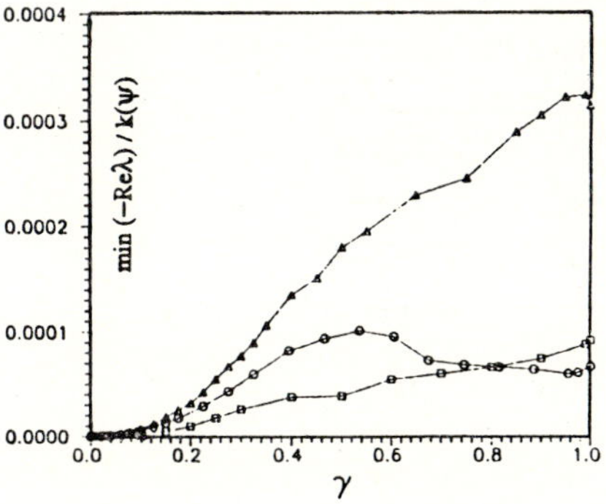

Figure 8 Convergence of Robustness Index

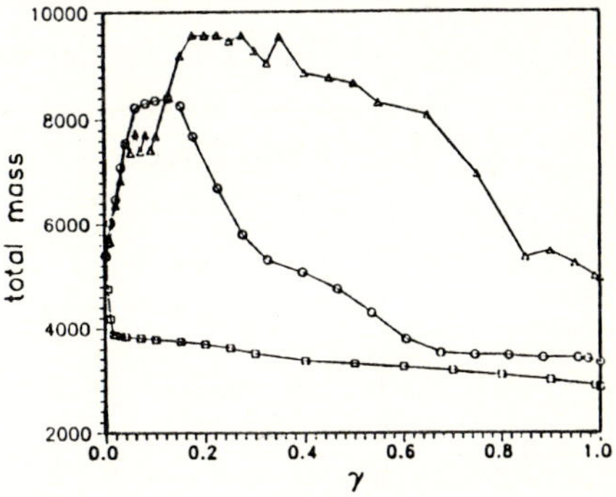

Figure 9 Convergence of Total Mass

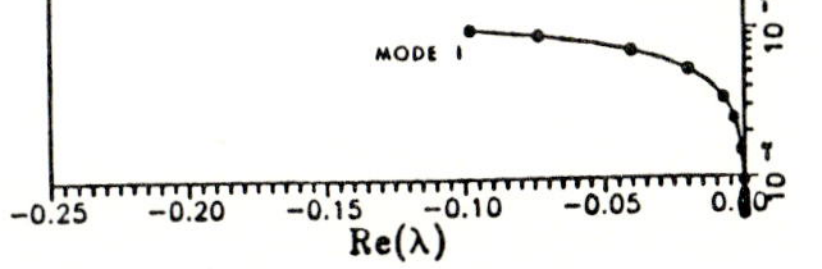

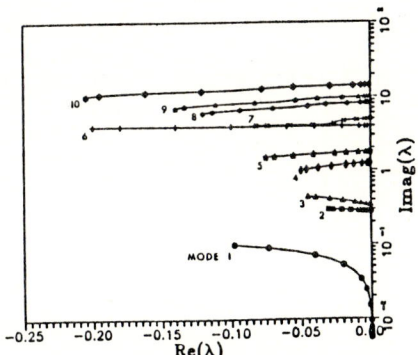

Figure 10 Closed Loop Eigenvalue Locus

Table 11. Converged Parameters for Three Designs

CASE I: Minimum Eigenvalue Sensitivity

CASE II: Maximum Stability Robustness Design

CASE III: Minimum Mass Design

CASE	a_1	a_2	a_3	k	t_F	E	ρ_R
I	4.924	7.878	14.72	1443	.0841	.1480E+9	51.22
II	4.797	8.924	15.48	1481	.0939	.1496E+9	262.1
III	4.087	9.795	14.75	2315	.0709	.1494E+9	50.0

Table 12. Actual Robustness of Three Designs

PARAMETER		PERCENT VARIATION CAUSING INSTABILITY		
		MASS DESIGN	*SENSITIVITY DESIGN*	*ROBUSTNESS DESIGN*
ACTUATOR LOCATIONS	a_1	6%	12%	16%
	a_2	1.5	9	8
	a_3	2.4	4	4
STRUCTURAL PARAMETERS	k	4	6	98
	t_F	3	5	10
	E	8	14	31
	ρ_R	6	24	28
SELECTED CONTROL GAINS	G(1,1)	9	9	67
	G(1,3)	12	50	28
	G(2,8)	43	31	91
	G(3,9)	38	51	25
	G(4,11)	24	8	31
55 PARAMETER AVERAGE		13%	18%	37%

Six pairs of position and velocity sensors and located on the structure, four linear motion pars are located on the beam (Table 7) available for re-design iterations. An angle and angular-velocity measuring sensor pair is fixed to the rigid body and another to the beam mid-span. Four actuators are assumed, one torquer acts on the rigid body, the other three torquers act on the beam with their stations available for optimization. There are a total of 55 design variables free for optimization (the 4x12=48 elements of the output feedback gain matrix, plus the 7 sensor, actuator, and model parameters of Table 6).

Table 8 shows the nominal system's open-loop eigenvalues for the first ten modes and some arbitrarily selected eigenspace constraints we wish to impose. Table 9 provides the additional set of inequality conditions imposed to establish Eq. (59). We considered the three performance measures of Eqs. (60) - (62) and established three separate design problems. For each case, we extremized one of the three indices, but calculated the other two for evaluation purposes.

Design I (minimum eigenvalue sensitivity) has a performance measure which involves a sum square of eigenvalue first partials. Obviously forming Eq. (64) requires eigenvalue second partials. We used the formulation of [2,3] to compute these by analytical means. As is evident in Table 10, the second partial derivative computation dominates the solution time, even in comparison to the cost of solving the linear program.

The results of the three optimizations are summarized with reference to Tables 11, 12 and Figures 6-11. Notice in Table 11 that the three sets of structural design and control gain parameters all satisfy the inequality bounds of Table 9, but the three designs are quite different. Figures 6-10 show the convergence of important quantities during the optimization/constraint enforcement process (as $\gamma \rightarrow 1$). Each point on these curves represent the converged results from the SLP algorithm with γ held fixed. In Figure 6, it is obvious that the minimum sensitivity design I converged to a point one to three orders of magnitude less sensitive (measured locally by Eq. (60)) than the sensitivity of the maximally robust (II) and minimum mass designs (III). In Figures 7 and 8, notice that the small condition number and large robustness index are well correlated, as one would expect. Notice for γ small that the minimum sensitivity designs and maximum robustness designs are very close together. This is because eigenvalue sensitivity obviously dominates robustness when some eigenvalues are near the imaginary axis (the eigenvalue trajectories begin, for $\gamma=0$, near open-loop positions on the imaginary axis). Notice, the obvious tradeoff between robustness and mass. The mass of the minimum mass design was 2853 kgs, compared to the nominal design of 5398 kgs, the minimum sensitivity design of 3336 kg and the maximum robustness design of 4952 kg. The robust design, not surprisingly, converged to the thickest beam, the most massive rigid body and the largest Young's modulus of the three designs.

A crucial question is now considered, is the Robust Design II *really* more robust? We investigate this question by sweeping the 55 parameters away from their respective converged values (for the three designs) and record the variation for which one of the eigenvalues loci first enters the right-half plane. Table 12 summarized the three sets of calculations. It is obvious that the actual robustness to these (highly structured) variations is most significantly increased. In [23,46], additional details are presented showing which modes were least stable (to the 55 parameter variations), consideration of a second Patal/Toda robustness measure, and a discussion of the computer implementation.

Concluding Remarks

In this chapter we have presented significant methods for design of robust full state feedback controllers for *fixed structures*. We have also given some results for simultaneous optimization of structures and output feedback controllers. Enhanced convergence of iterative processes is achieved through the use of homotopy methods. A multi-criterion approach is followed throughout to display the multidimensional nature of optimal control. Performance tradeoff surfaces are presented which provide an attractive means for generating a family of designs displaying inherent tradeoffs between robustness, small control errors, and small controller inputs. The favorable consequences of robustness optimization upon actual robustness is studied through generation of multi-parameter root locii. Numerical examples are given which provide a basis for optimism for future development and implementations.

Acknowledgements

This work has been the outgrowth of research sponsored by AFOSR (contract No. F49620-86-K-0014DEF), the support of A.K. Amos is greatly appreciated. Recent collaboration with Martin Marietta Aerospace (L. Morine, D. Wilks, et at.) have provided an educational opportunity to implement the robust eigenstructure algorithms to control the Rapid Retargeting and Precision Pointing experiment. We have also benefited significantly from on-going sharing of results and joint work with Dr. J.N. Juang (NASA Langley Research Center) and Dr. K.B. Lim (Kentron, Hampton, Va.).

References

[1] Meirovitch, L., Computational Method in Structural Dynamics, Sijthoff and Noordhoff, 1980, Rockville, Maryland, U.S.A.

[2] Plaut, R.H. and Huseyin, K., "Derivatives of Eigenvalues and Eigenvectors in Non-Self-Adjoint Systems," AIAA Journal, Vol. 11, No. 2, pp. 250-251, Feb, 1973.

[3] Bodden, D.S. and Junkins, J.L., "Eigenvalue Optimization Algorithms for Structure/Controller Design Iterations," AIAA Journal of Guidance, Control, and Dynamics.

[4] Srinathkumar, S., "Eigenvalue/Eigenvector Assignment Using Output Feedback," IEEE Transactions on Automatic Control, Vol. AC-23, No. 1, 1978, pp. 79-81.

[5] Brogan, W.L., Modern Control Theory, Quantum Publishers, Inc., New York, New York, 1974, pp. 311-315.

[6] Bhattacharyya, S.P. and deSouza,E., "Pole Assignment via Sylvester's Equation," Systems and Control Letters, Vol. 1, No. 4, Jan. 1982, pp. 261-263.

[7] Cavin III, R.K. and Bhattacharyya, S.P., "Robust and Well-Conditioned Eigenstructure Assignment via Sylvester's Equation," Journal of Optimal Control Applications and Methods, Vol. 4, 1983, pp. 205-212.

[8] Potter, B. and D'Azzo, J.J., "Algorithm for Closed-Loop Eigenstructure Assignment by State Feedback in Multivariable Linear Systems," Int. Journal of Control, Vol. 27, No.6, 1978, pp. 943-947.

[9] Moore, B.C., "On the Flexibility Offered by State Feedback in Multivariable Systems Beyond Closed-Loop Eigenvalue Assignments," IEEE Transaction on Automatic Control, vol. AC-21, 1976 pp. 689-692.

[10] Kautsky, J., Nichols, N.K. and Van Doorend, P., "Robust Pole Assignment in Linear State Feedback," Int. Journal of Control, Vol. 41, No. 5, 1985, pp. 1129-1155.

[11] Wonham, W.M., "On Pole Assignment in Multiinput, Controllable Linear Systems," IEEE Transactions on Automatic Control, Vol. AC-12, 1976, pp. 660-665.

[12] Juang, J-N, Lim, K.B., and Junkins, J.L., "Robust Eigensystem Assignment," preprint of a paper submitted to 1987

AIAA Guidance and Control Conference, Dec. 1986.

[13] Pontryagin, L.S., Boltyanskii, V.G., Gamkrelidze, R.V., and Mishchenko, E.F., The Mathematical Theory of Optimal Processes, Interscience Publishers, Inc., New York, 1962.

[14] Potter, J.E., "Matrix Quadratic Solution," SIAM J. of Applied Mathematics, Vol. 14, No. 3, May 1966, pp. 496-501.

[15] Laub, A.J., "A Schur Method for Solving Algebraic Riccati Equations," IEEE Transactions on Automatic Control, Vol. AC-24, No. 6, Dec. 1976, pp. 913-921.

[16] Rew, D.W., and Junkins, J.L., "Multi-Criterion Approaches to Optimization of Linear Regulators," AIAA Paper No., 86-2198-CP, J. of the Astronautical Sciences, to appear, Aug. 1986.

[17] Wilkinson, J.H., The Algebraic Eigenvalue Problem, Oxford University Press, Oxford, 1965.

[18] Dongarra, J.J., Moler, C.B., Bunch, J.R., and Stewart, G.W., LINPACK Users' Guide, SIAM Publication, Philadelphia, 1979.

[19] Rew, D.W. and Junkins, J.L., "Robust Eigenstructure Assignment by a Projection Method: Application to Multi--Criterion Optimization," preprint of a paper submitted to 1987 AAS/AIAA Astrodynamics Conference, Feb. 1987.

[20] Oz, H. and Meirovitch, L., "Optimal Modal-Space Control of Flexible Gyroscopic Systems," Journal of Guidance and Control, Vol. 3, Nov.-Dec. 1980, pp. 220-229.

[21] Junkins, J.L. and Dunyak,J.P., "Continuation Methods for Enhancement of Optimization Algorithms," Presented to 19th Annual Meeting, Society of Engineering Science, University of Missouri, Rolla, Oct. 1982.

[22] Freudenberg, J.S., Looze, D.P. and Cruz, J.B., "Robustness Analysis Using Singular Sensitivities," Int. Journal of Control, Vol. 35, No. 1, 1982, pp. 95-116.

[23] Lim, K.B., and Junkins, J.L., "Robustness Optimization of Structural and Controller Parameters," paper No. AIAA-87-0791-CP, presented at AIAA 28th SDM Conference, Monterey, CA, April 6-8, 1987.

[24] Balas, M.J., "Trends in LSS Control Theory: Fondest Hopes, Wildest Dreams," IEEE Transactions on Automatic Control, Vol. Ac-27, No. 3, June 1982, pp. 522-535.

[25] Bekey, I. and Naugle, J.E., "Just Over the Horizon in Space," Astronautics and Aeronautics, May 1980, pp. 64-76.

[26] Khot, N.S., et al., "Optimal Structural Modifications to Enhance the Optimal Active Vibration Control of Large Flexible Structures," 26th Structures, Structural Dyn., and Materials Conf. Orlando, FL, April 15-17,1985.

[27] Junkins, J.L., Bodden D.S. and Turner, J.D., "A Unified Approach to Structure and Control System Design Iterations," Fourth International Conference on Applied Numerical Modeling, Tainan, Taiwan, Dec. 27-29, 1984.

[28] Haftka, R.T., et al., "Sensitivity of Optimized Control Systems to Minor Structural Modifications," 26th Structures, Structural Dynamics and Materials Conference, Orlando, FL, April 15-17, 1985.

[29] Junkins, J.L. and Rew, D.W., "A Simultaneous Structure/Control Design Iteration Method," American Controls Conference, Boston, MA, June 1985.

[30] Dantzig, G.B., Linear Programming and Extensions, Princeton University Press, 1963.

[31] Hadley, G., Linear Programming, Addison-Wesley Publishing Co., Inc., Reading, MA, 1962.

[32] Horta, L.G., Juang, J-N and Junkins, J.L., "A Sequential Linear Optimization Approach for Controller Design," AIAA Paper 85-1971-CP, 1985.

[33] Lim, K.B. and Junkins, J.L., "Minimum Sensitivity Eigenvalue Placement via Sequential Linear Programming," Proceedings of the Mountain Lake Dynamics and Control Institute, ed. by J.L. Junkins, Mountain Lake, VA, June 9-11, 1985.

[34] Newsom, J.R. and Mukhopadhyay, "A Multiloop Robust Controller Design Study Using Singular Value Gradients," Journal of Guidance, Control, and Dynamics, Vol. 8, No. 4, July-Aug., 1985, pp. 514-519.

[35] Howze, J.W. and Cavin, R.K., "Regulator Design with Modal Insensivity," IEEE Transaction of Automatic Control, Vol. Ac-24, No. 3, June 1979, pp. 466-9.

[36] Raman, K.V., "Modal Insensitivity with Optimality," Ph.D. Dissertation, Drexel University, Philadelphia, PA, 1984.

[37] Raman, K.V. and Calise, A.J., "Design of an Optimal Output Feedback Control System with Modal Insensitivity," AIAA Paper 84-1940, AIAA Guidance and Control Conference, Seattle, WA, Aug. 20-22, 1984.

[38] Lim, K.B. and Junkins, J.L,, "Optimal Redesign of Dynamic Structures via Sequential Linear Programming," Fourth International Modal Analysis Conference, Los Angeles, CA, Feb. 3-6, 1986.

[39] Palacios-Gomez, R., Lasdon, L. and Engquist, M., "Nonlinear Optimization by Successive Linear Programming," Management Science, Vol. 28, No. 8, October, 1982, pp. 1106-1120.

[40] Palacios-Gomez, R., "The Solution of Nonlinear Optimization Problems Using Successive Linear Programming," Ph.D. Dissertation, The University of Texas, Austin, TX, 1980.

[41] Patel, R.B. and Toda, M., "Quantitative Measures of Robustness for Multivariable Systems," Proceedings of JACC, San Francisco, TP8-A, 1980.

[42] Noble, B. and Daniel, J.W., Applied Linear Algebra, Englewood Cliffs, NJ: Prentice-Hall, 1977.

[43] IMSL Reference Manual, International Mathematical and Statistical Library, Inc., 1982

[44] Chen, C.T., Introduction to Linear Systems Theory, Holt, Rinehart & Winston, Inc., 1970.

[45] Fleming, P., "Computer Aided Design of Regulators Using Multiobjective Optimization," preprint, University College of North Wales, November, 1986.

[46] Lim, K.B., "A Unified Approach to Structure and Controller Design Optimizations," Ph.D. Dissertation, Virginia Polytechnic Institute and State University, Blacksburg, VA, 1986.

[47] Sobel, K.M. and Shapiro, E.Y., "Application of Eigenstructure Assignment to Flight Control Design: Some Extension," AIAA J. of Guidance, Control, and Dynamics, Vol. 10, No. 1, Jan-Feb 1987, pp. 73-89.

An Integrated Approach to the Minimum Weight and Optimum Control Design of Space Structures

N. S. Khot
Structures Division
Flight Dynamics Laboratory, AFWAL/FIBRA
Wright-Patterson Air Force Base, Ohio 45433-6553, USA

ABSTRACT

This paper will describe two approaches to the design of a structure and its control system in an optimum manner. The primary objective would be to modify the structural stiffness by changing the cross-sectional areas of the members in order to achieve a minimum weight structure and the desired distribution of the closed-loop eigenvalues and structural frequencies. The application of both methods is illustrated on truss structures.

I. INTRODUCTION

A great deal of research has been conducted in developing efficient algorithms for the minimum weight design of structures and the design of control systems. However, the use of both of these disciplines in the design of an active controlled structure is essentially uncoupled. The control design is normally undertaken for a specified distribution of natural frequencies and vibration modes, and there is no interaction between the structural and control designers. Developing an integrated design approach would make the performance of the whole system more efficient and economical.

Literature research would reveal that different approaches have been proposed by various investigators, and, essentially, there is no consensus on the selection of objective functions and constraints. This paper contains a discussion of two of the approaches the author has investigated recently. The essential feature of these approaches is to systematically modify the structural stiffness in order to obtain a minimum weight structure and to enhance the closed-loop performance of the system. In the first approach the weight of the structure is the objective function, and constraints are imposed on the damping parameters and the distribution of the closed-loop eigenvalues of the linear quadratic regulator with constant gain feedback. The second approach consists of imposing constraints on the closed-loop eigenvalue distribution, the structural frequencies and the Frobenious norm of the required gains with weight as the objective function.

Springer Series in Computational Mechanics
S. N. Atluri, A. K. Amos (Eds.)
Large Space Structures: Dynamics and Control

II. PRELIMINARIES

The equation of motion can be written as

$$[\mathrm{M}]\{\ddot{\mathrm{U}}\} + [\mathrm{E}]\{\dot{\mathrm{U}}\} + [\mathrm{K}]\{\mathrm{U}\} = [\mathrm{D}]\{\mathrm{f}\} \tag{1}$$

where [M], [E] and [K] are $n \times n$ mass, damping and stiffness matrices, and [D] is a $n \times p$ input distribution matrix. $\{U\}$ and $\{f\}$ represent displacement and input vectors of n and p components. Using the coordinate transformation $\{U\} = [\phi]\{\eta\}$, where $[\phi]$ is the modal matrix and $\{\eta\}$ is the modal coordinates vector, Eq. 1 can be transformed into an uncoupled set of equations

$$[\overline{\mathrm{M}}]\{\ddot{\eta}\} + [\overline{\mathrm{E}}]\{\dot{\eta}\} + [\overline{\mathrm{K}}]\{\eta\} = [\phi]^{\mathrm{T}}[\mathrm{D}]\{\mathrm{F}\} \tag{2}$$

where $[\bar{M}]$, $[\bar{E}]$ and $[\bar{K}]$ are diagonal matrices. Eq. 2 can be converted into the state space representation,

$$\{\dot{x}\} = [\mathrm{A}]\{\mathrm{x}\} + [\mathrm{B}]\{\mathrm{f}\} \tag{3}$$

by using the transformation

$$\{x\}_{2n} = \begin{bmatrix} \eta \\ \dot{\eta} \end{bmatrix}_{2n} \tag{4}$$

In Eq. 3 [A] and [B] are the plant and input matrices and are given by

$$[\mathrm{A}] = \left[\begin{array}{c|c} 0 & \mathrm{I} \\ \hline -\omega^2 & -2\varsigma\omega \end{array}\right] \tag{5}$$

and

$$[\mathrm{B}] = \left(\begin{array}{c} 0 \\ \hline \phi^{\mathrm{T}}\mathrm{D} \end{array}\right) \tag{6}$$

where ω is the vector of structural frequencies and ς is the modal damping factor vector. In order to design a controller using a linear quadratic regulator, a performance index (PI) can be defined as

$$\mathrm{PI} = \int_0^t \left(\{x\}^{\mathrm{T}}[\mathrm{Q}]\{\mathrm{x}\} + \{\mathrm{f}\}^{\mathrm{T}}[\mathrm{R}]\{\mathrm{f}\}\right) \mathrm{dt} \tag{7}$$

where [Q] and [R] are the state and control weighting matrices which have to be positive semidefinite and positive definite respectively. For a linear state feedback law the control law is given by

$$\{f\} = -[\tilde{\mathrm{G}}]\{x\} \tag{8}$$

where

$$[\tilde{\mathrm{G}}] = [\mathrm{R}]^{-1}[\mathrm{B}]^{\mathrm{T}}[\tilde{\mathrm{P}}] \tag{9}$$

and $[\tilde{P}]$ is a Riccati matrix which is obtained by solving the algebraic Riccati equation,

$$[\mathrm{A}]^{\mathrm{T}}[\tilde{\mathrm{P}}] - [\tilde{\mathrm{P}}][\mathrm{B}][\mathrm{R}]^{-1}[\mathrm{B}]^{\mathrm{T}}[\tilde{\mathrm{P}}] + [\tilde{\mathrm{P}}][\mathrm{A}] + [\mathrm{Q}] = 0 \tag{10}$$

Using Eq. 8 the state input equation can be written as

$$\{\dot{x}\} = [\overline{\mathrm{A}}]\{x\} \tag{11}$$

where

$$[\bar{A}] = [A] - [X][\tilde{P}] \tag{12}$$

and

$$[X] = [B][R]^{-1}[B]^T \tag{13}$$

The eigenvalues of the closed-loop matrix $[\bar{A}]$ are a set of complex conjugate pairs written as

$$\lambda_i = \tilde{\sigma}_i \pm j\tilde{\omega}_i \tag{14}$$

The closed-loop dampng ratio ξ_i is given by

$$\xi_i = -\frac{\tilde{\sigma}_i}{(\tilde{\sigma}_i^2 + \tilde{\omega}_i^2)^{1/2}} \tag{15}$$

The active control energy or the work done by the controllers can be written as

$$W = \int_0^t f\dot{U}\, dt \tag{16}$$

The gradients of the closed-loop eigenvalues λ_i with the respect to the structural design variable A_ℓ are given by

$$\lambda_{i,l} = \{\beta\}_i^T[\bar{A}]_{,l}\{\alpha\}_i \tag{17}$$

where $\{\beta\}_i$ and $\{\alpha\}_i$ are the left-hand and right-hand eigenvectors of the closed-loop matrix. The partial derivative of the closed-loop matrix $[\bar{A}]$ can be obtained by differentiating Eq. 12. This gives

$$[\bar{A}]_{,l} = [A]_{,l} - [X]_{,l}[\tilde{P}] - [X][\tilde{P}]_{,l} \tag{18}$$

The plant matrix $[A]$ is a function of the circular frequency ω_j, and matrix $[X]$ is a function of the modal matrix $[\phi]$.

The sensitivity of the Riccati matrix can be found from the solution of the algebraic Lyapunov equation

$$[\bar{A}]^T[\tilde{P}]_{,l} + [\tilde{P}]_{,l}[\bar{A}] = [\tilde{B}] \tag{19}$$

where

$$[\tilde{B}] = -[A]_{,l}^T[\tilde{P}] - [\tilde{P}][A]_{,l} + [\tilde{P}][X]_{,l}[P] \tag{20}$$

which is obtained by differentiating Eq. 12.

Knowing the gradients of the closed-loop eigenvalues, the gradients of the damping ratio in Eq. 15 can be easily written.

III. MINIMUM WEIGHT DESIGN WITH CONSTRAINTS ON THE CLOSED-LOOP EIGENVALUES

In this case the optimization problem for the integrated design procedure can be stated as:[1,2]

$$\textit{minimize the weight}\ \mathrm{W} = \sum_{i=1}^{m} \rho_i \mathrm{A}_i l_i \tag{21}$$

$$\textit{subject to:}\quad g_j(\xi_i) = \xi_i - \bar{\xi}_i = 0$$

$$g_j(\tilde{\omega}_i) = \tilde{\omega}_i - \bar{\tilde{\omega}}_i = 0 \tag{22}$$

where ρ_i is the density of the material, ℓ_i is the length of the element, and A_i is the design variable which is the cross-sectional area of the ith member. $\bar{\xi}_i$ and $\bar{\tilde{\omega}}_i$ in Eq. 22 represent the desired values of the damping factors and the imaginary part of the closed-loop system for a linear quadratic regulator. The optimization problem was solved by using the NEWSUMT-A[3] optimization program which is based on an extended interior penalty method of constrained optimization and a modified Newton's method of unconstrained minimization. The major steps involved in the optimization are as follows:

1. Assign initial sizes to all the elements.
2. Determine the structural frequency distribution and the vibration modes.
3. Use the eigenvalues and eigenvectors to assemble the plant matrix and input matrix in Eqs. 5 and 6.
4. Evaluate the closed-loop eigenvalues (Eq. 14) and damping ratios (Eq. 15) and their derivatives with respect to the design variables A_i.
5. Use the optimization program to find the modified design variables.
6. Terminate the iterative procedure after an optimum design is obtained or go to step 2.

The structural model selected for this investigation is a tetrahedral truss[4] (Fig. 1) which models the feed tower in a generic class of large antenna applications. The apex of the structure, node 1, represents the antenna feed, and its motion in the X-Y plane has to be actively controlled. The coordinates of the node points are given in Table 1. The actuators and sensors are collocatd in the six bipods. The structure has twelve degrees of freedom with four masses of 2 units each attached at nodes 1 through 4. The dimensions of the structure are given in unspecified consistent units. The elastic modulus of the members was assumed to equal to 1.0, and the density of the structural material was set equal to 0.001. The initial design used for starting the optimization is denoted by Design A. The cross-sectional areas of the members for this design are equal to those assigned by the Charles Stark Draper Lab.[4] for their control design investigation. A comparison between the nominal Design A and the optimum Design B was made by comparing the weight, structural frequencies, closed-loop eigenvalues, and damping ratios. The weighting matrices [Q] and [R] in Eq. 7 were assumed to be identity matrices, and the modal damping parameter ς in Eq. 5 was set equal to zero, assuming that there is no passive damping.

The constraints imposed on the optimum design were as follows:

1. $\tilde{\omega}_1$ must be equal to 1.341, i.e., the same as that of Design A.
2. $\tilde{\omega}_2 \geq 1.5$
3. $\bar{\xi}_1 = \bar{\xi}_2 = \bar{\xi}_3 = \bar{\xi}_4 = 0.1093$. For Design A ξ_1 is equal to 0.05461.

The weight of the initial design was 43.69 units. The optimum design weighed 20.75 units. NEWSUMT-A required twenty-two iterations to attain the optimum design. The cross-sectional areas of the members, the structural frequencies, and the closed-loop damping parameters for the initial Design A and the optimum Design B are given in Tables 2 and 3. In the simulated transient response for and initial displacement of unit magnitude in the X-direction at node 2, the work done by the controllers (Eq. 16) for Design A and Design B was equal to 78.6 and 29.9 respectively.

IV. MINIMUM WEIGHT DESIGN WITH CONSTRAINTS ON THE CONTROL GAIN NORM

In this approach, instead of using the linear quadratic regulator to determine the gain matrix $[\tilde{G}]$, it was computed such that Eq. 11 satisfied the closed-loop eigenvalue constraints, and the Frobenious norm of the gain matrix $S_G = [\tilde{G}]^T[\tilde{R}][\tilde{G}]$ was minimized, where $[\tilde{R}]$ is the control gain weighting matrix. The optimization problem can be stated as[5]

$$\textit{minimize the weight } \mathrm{W} = \sum_{i=1}^{m} \rho_i \mathrm{A}_i l_i \tag{23}$$

subject to:

$$\begin{aligned} \text{Minimum} \quad \mathrm{S_G} &\leq \bar{\mathrm{S}}_\mathrm{G} \\ \tilde{\sigma}_j - \bar{\tilde{\sigma}}_j &\leq 0 \\ \tilde{\omega}_j - \bar{\tilde{\omega}}_j &\leq 0 \\ \omega_j - \bar{\omega}_j &\leq 0 \end{aligned} \tag{24}$$

where $\bar{S}_G$, $\bar{\tilde{\sigma}}_j$, $\bar{\tilde{\omega}}_j$ and $\bar{\omega}_j$ are the constraint values of the Frobenious norm and the real part and imaginary part of the closed-loop eigenvalues and structural frequencies respectively. The optimization procedure consists of first minimizing the control objective function, S_G, subject to the closed-loop eigenvalue constraints to obtain the optimal gains. Secondly, minimize the weight of the structure subject to constraints on the structural frequencies and the Frobenious norm. The optimization was carried out by using the NEWSUMT-A[3] program.

In order to present this optimization concept, a simple two bar truss shown in Fig. 2 was considered. A nonstructural mass of 2 units was placed at the vertex. The elastic modulus was assumed to be equal to 1.0 and the mass density $\rho_i = 0.001$. The minimum size constraint was 10 units. The specified equality constraints on the closed-loop eigenvalues were,

$$\tilde{\sigma}_1 = -0.0228 \quad \tilde{\sigma}_2 = -0.361$$

$$\tilde{\omega}_1 = 1.17 \quad \tilde{\omega}_2 = 4.81$$

The constraint on the gain norm $S_G = \sum_{i=1}^{4} g_i^2$ was

$$Min S_G \leq 1500$$

In addition to this the constraints on the structural frequencies were

$$\omega_1 = \tilde{\omega}_1 = 1.0$$

and

$$\omega_2 \geq \tilde{\omega}_2$$

In the different cases investigated[5], ω_2 was either free or constrained. The constraint on the Frobenious norm was either an equality or inequality constraint. Table 4 gives the details of the different optimum designs. It was found that with ω_2 unconstrained, different optimum designs were obtained depending on the initial design. For $S_G = 1500$, the main difference between the optimum designs was the magnitude of ω_2. No design with $\omega_2 \geq 6.457$ was obtained. Amongst different optimum designs, of particular interest were designs D_2 and D_6. Both these designs have the same weight, however, the gain norm for D_2 is 1500 and for D_6 is 245.4. This is achieved by switching the areas amongst the two members. The transient response was simulated for the optimum design for an initial displacement of unit magnitude in the X and Y direction. The conclusions reached by comparing the results were: 1) the settling time is reduced, not as the control effort increases but as the control effort per unit weight of the structure increases, 2) even though the two structures may be mechanically equivalent, from a control point of view they can be different, and 3) the heavier structure does not necessarily require a large control effort nor does a lighter structure imply a lower control effort.

V. CONCLUSIONS

Two integrated approaches to design an optimum structure and control system are presented with constraints on structural frequencies and closed-loop frequencies. The feasibility of the approaches has been illustrated on sample problems. It has been shown that different op illustrated on sample problems. It has been shown that different optimum solutions can be obtained depending on the constraint definition. The final choice amongst different local optimum designs would be a compromise between the weight of the structure, the settling time and the control effort.

VI. REFERENCES

1. Khot, N. S., Eastep, F. E. and Venkayya, V. B., "Optimal Structural Modifications to Enhance the Active Vibration Control of Flexible Structures," AIAA J., Vol. 24, No. 8, 1986.

2. Khot, N. S., "Structure/Control Optimization to Improve the Dynamic Response of Space Structures," To be published in the Computational Mechanics, An International Journal, Springer-Verlag.

3. Thareja, R. T. and Haftka, R., "NEWSUMT-A: A Modified Version of NEWSUMT for Inequality and Equality Constraints," VPI&SU Report, March 1986.

4. Strunce, R. R., et. al., "ACOSS-FOUR (Active Control of Space Structures) Theory Appendix," RADC-TR-80-78, Vol II, 1980.

5. Khot, N. S., Oz, H., Grandhi, R. V., Eastep, F. E. and Venkayya, V. B., "Optimal Structural Design with Control Gain Norm Constraint," (87-0019) Presented at AIAA 25th Aerospace Sciences Meeting, Reno, Nevada, Jan 12-15, 1987.

Table 1. Node Point Coordinates for ACOSS-FOUR

Node	X	Y	Z
1	0.0	0.0	10.165
2	-5.0	-2.887	2.00
3	5.0	-2.887	2.00
4	0.0	5.7735	2.00
5	-6.0	-1.1547	0.0
6	-4.0	-4.6188	0.0
7	4.0	-4.6188	0.0
8	6.0	-1.1547	0.0
9	-2.0	5.7735	0.0
10	2.0	5.7735	0.0

Table 2. Cross-Sectional Areas of the Members

Element No.	Design A Nominal	Design B IDESIGN
1 (1-2)	1000.	190.49
2 (2-3)	1000.	252.19
3 (1-3)	100.	379.03
4 (1-4)	100.	276.53
5 (2-4)	1000.	485.13
6 (3-4)	1000.	263.26
7 (2-5)	100.	347.73
8 (2-6)	100.	54.27
9 (3-7)	100.	64.45
10 (3-8)	100.	242.04
11 (4-9)	100.	38.15
12 (4-10)	100.	59.23
WEIGHT	43.69	20.75

Table 3. Structural Frequencies and Closed-Loop Damping Parameters of Acoss-Four

	Structural Frequencies (ω_j^2)		Modal Damping Parameters	
Mode	Design A	Design B	Design A	Design B
1	1.80	1.796	0.0546	0.1093
2	2.77	2.268	0.0653	0.1093
3	8.35	5.693	0.0737	0.1093
4	8.74	5.912	0.0801	0.1093
5	11.54	13.28	0.0839	0.09421
6	17.67	14.32	0.0864	0.0444
7	21.73	21.92	0.0760	0.06859
8	22.61	28.66	0.0723	0.04222
9	72.92	37.94	0.0341	0.04272
10	85.57	54.22	0.0298	0.03977
11	105.77	67.60	0.0207	0.02363
12	166.54	73.01	0.0064	0.03938

Table 4. Optimum Designs for the Two Bar Truss

DESIGN	A_1	A_2	W	$-g_1$	$-g_2$	$-g_3$	$-g_4$	S_G	ω_1	ω_2
D1	203.98	82.98	6.417	-16.73	-34.89	-1.39	-0.78	1500.	1.0	2.327
D2	72.70	691.83	17.095	14.36	-35.77	1.33	-3.50	1500.0	1.0	4.012
D3	71.29	1312.9	30.952	-16.84	34.62	1.71	-3.88	1500.0	1.0	5.473
D4	1838.1	70.9	42.687	15.55	35.41	-0.84	-1.33	1500.0	1.00	6.457
D5	371.53	75.70	10.0	-15.24	-26.24	-1.20	-0.97	923.5	1.0	3.00
D6	687.57	72.72	17.000	-8.86	-12.83	-1.02	-1.16	245.4	1.0	4.0

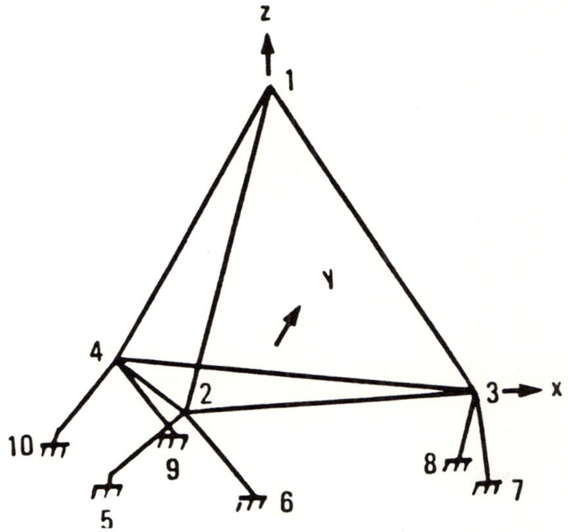

Fig. 1 Tetrahedral truss

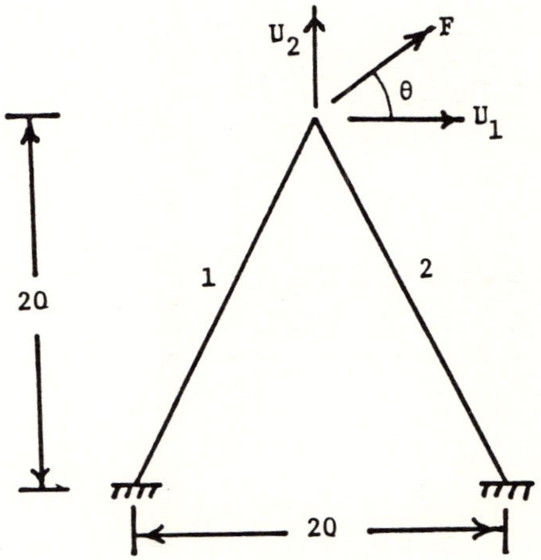

Fig. 2 Two-bar truss